CALCIUM ANTAGONISTS

PHARMACOLOGY AND CLINICAL RESEARCH

Medical Science Symposia Series

Volume 3

The titles published in this series are listed at the end of this volume.

Calcium Antagonists

Pharmacology and Clinical Research

Edited by

T. Godfraind

Laboratory of Pharmacology, Catholic University of Louvain,
Brussels, Belgium

S. Govoni

Institute of Pharmacological Sciences, University of Milan,
Milan, Italy

R. Paoletti

Institute of Pharmacological Sciences, University of Milan,
Milan, Italy

and

P. M. Vanhoutte

Center for Experimental Therapeutics, Baylor Institute of Medicine,
Houston, Texas, U.S.A.

SPRINGER-SCIENCE+BUSINESS MEDIA, B.V.

Library of Congress Cataloging-in-Publication Data

Calcium antagonists pharmacology and clinical research / edited by T.
   Godfraind ... [et al.].
       p.   cm. -- (Medical science symposia series ; v. 3)
     Includes index.

     ISBN 978-0-7923-2259-7   ISBN 978-94-011-1725-8 (eBook)
     DOI 10.1007/978-94-011-1725-8

     1. Calcium--Antagonists.   I. Godfraind, T. (Théophile)
   II. Series.
     [DNLM: 1. Calcium Channel Blockers--pharmacology--congresses.
   2. Calcium Channel Blockers--therapeutic use--congresses.   QV 150
   C14372 1993]
   RC684.C34C338   1993
   616.1'061--dc20
   DNLM/DLC
   for Library of Congress   93-15059

Printed on acid-free paper

This volume is dedicated to the memory of Dr. Albrecht Fleckenstein, in honor of his pioneering contribution to the field of cardiovascular science.

Albrecht Fleckenstein
(1917–1991)

CONTENTS

PREFACE

The importance of calcium (Ca^{2+}) for the maintenance of cardiac contractility has been recognized by Ringer as early as 1880, but the critical role of the ion in the contractile process in skeletal, cardiac, and smooth muscle has only been established in the last three decades. At the same time, the role of Ca^{2+} in secretory responses of various tissues and in the nervous system has become obvious, leading to the concept that Ca^{2+} is perhaps the most widespread second messenger in eucaryotic cells. During the 1960s the hypothesis of drugs as Ca^{2+} antagonists (Ca^{2+} channel inhibitors) was proposed by the research groups of A. Fleckenstein (who very unfortunately passed away since the symposium covered by these proceedings) and T. Godfraind. Since then, the number of newly discovered molecules acting at Ca^{2+} channels has increased exponentially. This has resulted in the need for reappraisal of the importance of these drugs as pharmacological tools and as therapeutic agents. As the complexity of the pharmacological actions of the Ca^{2+} channel inhibitors grows, there is a continued need to further clarify the inhibitors, both chemically and functionally. Indeed, new pharmacological properties of selected Ca^{2+} channel inhibitors have been described; notably, some of the activities are independent of their ability to block voltage-dependent Ca^{2+} channels.

Within this context, the goal of this volume is to provide an update of the field based on the work presented at the 5th International Symposium on **CALCIUM ANTAGONISTS: Pharmacology and Clinical Research.** The meeting provided an international forum for the discussion of recent progress in the knowledge of Ca^{2+} channels, as well as of the pharmacology of agents acting at these channels and their clinical applications.

In particular, the monograph reviews the current state of the molecular biology of Ca^{2+} channels, which is still a growing area. Several partial clones of dihydropyridine-sensitive Ca^{2+} channels are being sequenced. There is general agreement that the expression of the α_1 subunit alone is sufficient to yield functionally active, dihydropyridine-sensitive Ca^{2+} channels.

In the cardiovascular area, in addition to the now well-established clinical uses of Ca^{2+} channel inhibitors, which were reviewed during the meeting, exciting new work points to an application in atherosclerosis. Thus, some representatives of this class of drugs may simultaneously reduce arterial blood pressure and control the major cardiovascular risk factor, atheroma.

Both from the experimental and the clinical point of view, great attention has been paid to the possibility that Ca^{2+} antagonists may exert a significant action on the central nervous system. At the experimental level, likely mechanisms underlying the beneficial effects of Ca^{2+} antagonists in ischemic tissues are discussed, although it is apparent that under such conditions the ways of entry of the ion are multiple. Additional data are reported on the presence of both Ca^{2+}-antagonist-sensitive as well as on Ca^{2+}-antagonist-insensitive Ca^{2+} channels in neurons and in glia cells. Their regulation by other transmitters, G proteins, is reviewed, as well as their modulation under various physiopathological conditions, including aging. Besides the traditional use in neurological disorders such as stroke or subarachnoid hemorrhage, the possible treatment of age-related defects in memory or of dementia with Ca^{2+} antagonists is an exciting new development.

Important uses of Ca^{2+} antagonists in novel areas of interest are also included in the monograph. In particular, Ca^{2+} antagonists with increased selectivity for gastrointestinal smooth muscle cells may soon become available. Such agents might be useful for treating abnormal esophageal contractility, nonulcerogenic stomach motility, irritable bowel syndrome, and other conditions.

Of great interest also are the preliminary clinical findings showing a beneficial effect of Ca^{2+} antagonists on renal blood flow and graft rejection. Likewise, the reversal of multidrug resistance in the treatment of cancer, tumors, and malaria is puzzling, although it appears due to the binding of phenylalkylamine Ca^{2+} antagonists to structures other than calcium channels.

The organizers gratefully acknowledge the advice and the aid provided by the staff of the Fondazione Giovanni Lorenzini, which insured the smooth functioning of the meeting. We also are grateful for the financial support by the sponsors that made this exciting conference possible.

T. Godfraind
R. Paoletti
P.M. Vanhoutte

Co-chairmen,
II International Symposium on **CALCIUM ANTAGONISTS:**
Pharmacological and Clinical Research
September 5-8, 1991
Houston, Texas (USA)

List of Contributors

A. Anderson
Hypertension Clinic
Repatriation General Hospital
Heidelberg West 3081
Australia

C. Argentino
Dept. of Neurological Sciences
University "La Sapienza"
Viale dell'Università 30
00185 Rome
Italy

R. Bangalore
School of Pharmacy
State University of New York at Buffalo
C126 Cooke-Hochstetter Complex
Buffalo NY 14260
USA

F. Battaini
Institute of Pharmacological Sciences
Via Balzaretti 9
20133 Milan
Italy

D. J. Beech
Department of Pharmacology
and Clinical Pharmacology
St. Georges Hospital
Medical School
Cranmer Terrace
London SW17 ORE
U.K.

R. Bellazzi
Dipartimento di Informatica e Sistemistica
University of Pavia
27100 Pavia
Italy

S. Bellosta
Institute of Pharmacological Sciences
University of Milan
Via Balzaretti 9
20133 Milan
Italy

F. Bernini
Institute of Pharmacological Sciences
University of Milan
Via Balzaretti 9
20133 Milan
Italy

F. V. Bielen
University of Leuven
Campus Kortrijk
8500 Kortrijk
Belgium

T. B. Bolton
Department of Pharmacology and Clinical
Pharmacology
St. Georges Hospital
Medical School
Cranmer Terrace
London SW17 ORE
U.K.

A. Boraso
University of Brescia
Spedali Civili
P. le Spedali Civili 1
25231 Brescia
Italy

M. Borgers
Life Sciences
Janssen Research Foundation
Turnhoutseweg 30
2340 Beerse
Belgium

R. M. Brawley
Department of Pharmacology
Northwestern University
Medical School
303 E. Chicago Avenue
Chicago IL 60611
USA

L. J. Briggs
Department of Radiology Med. Biochemistry
and the Biomolecular Structure Analysis
Center
University of Connecticut
Health Center
Farmington CT 06032
USA

M. Bucay
Baylor College of Medicine and The
Methodist Hospital
Suite F-905
6535 Fannin Street
Houston TX 77030
USA

A. Cargnoni
University of Brescia
Spedali Civili
P. le Spedali Civili 1
25231 Brescia
Italy

C. Carpi
Glaxo Research Laboratories
Via A. Fleming 4
37135 Verona
Italy

C. F. Chang
Department of Pharmacology
Northwestern University
Medical School
303 E. Chicago Avenue
Chicago IL 60611
USA

M. Chen
Department of Physiology
and Biochemistry
Medical College of Pennsylvania
3300 Henry Avenue
Philadelphia PA 19129
USA

M. Chen
Georgetown University
School of Medicine
Dept. of Neurology
3800 Reservoir Road NW
Washington DC 20007-2197
USA

C. P. Cheng
Section of Cardiology
Dept. of Medicine
Bowman Gray School of Medicine
Wake Forest University
Medical Center Boulevard
Winston-Salem NC 27157-1045
USA

V. Chikvaidze
Institute of Physiology
Georgian Academy of Sciences
Gotua Street 14
380060 Tbilisi
Georgia

M. O. Christen
Laboratoires de Therapeutique
Moderne - L.T.M.
42 Rue Rouget-de-Lisle
B.P. 22
92151 Suresnes CEDEX
France

W. E. Code
Department of Pharmacology
and Anaesthesia
College of Medicine
University of Saskatchewan
Saskatoon Sask. S7N 0W0
Canada

S. L. Cohan
Georgetown University
School of Medicine
Dept. of Neurology
3800 Reservoir Road NW
Washington DC 20007-2197
USA

J. M. C. Connell
MRC Blood Pressure Unit
Gardiner Institute
Western Infirmary
Glasgow G 11 6NT
Scotland
U.K.

A. Corsini
Institute of Pharmacological Sciences
University of Milan
Via Balzaretti 9
20133 Milan
Italy

I. Dawidson
University of Texas
Southwestern Medical Center
Parkland Memorial Hospital
5323 Harry Hines Blvd.
General Surgery - E7.126
Dallas TX 75235-9031
USA

F. De Clerck
Dept. of Cardiovascular Pharmacology
Janssen Research Foundation
Turnhoutseweg 30
2340 Beerse
Belgium

M. De Ryck
Janssen Research Foundation
Turnhoutseweg 30
2340 Beerse
Belgium

R. Donnelly
Department of Medicine
and Therapeutics
Gardiner Institute
Western Infirmary
Glasgow G11 6NT
Scotland
U.K.

R. J. Dring
Dept. of Radiology Med. Biochemistry
and the Biomolecular Structure
Analysis Center
University of Connecticut
Health Center
Farmington CT 06032
USA

M. Epstein
Nephrology Section
Veterans Affairs Medical Center
1201 NW 16th Street
Miami FL 33125
USA

J. Ferrante
Laboratory of Biochemical Genetics
National Institute of Health
Bethesda MD 20892
USA

R. Ferrari
University of Brescia
Spedali Civili
P. le Spedali Civili 1
25231 Brescia
Italy

C. Fieschi
Dept. of Neurological Sciences
University "La Sapienza"
Viale dell'Università 30
00185 Rome
Italy

M. Fiorelli
Dept. of Neurological Sciences
University "La Sapienza"
Viale dell'Università 30
00185 Rome
Italy

A. Fleckenstein
Physiological Institute
University of Freiburg
Hermann-Herder Str. 7
W-7800 Freiburg
Germany

G. Fleckenstein-Grün
Physiological Institute
University of Freiburg
Hermann-Herder Str. 7
W-7800 Freiburg
Germany

J. D. Folts
Dept. of Medicine
Cardiology Section
Univ. of Wisconsin
Medical School
Madison WI 53792
USA

M. Frey
Physiological Institute
University of Freiburg
Hermann-Herder Str. 7
W-7800 Freiburg
Germany

M. Frontoni
Department of Neurological Sciences
University "La Sapienza"
Viale dell'Università 30
00185 Rome
Italy

P. Gaetani
Department of Surgery
Neurosurgery
IRCCS Policlinico S. Matteo
27100 Pavia
Italy

G. Gaviraghi
Glaxo Research Laboratories
Via A. Fleming 4
37135 Verona
Italy

H. Geerts
Life Sciences
Janssen Research Foundation
Turnhoutseweg 30
2340 Beerse
Belgium

M. Gentile
Dept. of Neurological Sciences
University "La Sapienza"
Viale dell'Università 30
00185 Rome
Italy

R. Getz
Georgetown University
School of Medicine
Dept. of Neurology
3800 Reservoir Road NW
Washington DC 20007-2197
USA

J. Gheuens
Janssen Research Foundation
Turnhoutseweg 30
2340 Beerse
Belgium

T. Godfraind
Laboratory of Pharmacology
UCL 7350
Catholic University of Louvain
1200 Brussels
Belgium

M. Gopalakrishnan
Department of Physiology
Baylor College of Medicine
One Baylor Plaza
Houston TX 77030
USA

S. Govoni
Institute of Pharmacological Sciences
University of Milan
Via Balzaretti 9
20133 Milan
Italy

L. M. Greenberger
Lederle Laboratories
American Cyanamid Company
Oncology and Immunology
Research Section
North Middle Pown Road
Pearl River NY 10965
USA

L. M. Gutierrez
Department of Pharmacology
Northwestern University
Medical School
303 E. Chicago Avenue
Chicago IL 60611
USA

J. W. Hamilton
Department of Medicine
Cardiology Section
University of Wisconsin
Medical School
Madison WI 53792
USA

M. Hawthorn
School of Pharmacy
State University of New York at Buffalo
C126 Cooke-Hochstetter Complex
Buffalo NY 14260
USA

P. D. Henry
Baylor College of Medicine and
The Methodist Hospital
Suite F-905
6535 Fannin Street
Houston TX 77030
USA

L. G. Herbette
Dept. of Radiology Med. Biochemistry and
the Biomolecular Structure Analysis Center
University of Connecticut
Health Center
Farmington CT 06032
USA

L. Hertz
Department of Pharmacology
and Anaesthesia
College of Medicine
University of Saskatchewan
Saskatoon Sask. S7N 0W0
Canada

M. M. Hosey
Department of Pharmacology
Northwestern University
Medical School
303 E. Chicago Avenue
Chicago IL 60611
USA

J. D. Huizinga
Intestinal Disease Res. Unit
Dept. of Biomedical Sciences
and Engineering Physics
McMaster University
1200 Main Street West
Hamilton Ontario L8N 3Z5
Canada

S. Illiano
Center for Experimental
Therapeutics
Baylor College of Medicine
One Baylor Plaza
Houston TX 77030
USA

W. Janssens
Dept. of Cardiovascular Pharmacology
Janssen Research Foundation
Turnhoutseweg 30
2340 Beerse
Belgium

O. Krizanova
Department of Pharmacology
and Cell Biophysics
University of Cincinnati
College of Medicine
231 Bethesda Avenue
Cincinnati, OH 45267-0575
USA

W. C. Little
Section of Cardiology
Dept. of Medicine
Bowman Gray School of Medicine
Wake Forest University
Medical Center Boulevard
Winston-Salem NC 27157-1045
USA

L. W. C. Liu
Intestinal Disease Res. Unit
Dept. of Biomedical Sciences
and Engineering Physics
McMaster University
1200 Main Street West
Hamilton Ontario L8N 3Z5
Canada

D. Lombardi
Department of Surgery
Neurosurgery
IRCCS Policlinico S. Matteo
27100 Pavia
Italy

P. Lory
Dept. of Pharmacology and
Cell Biophysics
University of Cincinnati
College of Medicine
231 Bethesda Avenue
Cincinnati OH 45267-0575
USA

C. Lu
University of Texas
Southwestern Medical Center
Parkland Memorial Hospital
5323 Harry Hines Blvd.
General Surgery-E7.126
Dallas TX 75235-9031
USA

J. Ma
Rush University
Medical College
1750 Congress Parkway
Chicago IL 60612
USA

A. Maggi
University of Brescia
Spedali Civili
P. le Spedali Civili 1
25231 Brescia
Italy

G. Mancia
Center of Clinical Physiology and Hypertension
Padiglione Sacco
Ospedale Maggiore
Via F. Sforza 35
20122 Milan
Italy

A. Martini
Centro di Studio per la
Biologia e Fisiopatologia
Muscolare del CNR
Istituto di Patologia Generale
Università di Padova
Via Trieste 75
35121 Padova
Italy

F. Marzatico
Institute of Pharmacology
University of Pavia
27100 Pavia
Italy

R. P. Mason
Dept. of Radiology
Biomolecular Structure
Analysis Center
The University of Connecticut
Health Center
263 Farmington Avenue
Farmington CT 06030
USA

M. Mazzotti
Institute of Pharmacological Sciences
University of Milan
Via Balzaretti 9
20133 Milan
Italy

R. W. McCallum
Division of Gastroenterology
Health Sciences Center
Dept. of Medicine
University of Virginia
Box 145
Charlottesville VA 22908
USA

F. B. Meyer
Department of Neurosurgery
Mayo Clinic
200 1st Street SW
Rochester MW 55905
USA

R. J. Miller
Dept. of Pharmacological
and Physiological Sciences
University of Chicago
947 E. 58th-St.
Chicago IL 60637
USA

N. Morel
Univ. Catholique du Louvain
Faculté de Medecine
Laboratory de Pharmacologie
UCL 7350
Avenue E. Mounier 73
1200 Bruxelles
Belgium

T. Morgan
University of Melbourne
Dept. of Physiology
Parkeville Victoria 3052
Australia

A. D. Morris
Dept. of Medicine
and Therapeutics
Gardiner Institute
Western Infirmary
Glasgow G11 6NT
Scotland
U.K.

C. Mundina-Weilenmann
Department of Pharmacology
Northwestern University
Medical School
303 E. Chicago Avenue
Chicago IL 60611
USA

T. Nagao
Center for Experimental
Therapeutics
Baylor College of Medicine
One Baylor Plaza
Houston TX 77030
USA

W. G. Nayler
Dept. of Medicine
University of Melbourne
Austin Hospital
Heidelberg Victoria 3084
Australia

A. Nori
Centro di Studio per la
Biologia e Fisiopatologia
Muscolare del CNR
Istituto di Patologia Generale
Università di Padova
Via Trieste 75
35121 Padova
Italy

B. Palmer
University of Texas
Southwestern Medical Center
Parkland Memorial Hospital
5323 Harry Hines Blvd.
General Surgery - E7.126
Dallas TX 75235-9031
USA

P. Paoletti
Department of Surgery
Neurosurgery
IRCCS Policlinico S. Matteo
27100 Pavia
Italy

R. Paoletti
Institute of Pharmacological Sciences
University of Milan
Via Balzaretti 9
20133 Milan
Italy

T. Peters
Janssen Research Foundation
Turnhoutseweg 30
2340 Beerse
Belgium

H. Porzig
Department of Pharmacology
University of Bern
Friedbuhlstrasse 49
3010 Bern
Switzerland

S. Quaglini
Dipartimento di Informatica e Sistemistica
University of Pavia
27100 Pavia
Italy

xxii

M. Raiteri
Institute of Pharmacological Sciences
University of Milan
Via Balzaretti 9
20133 Milan
Italy

B. F. X. Reber
Department of Pharmacology
University of Bern
Friedbuhlstrasse 49
3010 Bern
Switzerland

J. L. Reid
Dept. of Medicine
and Therapeutics
Gardiner Institute
Western Infirmary
Glasgow G11 6NT
Scotland
U.K.

H. Reuter
Department of Pharmacology
University of Bern
Friedbuhlstrasse 49
3010 Bern
Switzerland

E. Rios
Rush University
Medical College
1750 W Congress Parkway
Chicago IL 60612
USA

R. Risser
University of Texas
Southwestern Medical Center
Parkland Memorial Hospital
5323 Harry Hines Blvd.
General Surgery - E7.126
Dallas TX 75235-9031
USA

O. K. Rizanova
Dept. of Pharmacology and
Cell Biophysics
University of Cincinnati
College of Medicine
231 Bethesda Avenue
Cincinnati OH 45267-0575
USA

D. Rock
Department of Physiology
and Biochemistry
Medical College of Pennsylvania
3300 Henry Avenue
Philadelphia PA 19129
USA

R. Rodriguez y Baena
Department of Surgery
Neurosurgery
IRCCS Policlinico S. Matteo
27100 Pavia
Italy

P. Rooth
University of Texas
Southwestern Medical Center
Parkland Memorial Hospital
5323 Harry Hines Blvd.
General Surgery - E7.126
Dallas TX 75235-9031
USA

A. Rutledge
School of Pharmacy
State University of New York at
Buffalo
C126 Cooke-Hochstetter Complex
Buffalo NY 14260
USA

M. L. Sacchetti
Dept. of Neurological Sciences
University "La Sapienza"
Viale dell'Università 30
00185 Rome
Italy

A. Sagalowsky
University of Texas
Southwestern Medical Center
Parkland Memorial Hospital
5323 Harry Hines Blvd.
General Surgery - E7.126
Dallas TX 75235-9031
USA

Z. F. Sandor
University of Texas
Southwestern Medical Center
Parkland Memorial Hospital
5323 Harry Hines Blvd.
General Surgery - E7.126
Dallas TX 75235-9031
USA

V. B. Schini
Center for Experimental
Therapeutics
Baylor College of Medicine
One Baylor Plaza
Houston TX 77030
USA

K. P. Scholz
Dept. of Pharmacological
and Physiological Sciences
University of Chicago
947 E. 58th-St.
Chicago IL 60637
USA

A. Schwartz
Dept. of Pharmacology and
Cell Biophysics
University of Cincinnati
College of Medicine
231 Bethesda Avenue
Cincinnati OH 45267-0575
USA

A. Scriabine
Miles Inc.
400 Morgan Lane
West Haven CT 06516
USA

M. Soma
Institute of Pharmacological Sciences
University of Milan
Via Balzaretti 9
20133 Milan
Italy

D. Stepp
Department of Physiology and Biochemistry
Medical College of Pennsylvania
3300 Henry Avenue
Philadelphia PA 19129
USA

C. K. Stone
Dept. of Medicine
Cardiology Section
Univ. of Wisconsin
Medical School
Madison WI 53792
USA

H. Tasaki
Department of Physiology
and Biochemistry
Medical College of Pennsylvania
3300 Henry Avenue
Philadelphia PA 19129
USA

F. Thimm
Physiological Institute
University of Freiburg
Hermann-Herder Str. 7
W-7800 Freiburg
Germany

D. Toni
Dept. of Neurological Sciences
University "La Sapienza"
Viale dell'Università 30
00185 Rome
Italy

M. Trabucchi
Dipartimento di Medicina Spearimentale e
Scienze Biochimiche
II Università di Roma
Via Raimondi 6
20156 ROMA
Italy

D. J. Triggle
School of Pharmacy
State University of New York at
Buffalo
C126 Cooke-Hochstetter Complex
Buffalo NY 14260
USA

T. N. Tulenko
Department of Physiology
and Biochemistry
Medical College of Pennsylvania
3300 Henry Avenue
Philadelphia PA 19129
USA

M. M. Usowicz
Department of Pharmacology
University of Bern
Friedbuhlstrasse 49
3010 Bern
Switzerland

C. Van Dyke
Department of Pharmacology and Toxicol-
ogy
West Virginia University
Health Sciences North
Morgantown WV 26506
USA

K. Van Dyke
Department of Pharmacology
and Toxicology
West Virginia University
Health Sciences North
Morgantown WV 26506
USA

J. Van Reempts
Janssen Research Foundation
Turnhoutseweg 30
2340 Beerse
Belgium

P. A. van Zwieten
Departments of Pharmacotherapy and
Cardiology
Academic Medical Center
University of Amsterdam
Meibergdreef 15
1105 AZ Amsterdam
The Netherlands

P. M. Vanhoutte
Center for Experimental
Therapeutics
Baylor College of Medicine
One Baylor Plaza
Houston TX 77030
USA

L. Ver Donck
Life Sciences
Janssen Research Foundation
Turnhoutseweg 30
2340 Beerse
Belgium

F. Verdonck
University of Leuven
Campus Kortrijk
8500 Kortrijk
Belgium

F. Verheyen
Dept. of Cardiovascular Pharmacology
Janssen Research Foundation
Turnhoutseweg 30
2340 Beerse
Belgium

O. Visioli
University of Brescia
Spedali Civili
P. le Spedali Civili 1
25231 Brescia
Italy

P. Volpe
Centro di Studio per la Biologia e
Fisiopatologia
Muscolare del CNR
Istituto di Patologia Generale
Università di Padova
Via Trieste 75
35121 Padova
Italy

M. Wibo
Univ. Catholique de Louvain
Lab. de Pharmacologie
UCL 7350
Avenue E. Mounier 73
1200 Bruxelles
Belgium

Z. Ye
Department of Pharmacology
and Toxicology
West Virginia University
Health Sciences North
Morgantown WV 26506
USA

W. Zheng
School of Pharmacy
State University of New York at Buffalo
C126 Cooke-Hochstetter Complex
Buffalo NY 14260
USA

STRUCTURE-FUNCTION STUDIES OF THE
VOLTAGE-DEPENDENT CALCIUM CHANNELS

Olga Krizanova, Philippe Lory, and Arnold Schwartz

Department of Pharmacology and Cell Biophysics
University of Cincinnati, College of Medicine
231 Bethesda Avenue
Cincinnati, Ohio 45267-0575

Voltage-dependent calcium channels (VDCCs) form the basis for fundamental cellular activity and regulation. The importance of these calcium channels lies in their ability to link electrical activity of membranes to biological effects.

VDCCs are multipeptide structures that conduct calcium upon depolarization of the membrane and are affected by special classes of drugs or toxins (calcium antagonists, ω-conotoxins). The bulk of our knowledge about these channels comes from indepth studies of the L-type VDCCs, so named because of its specific electrophysiological characteristics and sensitivity to organic calcium antagonists. The role of the calcium channel subunits is studied by coexpression of various combinations of the α_1, α_2/δ, ß, and γ subunits that comprise the L-type VDCCs. Structural characteristics of these subunits provide important functional information concerning channel activity.

EXPRESSION SYSTEMS

Although structure-function studies are still in the preliminary stages of investigation, they have already revealed exciting fundamental information. Several expression systems have been established. Among these are *Xenopus laevis* oocytes, primary cell cultures from dysgenic mice, and certain mammalian cell lines.

Of the systems available, the *Xenopus laevis* oocyte system is the most widely used for transient expression of mRNA (1). After mRNA injection, oocytes correctly perform many posttranslational modifications, such as proteolytic modification, glycosylation or acetylation, and the expressed proteins can be studied with a variety of electrophysiological and biochemical techniques. Recently using vaccinia virus, cDNA has been used instead of mRNA (2). This modification can markedly facilitate expression experiments. The α_1 subunits of L-type VDCCs from brain (3), smooth muscle (4), and heart (5) have been expressed in

1

T. Godfraind et al. (eds.), Calcium Antagonists, 1–8.
© 1993 *Kluwer Academic Publishers and Fondazione Giovanni Lorenzini.*

oocytes. It is curious that functional expression of the skeletal muscle α_1 subunit has thus far been unsuccessful.

The primary mouse dysgenic (*mdg*) cell culture is derived from mouse dysgenic skeletal muscle. Such muscle is not able to contract because it lacks the α_1 subunit of the L-type VDCCs. All other channel subunits (α_2/δ, ß, γ) however, have been identified in these cells. These cells are being used to investigate the important parts of the α_1 subunit (6,7).

The *mdg* cell culture can be easily contaminated with fibroblasts. For this and other reasons, an *mdg* cell line has been created. The advantage of the cell line is that it can be grown indefinitely as a single cell type. The immortalized *mdg* cells express the same properties as their primary culture counterparts, viz., a lack of contraction, slow Ca^{2+} current, and triadic differentiation (8).

Stable transfection in mammalian cell lines has been used for studies of the skeletal muscle α_1 subunit. For this purpose, mouse L-cells have been established as a convenient system, because of a lack of all calcium channel subunits (9,10,11). Transfection of L-cells with the α_1 subunit of skeletal muscle results in expression of 600-1000 dihydropyridine binding sites per cell (9). Accordingly, this system forms a basis for biochemical and electrophysiological studies.

THE ROLE OF SPECIFIC REGIONS OF THE α_1 SUBUNIT

For evaluation of the most important regions on α_1 which may be involved in tissue-specific difference of function, the construction of chimeras has been established, in a series of elegant experiments (6,7). These experiments were carried out in primary cultures from mdg myotubes, which, as mentioned, represent a convenient "tool" for investigating the α_1 subunit of L-type VDCCs. The first chimeras were constructed between the skeletal and cardiac muscle α_1. Two major results were obtained:

a)   The intracellular loop between motifs II and III is a major determinant of skeletal muscle E-C coupling (6). The replacement of this region with the corresponding region from the heart α_1 abolished skeletal muscle E-C coupling. Replacement of the loop between motifs I and II is possibly less important and the amino and carboxyl terminal regions are probably unimportant in skeletal muscle E-C coupling. It is possible that the I-II loop is more functionally interchangeable between the cardiac and skeletal muscle α_1 subunits than is the II-III loop.

b)   Motif I is responsible for Ca^{2+} channel activation kinetics. This region determines whether the chimeric Ca^{2+} channel exhibits slow (skeletal muscle-like) or rapid (cardiac-like) activation (7). It remains to be determined which specific areas of motif I are responsible for the difference in activation kinetics between skeletal and cardiac muscle L-channels (fig. 1).

A typical feature of the α_1 subunit of the L-type VDCCs is the binding of Ca^{2+} antagonists. Because the calcium antagonists differ widely in their structure, it is likely that the binding sites for the different classes of Ca^{2+} antagonists are not the same. In recent papers, using azido-labeled ligands, a binding site for phenylalkylamines (12) was found on the intracellular side of the membrane, close to the carboxy terminus of the α_1 subunit. Controversial results have been reported thus far concerning the DHP binding site (13,14), not allowing a precise localization of the site at this time. Further experiments using site-directed mutagenesis, chimeras, and point mutations are required for evaluating all of the drug receptor sites.

THE ROLE OF OTEHR CALCIUM CHANNEL SUBUNITS

As stated above, the L-type VDCCs is a multisubunit complex composed of 5 subunits: α_1, α_2/δ, ß, and γ (fig. 1). The α_1 subunit contains the basic properties of a functional calcium channel, whereas the other subunits have modulatory functions.

The Role of the α_2/δ Subunit Complex

A possible role for the α_2/δ subunit complex has been shown by heterologous and homologous coexpression with the α_1 subunit. For example, heterologous coexpression of the skeletal muscle α_2 with the cardiac α_1 (5) or a brain DHP-insensitive α_1 (3) in *Xenopus* oocytes increased the amplitude without changes in kinetics of the Ca^{2+} channel current. Homologous coexpression of the α_1 and α_2/δ subunits from skeletal muscle, which was done in our laboratory using L-cells, resulted in a slight increase in maximal binding capacity of PN 200-110 (isradipine) without a significant change in its affinity. In addition, the peak currents were slightly higher than that observed with the α_1 subunit alone (10). Although the exact role of the α_2/δ subunit complex is not known, it is likely that it facilitates incorporation of the α_1 subunit into the membrane. This is analogous to a proposed role for the ß subunit of the Na^+/K^+ ATPase. In this system, the ß subunit is a glycosylated protein with a single membrane-spanning region, which may be required for the insertion of the α subunit of Na^+/K^+ ATPase into the membrane (15). Interestingly, recent studies have shown that the α_2/δ complex may have only one membrane-spanning region (fig. 1)(16).

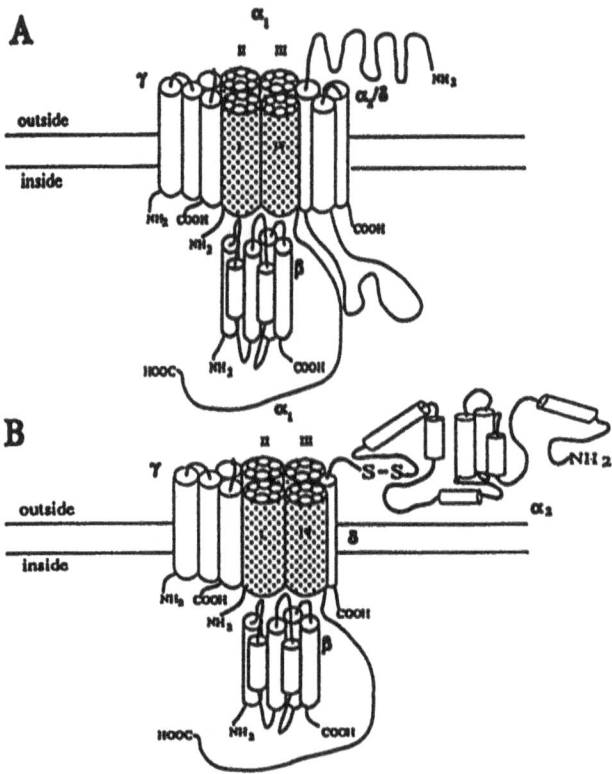

FIG. 1. Two models of the L-type VDCCs. The α_1 subunit, the major functional component of the L-type VDCCs is a highly hydrophobic protein, composed of 24 transmembrane segments grouped into four motifs. The organization of the motifs is not yet known, although it is likely that they are organized like a four-leaf clover. The γ and the α_2/δ subunits probably serve as stabilizing components for the α_1 subunit. The γ subunit is a transmembrane protein containing four hydrophobic segments and probably is directly attached to the α_1 subunit. The α_2/δ subunit complex was originally proposed to have three membrane spanning regions according to the amino acid hydropathy plot (A). Recently it has been suggested that only the δ subunit possesses a membrane spanning region. The α_2 subunit is probably localized on the extracellular side of the membrane and attached to the δ-subunit by disulphide bonds (B). The ß subunit is hydrophilic and is located on the cytoplasmic side of the membrane. The exact localization is not yet determined, although it is attractive to suggest that it is attached to motifs I and IV, as well as to the carboxy terminal region of the α_1. The regulatory role of this subunit has been proven (3,10,11).

The Role of the ß Subunit

The ß subunit has been shown by several laboratories to modulate the activity of the α_1 subunit not only in skeletal muscle (10,11), but also in brain (3) and heart (5). Homologous coexpression of the skeletal L-type VDCCs α_1 subunit with the ß subunit from the same tissue, resulted in a significant increase in DHP binding, whereas the affinity of the DHP binding site remained unchanged (10,11). Coexpression of the α_1 and ß subunits together with the γ or the γ and α_2 revealed a qualitatively similar effect to the α_1ß construct, although not as dramatic as the former.

The pharmacological response of the α_1ß-related Ca channel current to Bay K 8644 was either markedly attenuated or absent. Peak currents were more than two times lower in the cells coexpressing α_1ß compared to the α_1 alone. Of considerable interest, the ß subunit dramatically accelerated activation and inactivation kinetics of the calcium current (10). The results on activation kinetics have been confirmed by Lacerda and coworkers (11), although the data are somewhat difficult to interpret since the experiments presented were carried out in the presence of Bay K 8644. Taken together, these results suggest that the cloned skeletal calcium channels can be engineered to resemble the native skeletal muscle channel (10,11). It should be remembered that although fast skeletal muscle contains an abundance of L-type VDCCs and calcium antagonist binding sites, even high doses of these drugs produce no effect in animals or clinically.

Interestingly, the skeletal ß subunit is also able to modulate the α_1 current from a brain DHP-insensitive VDCCs by increasing the peak of I_{Ba}. A further increase was observed after coinjection of α_1, α_2 and ß together (3). All of the results discussed here stress the importance of the ß subunit in regulating the α_1 subunit of calcium channels.

The Role of the γ Subunit

It has been suggested by recent experiments carried out in the *Xenopus* oocyte system (17), that the skeletal γ subunit may regulate the inactivation process of a cloned rabbit cardiac α_1 subunit, making it faster and perhaps more sensitive to voltage. This appears to be a voltage- rather than current-dependent phenomenon, since in the presence of the γ subunit barium current inactivated at voltages that do not cause any current flow through the channel (17).

ALTERNATIVE SPLICING

Comparison of the cloned α_1 subunits has recently revealed considerable diversity, which arises from both multiple genes and

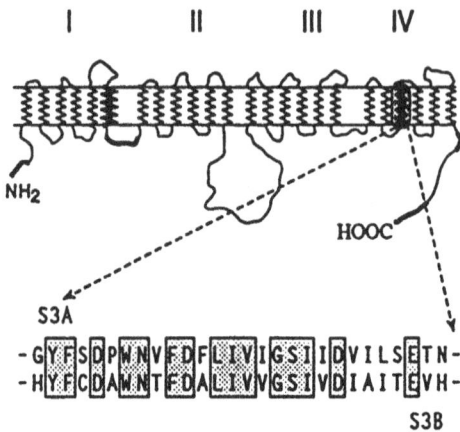

FIG. 2. A model of the α_1 subunit based on hydropathy analysis, showing regions of alternative splicing. Several regions of the α_1 subunit have been found to be alternatively spliced. The variable sequences have been found at the amino and carboxyl terminal ends, on the loop between motifs I and II as well as IS6 and IVS3 segments (bold line). Mutually exclusive alternative splicing of IVS3 region illustrates the presence of a developmentally regulated switch in cardiac tissue. Comparison of the IVS3 sequence found in heart from adult rats (S3B) shows some identities with the form from newborn rat heart (S3A). Amino acids asparagine (D) and glutamine (E) are highly conserved in both fetal (S3A) and adult (S3B) form. Although the function of IVS3 region is not yet known, these charged amino acids can participate in the function of the voltage-dependent calcium channels.

alternative splicing (fig. 2). The regions of interest include the amino and carboxyl terminal ends, the loop between motifs I and II and IVS3 and IS6 segments. The physiological relevance of these alternative spliced products is at present unknown. One of the more interesting spliced regions is the third membrane-spanning region of IV motif (IVS3), which is found in heart (18), smooth muscle (19), and brain (3). Recent experiments in our laboratory using S1 nuclease protection analysis have revealed that both variant forms of the IVS3 are equally expressed in the newborn and fetal rat heart, while only a single isoform is expressed in the adult rat heart. The phenomenon is the product of mutually exclusive alternative splicing and illustrates the presence of a developmentally regulated switch in cardiac tissue (fig. 2) (18). The proof of the splice mechanism was revealed by cloning and sequencing the exons involved as well as the up and downstream regions (18,20). The cartoon illustrated in fig. 2, present our current state of information concerning the regions of the α_1 subunit that are divergent, perhaps giving rise to tissue specific functions.

CONCLUSIONS

The rapid progress in calcium channel research in the past few years has resulted in an explosion of information. Current research is now concentrated on the connection between structure and function. The result of these experiments should reveal specific roles for various regions of the channel proteins as well as for the function of the subunits and how the channels are regulated. Additional work on this and other aspects should reveal the interesting and unusual interplay of subunits in controlling the activity of calcium channels. In the future we should be able to localize the exact binding domains of all types of Ca^{2+} modulator drugs, and hopefully, provide directions for new chemical entities that can be designed specifically for a tissue-specific isoform.

ACKNOWLEDGEMENTS

Authors wish to thank Dr. Gyula Varadi and David Schultz for reading the manuscript and providing valuable comments, L. Wendelmoot for help with the manuscript as well as to G. Kraft for the help with figures. Supported in part by National Institutes of Health Grants PO1-HL22619-14 (A.S.); T32HL07382; 1R37HL43231-01 (A.S.) P.L. was supported by a Postdoctraral Fellowship from INSERM, France.

REFERENCES

1. Lory P., Rassendren F.A., Richard S., Tiaho F. and Nargeot J. (1990): *J. Physiol. Lond.*, 429: 95-112.
2. Yang X.Ch., Karschin A., Labarca C., Elroy-Stein O., Moss B., Davidson N. and Lester H.A. (1991): *FASEB J.,* 5: 2209-2216.
3. Mori Y., Friedrich T., Kim M.S., Mikami A., Nakai J., Ruth P., Bosse E., Hofmann F., Flockerzi V., Furuichi T., Mikoshiba K., Imoto K., Tanabe T. and Numa S. (1991): *Nature,* 350: 398-402.

4. Biel M., Hullin R., Freudner S., Singer D., Dascal N., Flockerzi V. and Hofmann F. (1991): *Eur. J. Biochem.,* 200: 81-88.

5. Mikami N., Imoto K., Tanabe T., Niidome T., Mori Y, Takeshima H., Narumiya S. and Numa S. (1989): *Nature,* 340: 230-233.

6. Tanabe T., Beam K.G., Adams, B.A., Niidome T. and Numa S. (1990): *Nature,* 346: 567-569.

7. Tanabe T., Adams B.A., Numa S. and Beam K.G. (1991): *Nature,* 352: 800-803.

8. Pincon-Raymond M., Vicart P., Bois P., Chassande O., Romey G., Varadi G., Li Z.L., Lazdunski M., Rieger F. and Paulin D. (1991): *Developmental Biology,* 148: 517-528.

9. Perez-Reyes E., Kim H.S., Lacerda A.E., Horne W., Wei X., Rampe D., Campbell K.P., Brown A.M. and Birnbaumer L. (1989): *Nature,* 340: 233-236.

10. Varadi G., Lory P., Schultz D., Varadi M. and Schwartz A. (1991): *Nature,* 352: 159-162.

11. Lacerda A.E., Kim S.H., Ruth P., Peres-Reyes E., Flockerzi V., Hofmann F., Birnbaumer L. and Brown A.M. (1991): *Nature,* 52: 527-530.

12. Striessnig J., Glossmann H. and Catterall W.A. (1990): *Proc. Natl. Acad. Sci. USA,* 87: 9108-9112.

13. Regulla S., Schneider T., Nastainczyk W., Meyer H.E. and Hofmann F. (1991): *The EMBO Journal,* 10: 45-49.

14. Nakayama H., Taki M., Striessnig J., Glossmann H., Catterall W.A. and Kanaoka Y. (1991): *Proc. Natl. Acad. Sci, USA,* 88: 9203-9207.

15. Noguchi S., Mishina M., Kawamura M. and Numa S. (1987): *FEBS Lett.,* 225: 27-32.

16. Jay S.D., Sharp A.H., Kahl S.D., Vedvick T.S., Harpold M.M. and Campbell K.P. (1991): *J. Biol. Chem.,* 266: 3287-3293.

17. Singer D., Biel M., Lotan I., Flockerzi V., Hofmann F. and Dascal N. (1991): *Science,* 253: 1553-1557.

18. Diebold R., Koch W.J., Ellinor P.T., Wang J.J., Muthuchamy M., Wieczorek D.F. and Schwartz A. (1991): *Proc. Natl. Acad. Sci. USA.,* (in press).

19. Koch W.J., Ellinor P.T. and Schwartz A. (1990): *.J Biol. Chem.,* 265: 17786-17791.

20. Snutch T.P., Tomlison W.J., Leonard J.P. and Gilbert M.M. (1991): *Neuron,* 7: 45-57.

REGULATION OF CALCIUM CHANNELS BY PROTEIN PHOSPHORYLATION

M.M. Hosey, R.M. Brawley, C.F. Chang, L.M. Gutierrez, J. Ma*,
E. Rios*, and C. Mundina-Weilenmann

Northwestern University Medical School
303 E. Chicago Avenue, Chicago, Illinois 60611

*Rush University Medical College
1750 W. Congress Parkway, Chicago, Illinois 60612

There are many types of voltage-activated Ca^{2+} channels that are found in a wide variety of excitable and nonexcitable cells. While voltage is the primary factor that determines whether these Ca^{2+} channels are opened or closed, many Ca^{2+} channels are regulated by receptor-dependent events (1-3). An example of this type of regulation can be seen with the dihydropyridine (DHP)-sensitive Ca^{2+} channels. In cardiac cells, activation of ß-adrenergic receptors leads to an increase in the probability that these channels will open in response to a given membrane depolarization (1-3). This "activation" of the channels is believed to occur by two distinct regulatory processes. The first and most well understood is thought to involve regulation of the channels by protein phosphorylation. In this case, ß-adrenergic receptor-mediated increases in cyclic AMP (cAMP) and subsequent activation of cAMP-dependent protein kinase (PKA) is thought to result in phosphorylation of either the channel proteins or associated regulatory proteins (1-3). The phosphorylation event is believed to result in the increased probability of channel opening. The second receptor-dependent regulatory mechanism involves receptor-mediated activation of stimulatory GTP binding regulatory proteins (G proteins) which in turn are believed to directly interact with and activate the channels (4,5). Ca^{2+} channels in other types of cells are also believed to be regulated by a variety of G proteins and by protein phosphorylation catalyzed by a variety of protein kinases (1-4).

Most of the evidence that supports both the phosphorylation and the G protein hypotheses comes from electrophysiological studies. We wished to provide further insight into these processes by defining the biochemical and molecular events that are responsible for receptor-mediated regulation of Ca^{2+} channels by protein phosphorylation. This chapter summarizes our findings.

T. Godfraind et al. (eds.), Calcium Antagonists, 9–14.

The most widely used model for biochemical studies of Ca^{2+} channels is the dihydropyridine (DHP)-sensitive Ca^{2+} channel from skeletal muscle. This model has been the most well characterized because this class of Ca^{2+} channels can be biochemically identified as high-affinity DHP receptors, and the transverse tubule (T-tubule) membranes of skeletal muscle contain the highest density of these receptors. The DHP-sensitive Ca^{2+} channels from skeletal muscle are multisubunit proteins (3,6). The 165 kDa α_1 subunit is the channel-forming pore and contains the receptors for dihydropyridines and other Ca^{2+} channel modulators. The accessory subunits include the 175 kDa α_2/δ subunits, the 52 kDa ß subunit and a 32 kDa γ subunit (3,6). We have performed biochemical experiments to determine which subunits are the preferential targets of various protein kinases when the channels are phosphorylated *in vitro* and in intact cells, as well as biochemical and electrophysiological reconstitution studies to determine the consequences of phosphorylation.

WHICH CHANNEL SUBUNITS ARE SUBSTRATES FOR CYCLIC AMP-DEPENDENT PROTEIN KINASE?

The first approach was to phosphorylate the skeletal muscle DHP-sensitive channels in the T-tubule membranes *in vitro* with the purified catalytic subunit of PKA and to analyze for phosphorylation of channel subunits. The results of these studies established that the α_1 subunit is an excellent substrate for PKA *in vitro* (7,8). The rate of phosphorylation of the α_1 subunit by PKA is similar to that for well-characterized substrates of this kinase (7) and the extent of phosphorylation is 2-2.5 mol phosphate/mol protein (7,8). The ß subunit is also a substrate for PKA but under the conditions utilized we found (8) that the extent of phosphorylation was much lower ($\approx$0.5 mol phosphate/mol protein). Phosphorylation of the α_1 subunit by PKA occurs on serine (90%) and threonine (10%) residues and phosphopeptide mapping indicates that multiple sites are phosphorylated by this kinase (7).

FUNCTIONAL CONSEQUENCES OF PHOSPHORYLATION: BIOCHEMICAL STUDIES

Biochemical reconstitution studies

We determined how the observed phosphorylation catalyzed by PKA affected channel function by using *in vitro* reconstitution strategies. In one approach we developed a biochemical assay of reconstituted channels that allowed us to observe "macroscopic" properties of the channels (9,10).

This assay relies on the reconstitution of purified channels into liposomes containing the Ca^{2+}-sensitive dye fluo-3 (9,10). The channels reconstituted with this approach exhibit Ca^{2+} influx that is sensitive to membrane potential created by valinomycin and K^+ gradient (9,10). The reconstituted channels display their expected pharmacological properties: they are activated by the DHP agonist Bay K 8644 and inhibited by the DHP blocker (+)PN 200-110 (izradipine), as well as by verapamil and diltiazem (10). In addition, the reconstituted channels retain the proper reactivity to the stereoisomers of Sandoz-202-791: (+)202-791 activates the reconstituted channels with a K_D similar to that observed in intact cell studies, and (-)202-791 inhibits the channels with a K_i similar to that observed with native channels (10). Thus this biochemical reconstitution system provides a convenient assay of channels that appear to retain the properties observed with channels in native membranes.

Roles of subunits

The cDHN for the α_1 subunit of the skeletal muscle Ca^{2+} channel predicts that it should have a molecular weight of 210 kDa (11). Can a 165 kDa α_1 subunit support channel activity? We used site-specific antipeptide antibodies raised against portions of the α_1 subunit to characterize the size of the α_1 subunit. The channels used in reconstitution studies possessed an α_1 subunit of 165 kDa and did not appear to contain a larger α_1 subunit of $\approx$210 kDa (9), suggesting that channel activity with properties similar to those observed in intact cells can be supported with an α_1 subunit of 165 kDa. We also determined if fully functional channels could be obtained if the α_2/δ subunits were removed. Reconstituted channels that are 85% depleted in the α_2/δ subunits show a significant decrease in the initial rate of Ca^{2+} influx induced by valinomycin but retain responsiveness to Bay K 8644 and (+)PN 200-110 (9). When the separated α_2 and δ subunits are added back to the α_1 subunit-containing preparation, the channels exhibit their normal rate of Ca^{2+} influx (9). These results demonstrate that the DHP-sensitive Ca^{2+} channels from skeletal muscle require the presence of the α_2/δ complex in stoichiometric amounts to exhibit full activity.

Effects of PKA on the activity of
the reconstituted channels in liposomes

We combined assays of subunit phosphorylation and reconstitution to determine how phosphorylation of the channels by PKA affects channel properties (8). We observed that phosphorylation by PKA resulted in a $\approx$2-fold increase in the rate and extent of Ca^{2+} influx through the reconstituted channels (8). In parallel phosphorylation studies we observed that this activation occurs under conditions in which the α_1 subunit is the preferred target of PKA, and the ß subunit is less extensively phosphorylated (8).

ELECTROPHYSIOLOGICAL ANALYSIS OF EFFECTS OF PKA

Electrophysiological reconstitution studies

The biochemical reconstitution studies provide a convenient assessment of the "macroscopic" properties of the channels, but in order to provide more insight into how phosphorylation by PKA activates the channels we reconstituted the T-tubule membranes containing the channels into planar lipid bilayers and used an electrophysiological approach to characterize channel properties (12,13). The solutions used for these studies were designed to resemble those used in whole cell recordings and contained MgATP, Ba^{2+} as the charge carrier in the trans chamber so that inward currents could be recorded, and the lowest concentration of Bay K 8644 (0.1 μM) that allowed for consistent detection of channels (12).

The reconstituted channels exhibit a voltage-dependence of activation (12). Using a holding potential of -80 mV and stepping to more positive potential we found that control, nonphosphorylated channels exhibit a threshold of activation of $\approx$-40 mV and half maximal activation is observed at -27 mV (12). The reconstituted channels exhibit a very low open probability (0.08) even when the channels are maximally activated by voltage. This finding has important implications concerning the prior suggestion that only 5% of the DHP receptors may be functional Ca^{2+} channels in skeletal muscle (14). This calculation assumed that the maximal open probability is much higher than that which we have observed. However, if the channels that are maximally activated by voltage show a very low tendency to open, then the interpretation is different and suggests that a much higher percentage of skeletal muscle dihydropyridine receptors can be function as Ca^{2+} channels.

Effects of PKA on channels in bilayers

The addition of PKA to the cis side of the bilayer chamber resulted in several changes in channel properties: a) open probability increased by 50% at all voltages more positive than -50 mV; b) there was a leftward shift by -7 mV in the voltage dependence of activation, i.e., the phosphorylated channels were activated at more negative voltages; c) there was a two-fold decrease in the rate of inactivation; and d) phosphorylation increased the availability of the channels (13). These results demonstrate that activation of the channels by PKA-mediated phosphorylation involves several molecular transitions and the overall result is an increased readiness of the channels to open in response to depolarization (13).

PHOSPHORYLATION OF CHANNELS IN INTACT CELLS

The results described above were obtained in *in vitro* studies of phosphorylation. In order to determine if similar events occur in intact cells, we analyzed the effects of cAMP-elevating agents on the phosphorylation of DHP-sensitive Ca^{2+} channels in newborn chick skeletal muscle (10). In situ treatment with the ß-adrenergic receptor agonist isoproterenol resulted in the phosphorylation of the 165 kDA α_1 subunit in the intact cells, as evidenced by a marked decrease in the ability of the α_1 polypeptide to serve as a substrate in *in vitro* back phosphorylation reactions with $[\gamma\text{-}^{32}P]ATP$ and the purified catalytic subunit of cAMP-dependent protein kinase (10). The phosphorylation of the 52 kDa ß subunit was not affected by the isoproterenol treatment (10). The effects of isoproterenol were time- and concentration-dependent and were mimicked by other cAMP-elevating agents. To test for functional effects of the observed phosphorylation, purified channels were reconstituted into liposomes containing entrapped fluo-3, and depolarization-sensitive and dihydropyridine-sensitive Ca^{2+} influx was measured. Channels from isoproterenol-treated muscle exhibited an increased rate and extent of Ca^{2+} influx compared to control preparations (10). The effects of isoproterenol pretreatment could be mimicked by phosphorylating the channels with cAMP-dependent protein kinase *in vitro* (10). these results demonstrate that the α_1 subunit of the DHP-sensitive Ca^{2+}-channels is the primary target of cAMP-dependent phosphorylation in intact muscle, and that the phosphorylation of this protein leads to activation of channel activity.

EFFECTS OF OTHER PROTEIN KINASES

We also determined which subunits are the preferential targets of other protein kinases that have been suggested to regulate channel activity (8,15). Using as substrates channels present in purified T-tubule membranes, we found that protein kinase C (PKC) (8,15) and a multifunctional Ca^{2+}/calmodulin-dependent protein kinase (CaMK) (7,8) preferentially phosphorylated the 165 kDa α_1 subunit *in vitro* to an extent that was 2-3 times greater than the 52 kDa ß subunit. A protein kinase endogenous to the T-tubule membranes preferentially phosphorylated the ß peptide and showed little activity toward the α_1 subunit, however, the extent of the phosphorylation was low (8). Reconstitution of partially purified channels into liposomes was used to determine the functional consequences of phosphorylation by these kinases. Phosphorylation of channels by PKC resulted in an activation of the channels that was observed as a ≈2-fold increase in both the rat and extent of Ca^{2+} influx (8). However, phosphorylation of channels by either the CaMK or the endogenous kinase in T-tubule membranes had no effect (8). Phosphorylation did not affect the sensitivities of the channels toward the DHPs. Taken together with the results obtained with PKA, the results demonstrate that the α_1 subunit is the preferred substrate of PKA, PKC, and CaMK when the channels are phosphorylated in the membrane-bound state and that phosphorylation of the channels by PKA and PKC, but not by CaMK or an endogenous T-tubule membrane protein kinase, results in activation of the DHP-sensitive Ca^{2+} channels from skeletal muscle.

REFERENCES

1. Hess P. (1990): *Annu. Rev. Neurosci.*, 13: 337-356.
2. Tsein R.W., Bean B.P., Hess P., Lansman J.B., Nilius B. and Nowycky M.C. (1986): *J. Mol. Cell Cardiol.*, 18: 691-710.
3. Hosey M.M. and Lazdunski M. (1988): *J. Memb. Biol.*, 104: 81-105.
4. Brown A.M. and Birnbaumer L. (1990): *Annu. Rev. Physiol.*, 52: 197-213.
5. Brown A.M. (1991): *5th International Symposium on Calcium Antagonists: Pharmacology and Clinical Research,* September 25-28, 1991, Houston (abstract).
6. Catterall W.A. (1991): *Science,* 253: 1499-1500.
7. O'Callahan C.M. and Hosey M.M. (1988): *Biochem.,* 27: 6071-6077.
8. Chang C.F., Gutierrez L.M., Mundina-Weilenmann C. and Hosey M.M. (1991): *J. Biol. Chem.,* 266: 16395-16400.
9. Gutierrex L.M., Brawley R.M. and Hosey M.M. (1991): *J. Biol. Chem.,* 266: 16387-16394.
10. Mundina-Weilenmann C., Chang C.F., Gutierrex L.M. and Hosey M.M. (1991): *J. Biol. Chem.,* 266: 4067-4073.
11. Tanabe T., Takeshima H., Mikami A., Flockerzi V., Takahashi H., Kangawa K., Kojima M., Matsuo H., Hirose T. and Numa S. (1987): *Nature,* 328: 313-318.
12. Ma J., Mundina-Weilenmann C., Hosey M.M. and Rios E. (1991): *Biophys. J.,* 60: 890-901.
13. Mundina-Weilenmann C., Ma J., Rios E. and Hosey M.M. (1991): *Biophys. J.,* 60: 902-909.
14. Schwartz L.M., McCleskey E.W. and Almers W. (1985): *Nature,* 314: 747-751.
15. O'Callahan C.M., Ptasienski J. and Hosey M.M. (1989): *J. Biol. Chem.,* 263: 17342-17349.

INTERACTIONS OF LACIDIPINE AND OTHER CALCIUM CHANNEL DRUGS WITH BIOLOGICAL MEMBRANES: A STRUCTURAL MODEL FOR RECEPTOR/DRUG BINDING UTILIZING THE MEMBRANE BILAYER

Leo G. Herbette, Giovanni Gaviraghi*, Robert J. Dring, Lawrence J. Briggs, and R. Preston Mason

Departments of Radiology, Medicine, Biochemistry, and the Biomolecular Structure Analysis Center University of Connecticut Health Center Farmington, Connecticut 06032 *Glaxo Research Laboratories, Verona, Italy

The analysis of calcium channel drugs that bind to membrane-associated receptors must take into account the local environment where the binding event occurs. The partitioning of these drugs in an isotropic two-phase bulk solvent system, such as octanol/buffer, apparently is not a good model for their partitioning into model or in native membranes where they exert their effects. Knowledge of these membrane-based partition coefficients then necessitates reanalysis of other physical, chemical, and functional parameters.

MOLECULAR BASIS FOR DRUG BINDING TO MEMBRANE RECEPTORS

In contrast to ligand binding directly from the aqueous, extracellular environment, there is experimental support for highly lipophilic drugs, such as the calcium channel drugs to bind to their receptors via the membrane bilayer (1). A "membrane bilayer pathway" has been postulated for the binding of 1,4-dihydropyridine (DHP) Ca^{+2} channel agonists and antagonists to voltage-dependent Ca^{+2} channel blockers in cardiac and smooth muscle sarcolemma. This would occur in a two-step process. First, the drug molecule must partition to a well-defined, energetically favorable location, orientation, and conformation in the membrane bilayer before laterally diffusing to an intrabilayer receptor binding site (2).

The probability that DHP interacts with the bulk lipid phase in the cardiac sarcolemma is high in light of high partition coefficients measured for several DPHs ($Kp > 10^3$) and the very low receptor density of approximately one receptor site per square micron in the cardiac

15

T. Godfraind et al. (eds.), Calcium Antagonists, 15–23.

sarcolemmal membrane (3). Diffusion-limited rates calculated for a membrane pathway are approximately 3 orders of magnitude greater that those for an aqueous approach in which the drug reaches the receptor by diffusion through the bulk solvent (2). The two-dimensional component of this process, lateral diffusion through the bilayer, has a significant rate advantage if the ligand has the appropriate location and orientation for binding to the receptor site (4).

Experimental support for the first step of this pathway, namely DHP partitioning to a discrete, time-averaged location in the membrane bilayer, has been shown using small-angle x-ray and neutron diffraction (1). The second step of the membrane bilayer pathway, namely DPH lateral diffusion through the membrane, was measured using florescence redistribution after photobleaching (FRAP). Using an active rhodamine-labeled DHP analog, the microscopic rate of drug lateral diffusion was measured in canine cardiac sarcolemmal lipid multilayers over a wide range of relative humidities. At the highest relative humidity, the rate of lateral diffusion for the DHP was identical to that measured for phospholipid analogs, namely 3.8×10^{-8} cm^2/sec (5). These rapid rates of diffusion suggest that the overall binding rate by a membrane bilayer pathway is generally not rate-limited by the drug's diffusion through the membrane.

DRUG PARTITION COEFFICIENTS INTO BIOLOGICAL MEMBRANES

Data in table 1 highlights the fact that DHP interactions with both model and biological membranes are complex and cannot be mimicked by isotropic model systems, e.g. octanol/buffer. The charged DHP Ca^{+2} channel antagonist, amlodipine, is a case in point. The partition coefficient measured in octanol/buffer, $K_{P[iso]}$, for amlodipine was nearly an order of magnitude lower than that of the uncharged DHP, nimodipine. By contrast, its partition coefficient K_p in a biological membrane, $K_P[mem]$, is over three-fold higher than that of nimodipine.

TABLE 1.   1,4-Dihydropyridine Partition
               Coefficients into Biological Membranes
                       and Octanol/Buffer

Drug	Biological Membranes[a] (Sarcoplasmic Reticulum)	Octanol/Buffer
Bay P 8857	125,000	40
Iodipine	26,000	--
Amlodipine	19,000	30
Nisoldipine	13,000	40
Bay K 86444	11,000	290
Nimodipine	6,300	260
Nifedipine	3,000	--

[a]Similar values were obtained with sarcoplasmic reticulum lipid extracts, indicating a primary interaction of the drug with the membrane bilayer component of these biological membranes. Data from reference 1.

EFFECTS OF CHOLESTEROL ON MEMBRANE PARTITIONING OF LACIDIPINE

Table 2 provides the membrane partition coefficients for lacidipine as a function of cholesterol content in multilamellar membrane vesicles (MLVs). The membrane partition coefficient for lacidipine decreases as a function of increasing cholesterol, suggesting that lacidipine might be interacting within the hydrocarbon core domain of the membrane bilayer.

TABLE 2.   Lacidipine Membrane Partition
               Coefficients As a Function of Cholesterol Content

Cholesterol/Lipid Mole Ratio	$K_{P[mem]}$[a]
0:1	600,000
0.3:1	300,000
0.6:1	100,000
0.9:1	25,000

[a]Partition coefficients were obtained in MLVs; preliminary experiments with unilamellar membrane vesicles indicate lower partition coefficients.

LACIDIPINE ASSOCIATION AND DISSOCIATION
KINETICS WITH MEMBRANES

The kinetics of lacidipine partitioning into MLVs were measured using radioligand binding assays. It was shown that the membrane association rate of lacidipine with MLVs is slow (fig. 1). This is in contrast to other DHPs which come to equilibrium very rapidly (< 2 minutes). In addition, the membrane dissociation of lacidipine in MLVs was shown to be a slow process with $t\frac{1}{2} \approx 1$ hour (fig. 2). Owing to the very high $K_{P[mem]}$ for lacidipine, even after 4 hours there is still a substantial amount of lacidipine in the membrane compared to other DHP calcium channel drugs which explains its long-acting activity.

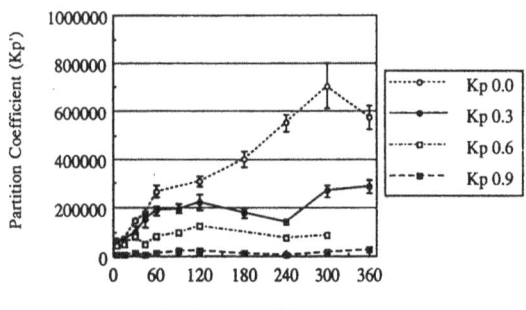

FIG. 1. Time-Dependent Partitioning of Lacidipine into Bovine Myocardial Phosphatidylcholine Multilamellar Vesicles at Cholesterol: BCPC Mole Ratios of 0:1, 0.3:1, 0.6:1 and 0.9:1.

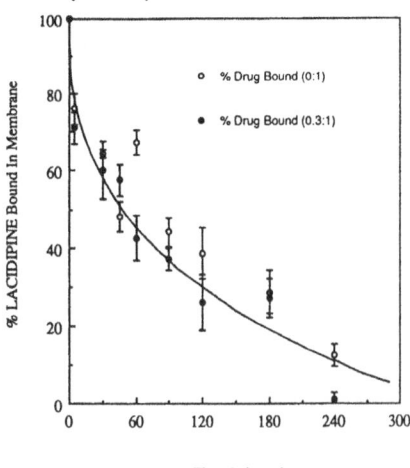

FIG. 2. Lacidipine Dissociation from Multilamellar Vesicles.

THE EFFECT OF CHOLESTEROL ON THE RATE OF
LACIDIPINE BINDING TO MEMBRANES

Partitioning experiments of radioactively labeled (^{3}H)lacidipine into MLVs were performed over a range of concentrations of cholesterol. Mole ratios of 0:1, 0.3:1, 0.6:1 and 0.9:1 [cholesterol:bovine cardiac phosphatidylcholine (BCPC)] were studied. The partition coefficient (Kp') was calculated over a range of time points from 0 to 6 hours. Kp' is not an equilibrium expression as is $K_{P[mem]}$; rather, it reflects, at a given point in time, the ratio of grams of drug in lipid per gram lipid to grams of drug in buffer per gram buffer.

Cholesterol lowered the initial rate of change of Kp'(ΔKp'/Δt) with increasing cholesterol concentration. Fig. 3 shows a least-squares linear fit of the data at each cholesterol concentration for the initial time points from 0 to 150 minutes. Fig. 4 shows the slopes of those lines at each cholesterol concentration, with the error bars representing the standard error of the slope from a least-squares linear regression analysis.

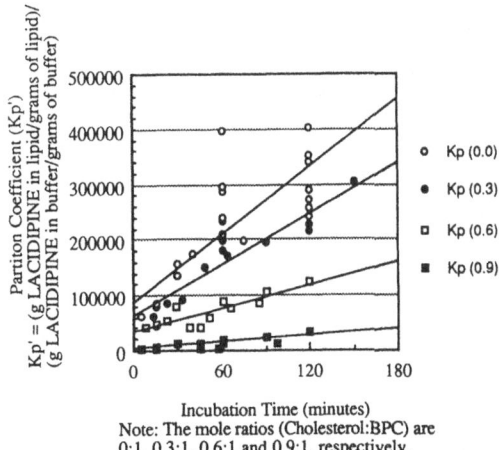

FIG. 3. Initial Partitioning of Lacidipine into Bovine Myocardial Phosphatidycholine Multilamellar Vesicles (Model Membrane). Incubation Times: 0 to 150 Minutes.

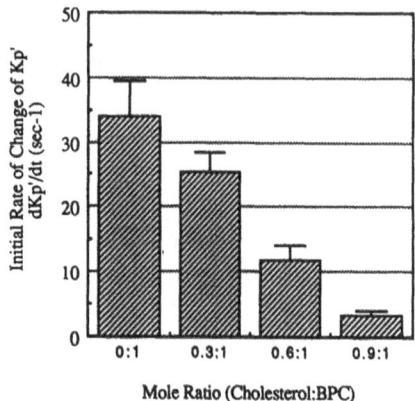

FIG. 4. Initial Rate of Change of Partitioning Coefficient of Lacidipine into Bovine Myocardial Phosphatidylcholine Multilamellar Vesicles (Model Membrane).

LOCATION OF LACIDIPINE IN THE MEMBRANE BILAYER

The location of lacidipine in cardiac model membrane bilayers with and without cholesterol was determined utilizing small-angle x-ray scattering. Lacidipine was shown to be 7 Å from the center of the membrane bilayer (fig. 5). This location for lacidipine is in marked contrast to that determined for nitrendipine and other DHP calcium channel drugs which are typically 12-16 Å from the center of the membrane bilayer as shown in fig. 5.

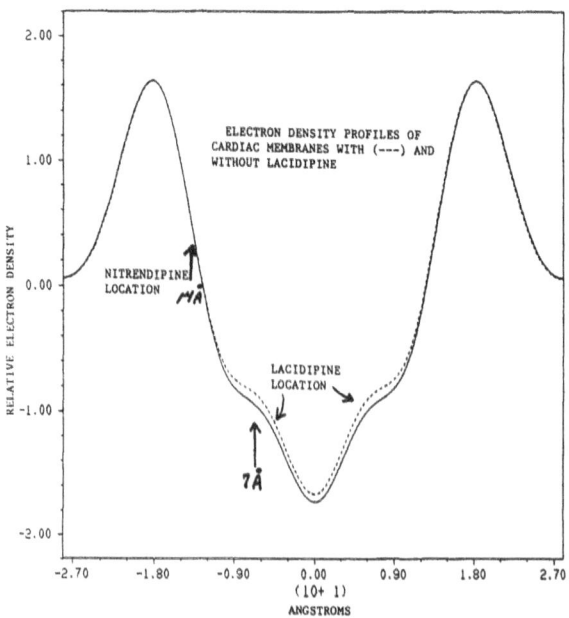

FIG. 5.

SIGNIFICANCE OF DETERMINING MOLECULAR INFORMATION FOR DRUG DESIGN

The combination of x-ray diffraction with physical chemical measurements provides valuable information regarding the molecular interactions of drugs with membranes. Apart from the significance in determining mechanisms for drug membrane and drug receptor actions, these data can be used to define the activities and selectivities of drug substances. A drug that does not penetrate to the proper depth within the membrane bilayer (drug y in fig. 6b) would not be as active as drug x, which at the proper depth within the membrane bilayer can diffuse to the receptor site and participate in a successful interaction. This feature of the model may also accommodate a selection for activity, whereby only a portion of the drug structure is the "active site," so that a hydrophobic (that portion of the drug within the hydrocarbon core) or a hydrophilic (that portion of the drug within the aqueous region of the membrane bilayer) interaction may occur. The limited amount of information to date indicates that cardiovascular agents have precise locations (i.e., depth of penetration) within the membrane bilayer, and this finding should not be ignored. Finally, the model provides another way of selecting the proper drug by its orientation within the membrane. Drug y in fig. 6c will not be active, because its orientation does not match that of the receptor site, whereas drug x in fig. 6c does. In addition, conformation may be important as demonstrated in fig. 6d. The lipid bilayer may constrain the flexible portions of drug structures in a given conformation relative to the inflexible regions of the drug molecule, thus setting up different overall drug conformations that are dependent on the lipid bilayer structure and composition. When the drug leaves the lipid phase and first makes contact with the protein receptor site in direct contact with this lipid phase, the drug conformation as mandated by the lipid bilayer may now be optimized for binding to the protein receptor site. The limited amount of structure data to date also points to precise orientations and conformations of these cardiovascular agents within the membrane bilayer, and this probably should not be ignored.

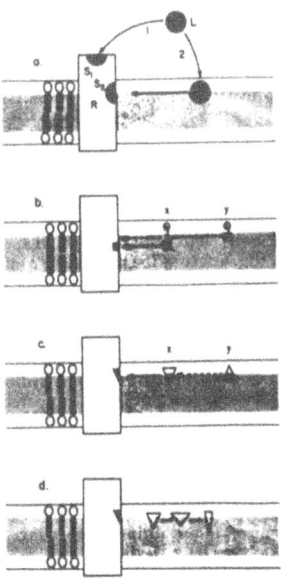

FIG. 6.

These structure tools may become part of an overall integrated system composed of three parts that could allow new breakthroughs in drug design. Single crystal studies of drug structures (obtained by both x-ray and neutron diffraction) could be combined with drug structures (again obtained by x-ray and neutron diffraction) determined directly in biological membranes. These experimentally determined structures for the drug in crystal and membrane-bound forms could be put into molecular modeling schemes. Based on studies of several drug substances in a particular class, similarities, and/or differences of the drug structures in the crystal and membrane-bound forms, in addition to their structures when bound at a protein receptor site, will allow calculations of potential site reactivity and selectivity. Optimized drug structures calculated in this manner could then be synthesized, and both their structures and biological activities could then be determined, completing the cycle. This cycle scheme of combining experimentally determined drug structures (crystalline versus membrane-bound form) with molecular modeling and predicting new structures should allow the optimization of therapeutic agents within a particular class of drugs and possibly predicting new drug classes with clinical potential.

CONCLUSION

Lacidipine is located deeper in the membrane bilayer compared to other DHP calcium channel drugs which may be the molecular basis for its unique pharmacokinetic properties as depicted in fig. 7. The novel side chain on the phenyl ring may be the key chemical moiety which results in unique interactions with the cell membrane to both determine lacidipine's location and proper orientation for binding to the calcium channel receptor.

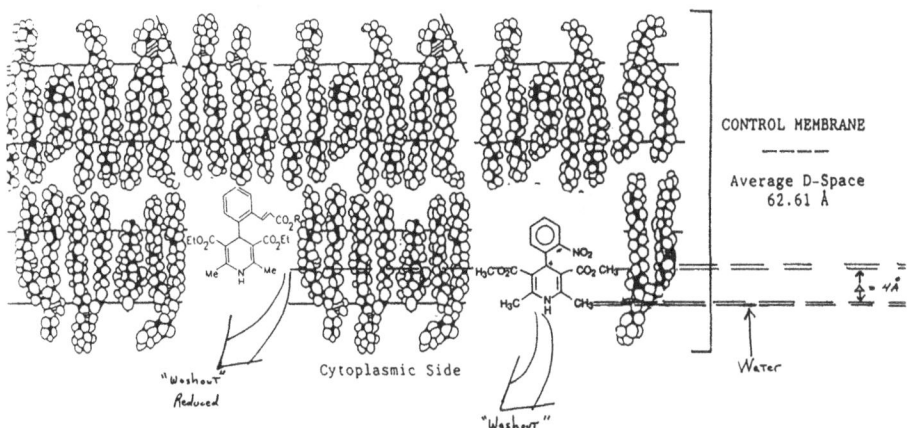

FIG. 7.

ACKNOWLEDGMENTS

This work was carried out in the Biomolecular Structure Analysis Center at the University of Connecticut Health Center. This project was supported by a research grant from Glaxo Research Laboratories, Verona, Italy. LGH was an Established Investigator of the American Heart Association. The Biomolecular Structure Analysis Center acknowledges support from the Patterson Trust Foundation and the State of Connecticut Department of Higher Education's High Technology Programs.

REFERENCES

1.   Herbette L.G., Rhodes D.G. and Mason R.P. (1991): *Drug Design and Delivery*, 7: 75-118.
2.   Rhodes D.G., Sarmiento J.G. and Herbette L.G. (1985): *Mol. Pharmacol.*, 27: 612-623.
3.   Colvin R.A., Ashavaid T.F. and Herbette L.G. (1985): *Biochim. Biophys. Acta*, 812: 601-608.
4.   McCloskey M. and Poo M.-M. (1986): *J. Cell Biol.*, 102: 88-96.
5.   Chester D.W., Herbette L.G., Mason R.P., Joslyn A.F., Triggle D.J. and Koppel D.E. (1987): *Biophys. J.*, 52: 1021-1030.

REGULATION OF CALCIUM CHANNELS IN VASCULAR SMOOTH MUSCLE

Nicole Morel and Theophile Godfraind

Laboratoire de Pharmacologie, UCL 7350,
Avenue E. Mounier, 73
B1200 Bruxelles, Belgium

Most contractile responses of vascular smooth muscle are dependent on Ca entry, as indicated by the absence or the reduction of the responses when tissues are bathed in solution devoid of Ca. At rest, Ca influx across the plasma membrane is about 20 nmol Ca/g tissue per minute. Opening of Ca channels in the plasma membrane causes a rapid increase of Ca influx resulting from the large electrochemical gradient existing between the extra- and intracellular compartments (1). Different molecular structures appear to ensure Ca movement across the plasma membrane. In addition to their typical electrical properties, the L subtype of voltage-operated Ca channels (VOCs) can be distinguished from other types of Ca channels by their sensitivity to Ca agonists and antagonist dihydropyridines. We have investigated whether, besides the control exerted by membrane potential, VOC activity could be regulated by agonists and intracellular messengers in vascular smooth muscle cells.

CA INFLUX EVOKED BY HIGH-KCL DEPOLARIZING SOLUTION

Increase of the KCl concentration in the bathing solution is known to depolarize cell membrane. ^{3}H-tetraphenyl-phosphonium (^{3}H-TPP$^+$) is a lipophilic cation which distributes across the membranes as a function of their potential (2). When the KCl concentration in the bath is enhanced from 6 up to 130 mM, ^{3}H-TPP$^+$ uptake in rat aorta dose-dependently decreases, indicating that the tissue progressively depolarizes (3). As a consequence of KCl-induced depolarization, ^{45}Ca influx in the lanthanum-resistant compartment of rat aorta increases and muscle contraction occurs. Both responses, ^{45}Ca influx and contraction, are blocked by Ca antagonist dihydropyridines (4). Inhibitory potency of some dihydropyridines is markedly increased after equilibration of the tissue with the drug in high KCl-depolarizing solution (3,5). Binding studies performed in intact arteries or microarterioles using ^{3}H-(+)isradipine as the Ca channel ligand, indicated that voltage-dependency of

T. Godfraind et al. (eds.), Calcium Antagonists, 25–30.
© 1993 *Kluwer Academic Publishers and Fondazione Giovanni Lorenzini.*

dihydropyridine inhibitory potency is related to the existence of voltage-dependent binding (3,6,7). In the different tissues tested, one population of specific binding sites for ^{3}H-(+)isradipine is detected. When the KCl concentration in the bath is enhanced to produce membrane depolarization, binding of ^{3}H-(+)isradipine increases (fig. 1). Saturation studies showed that the number of binding sites is not changed by depolarization but that the affinity of binding is increased in such a way that high-affinity binding is associated with the inactivated state of Ca channels in depolarized tissue and low-affinity binding is associated with resting channels.

AGONIST-EVOKED OPENING OF CALCIUM CHANNELS

Exposure of aortic segments to noradrenaline (10 μM) increases ^{45}Ca influx in the lanthanum-resistant compartment by 38 ± 1.5 nmol g^{-1} min^{-1}. Thirty-eight percent of this response is resistant to a maximally active concentration of nisoldipine (10 nM), the other part is dose-dependently depressed by dihydropyridine Ca antagonists. Exposure of rat aortic rings to noradrenaline also produces a contractile response that is partly blocked by dihydropyridine derivatives (3). This suggests that, although the response to α_1-adrenergic stimulation involves activation of VOC, another Ca entry pathway has to be considered to account for the dihydropyridine-resistant Ca entry. In many tissues it has been observed that agonist-evoked responses are less sensitive to dihydropyridine inhibition than are the responses induced by 100 mM KCl solutions (4). Also in rat aorta, concentrations producing 50% inhibition of the contraction induced by noradrenaline (10μM) and 100 mM KCl solution are (with their 95% confidence limits) 170 (124-233) pM and 49 (44-56) pM, respectively. To investigate the reason for the lower potency of Ca antagonists against noradrenaline-evoked responses, we measured the specific binding of ^{3}H-(+)isradipine in rat aortic segments exposed to noradrenaline (10μM). In the presence of the agonist, one population of specific binding sites for ^{3}H-(+)isradipine was detected (3). Density of binding sites is not different from density measured in the absence of agonist. However, binding affinity (K$_D$: 126 pM) increases when it is compared to the value measured in physiological solution (K$_{Dapp}$: 550 pM) but it is lower than the value measured in fully depolarized preparations (K$_D$: 68 pM). High concentrations of noradrenaline have been reported to produce membrane depolarization (8). To estimate the depolarization that could be produced by noradrenaline in rat aorta, we tested the effect of exposure of aortic segments to noradrenaline on the uptake of ^{3}H-TPP$^+$. Fig. 1 shows that noradrenaline (10^{-7}-10^{-5} M) dose-dependently decreases the uptake of ^{3}H-TPP$^+$ but that the maximum effect produced by noradrenaline corresponds to the effect of a solution containing 25 mM of KCl.

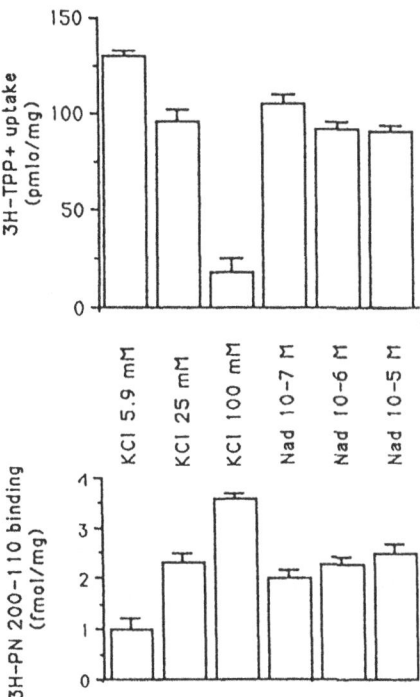

FIG. 1. Effect of KCl concentration and of noradrenaline on ^{3}H-tetraphenylphosphonium uptake (upper panel) and ^{3}H-(+)isradipine specific binding (lower panel) in segments of rat aorta (from reference 3, modified).

Comparison of data from ^{3}H-TPP$^+$ uptake and from ^{3}H-(+)isradipine binding (fig. 1) indicates that the binding of ^{3}H-(+)isradipine is directly related to the amplitude of the depolarization produced by noradrenaline. In agreement with these binding data, inhibitory potency of voltage-dependent dihydropyridines can be enhanced by preequilibrating the tissue with the drug in depolarizing solution (fig. 2). This procedure has the same effect on the inhibition of short contractions evoked by high KCl solution or by noradrenaline. These results suggest that high concentrations of noradrenaline produce membrane depolarization and activate VOCs, and that the potency of voltage-dependent dihydropyridines in inhibiting noradrenaline-evoked Ca influx is dependent on the degree of depolarization produced by the agonist.

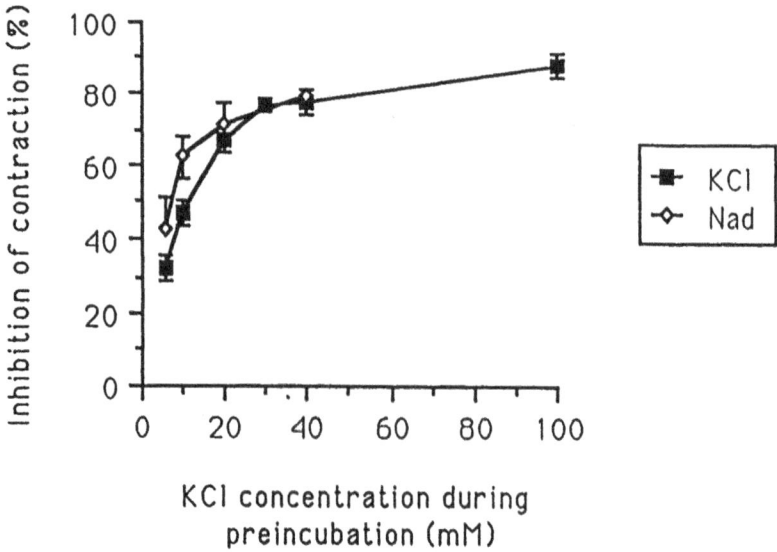

FIG. 2. Influence of the KCl concentration during preincubation of rat mesenteric artery segments with (+)isradipine (100 pM) on the inhibition of contractile response evoked by 2-minute exposure to noradrenaline (10μM) or to 100 mM KCl solution. Data are means ± SEM from 6 determinations.

MODULATION OF CA INFLUX BY PROTEIN KINASE C

Protein kinases C (PKC) are known to phosphorylate a large number of proteins and thereby to modulate a variety of cell functions and to interact with ionic channels in different ways (9). In view of the different roles that PKC has been suggested to play in the mediation of α_1-adrenergic responses in vascular smooth muscle, we investigated whether this enzyme could be involved in the activation of Ca entry by noradrenaline. In rat aortic segments, staurosporine (10 nM), a PKC inhibitor (10), inhibits the nisoldipine-sensitive ^{45}Ca entry evoked by noradrenaline while ^{45}Ca influx stimulated by high-KCl solution is unaffected (fig. 3). Responses evoked by low concentrations of noradrenaline, which depolarize smooth muscle cells by only a few millivolts (8), are markedly inhibited by staurosporine so that it may be suggested that VOC activation by nondepolarizing concentrations of noradrenaline could involve a PKC-mediated phosphorylation. In agreement with this hypothesis, phorbol 12,13-dibutyrate (100 nM), which directly activates PKC, produces an increase of ^{45}Ca influx that is completely blocked by nisoldipine (data not shown).

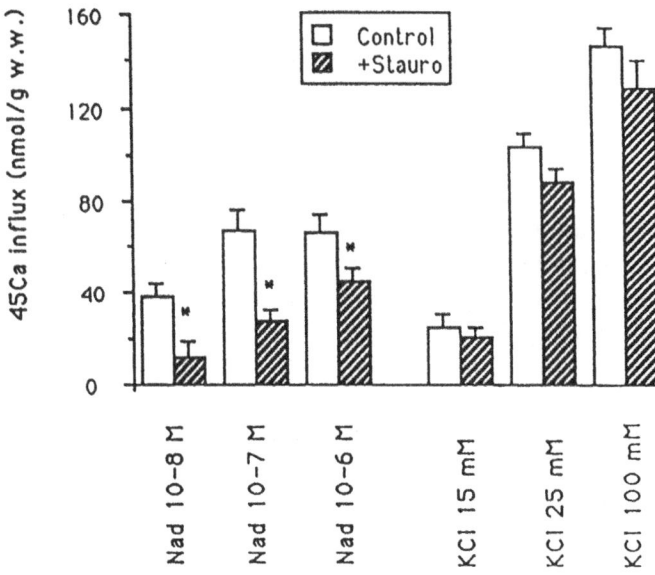

FIG. 3. Effect of staurosporine (10 nM) on nisoldipine-sensitive ^{45}Ca influx evoked by 2-minute exposure to noradrenaline or to high-KCl solution in rat aorta. Data are means ± SEM from 12 determinations. * indicates value significantly different from control.

MODULATION OF CA INFLUX BY EDRF AND cGMP

EDRF is known to modulate the contractile behavior of vascular smooth muscle by increasing cGMP production in vascular smooth muscle cells. It has been observed that the presence of a functional endothelium influences the activity of Ca channels in noradrenaline-stimulated preparation (11). Also in resting tissue, ^{45}Ca influx is significantly lower in the presence, than in the absence, of functional endothelium (35 ± 0.85 and 42 ± 4.4 nmol g^{-1} min^{-1} ($n=30$; $P<0.01$, respectively). This difference is abolished in the presence of nisoldipine (100 nM), or by the addition of 8Br-cGMP in preparations without endothelium, suggesting that VOCs are activated to a lower degree in the presence of endothelium. We investigated to what extent the effect of endothelium on ^{45}Ca influx could be related to change of membrane potential (12). The results obtained by microelectrode technique and by the measurement of ^{3}H-TPP$^+$ uptake indicated that EDRF produces a permanent hyperpolarization of the membrane of vascular smooth muscle cells of rat aorta of 4-10 mV. Addition of substances that mimic the effect of EDRF (SIN-1 or 8Br-cGMP) in preparation without functional endothelium, leads to a repolarization of the membrane potential. Thus the effect of cGMP on Ca influx is at least partly mediated by a modulation of membrane potential.

CONCLUSION

In vascular smooth muscle cells, the main determinant of VOC activity is the membrane potential which can be modulated by EDRF or agonists. However, opening of VOC by non-depolarizing concentration of agonist could involve PKC-mediated phosphorylation. The target of this phosphorylation remains to be identified.

REFERENCES

1.   Godfraind-De Becker A. and Godfraind T. (1980): *Int. Rev. Cytol.*, 67: 141-170.
2.   Lichtshtein D., Kaback H.R. and Blume A.J. (1979): *Proc. Natl. Acad. Sci.* USA, 76: 383-396.
3.   Morel N. and Godfraind T. (1991): *Br. J. Pharmacol.*, 102: 467-477.
4.   Godfraind T., Miller R. and Wibo M. (1986): *Pharmacol. Rev.*, 38: 321-415.
5.   Wibo M., Deroth L. and Godfraind T. (1988): *Circ. Res.*, 62: 91-96.
6.   Morel N. and Godfraind T. (1987): *J. Pharmacol. Exp. Ther.*, 243: 711-715.
7.   Morel N. and Godfraind T. (1989): *Naunyn-Schmiedeberg's Arch. Pharmacol.*, 340: 442-451.
8.   Bolton T.B. and Large W.A. (1986): *QJ Exp. Physiol.*, 71: 1-28.
9.   Nishizuka Y. (1986): *Science*, 233: 305-312.
10. Ruëgg U.T. and Burgess G.M. (1989): *Trends Pharmacol. Sci.*, 10: 216-220.
11. Godfraind T. (1986): *Eur. J. Pharmacol.*, 126: 341-343.
12. Krippeit-Drews P., Morel N. and Godfraind T. (1991): *Pflügers Arch.*, 419: R128.

CALCIUM AND THE PRODUCTION
OF ENDOTHELIUM-DERIVED VASOACTIVE FACTORS

V.B. Schini, T. Nagao*, S. Illiano, and P.M. Vanhoutte

Center for Experimental Therapeutics, Baylor College of Medicine
One Baylor Plaza, Houston, Texas 77030
*Current address: 2nd Department of Internal Medicine
Kyushu University, Maidash 3-1-1 Highashi-ku
Fukuoka 812 Japan

Endothelial cells play a major role in the local control of the vascular system. They form a semipermeable barrier which regulates the exchange of nutrients and waste products between the blood and the underlying smooth muscle. These cells also influence the tone of the blood vessel wall, in part, by releasing potent and short-lasting relaxing factors such as prostanoids, nitric oxide and endothelium-derived hyperpolarizing factor, as well as contractile factors such as thromboxane/endoperoxides, superoxide anions, and endothelin(s). The present review will discuss the biological properties of the vasoactive factors and emphasize the importance of calcium ions for both the production of the factors by the endothelial cells and the modulation of the vascular tone that they achieve.

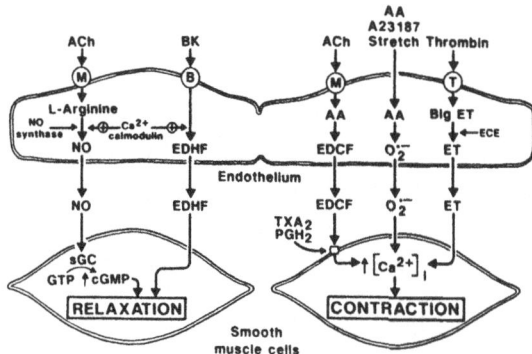

FIG. 1. Endothelial cells control vascular tone by producing relaxing factors such as nitric oxide (NO) and endothelium-derived hyperpolarizing factor (EDHF). They also produce endothelium-derived contracting factors (EDCF) such as an unidentified metabolite of arachidonic acid (AA), superoxide anions (O_2^-) and endothelin (ET). ACh: acetylcholine; BK: bradykinin; sGC: soluble guanylate cyclase; cGMP: guanosine 3',5'-cyclic monophosphate; ECE: endothelin converting enzyme; TXA2/PGH2: thromboxane A_2/endoperoxide PGH_2.

T. Godfraind et al. (eds.), Calcium Antagonists, 31–36.

ENDOTHELIUM-DERIVED RELAXING FACTORS

Nitric Oxide

In 1980 Furchgott and Zawadzki (1) discovered that acetylcholine evoked a concentration-dependent relaxation of a contracted ring of rabbit thoracic aorta only in the presence of a functional endothelium. Thereafter, the pivotal role of the endothelial cells in mediating relaxations of isolated blood vessels has been demonstrated also for kinins, thrombin, 5-hydroxytryptamine, purines, α_2-adrenergic agonists, shear and pulsatile stress as well as the calcium ionophore A23187 (2,3).

The endothelium-dependent relaxation of isolated blood vessels requires calcium in the incubation medium (4,5) and is due to the release of a short-lived relaxing factor which has been identified as nitric oxide (6). Nitric oxide is produced from the terminal guanidino nitrogen atom(s) of the amino acid L-arginine by nitric oxide synthases (7). The enzymatic production of nitric oxide is associated with both the cytosolic and the particulate fractions of homogenates of endothelial cells (8). Both enzyme activities require L-arginine, NADPH, and calcium (8-10). The production of nitric oxide from L-arginine is inhibited by structural analogues of L-arginine such as monomethyl L-arginine (7). The stimulating effect of calcium on the enzymatic production of nitric oxide by endothelial cells is mediated by the calcium binding protein, calmodulin (8,10). The cofactor role of the complex calcium-calmodulin is supported also by the fact that calmodulin inhibitors, but not calcium channel antagonists, impair endothelium-dependent relaxations of isolated blood vessels (11,12).

The relaxation evoked by nitric oxide is due to the production of cyclic GMP following the binding of the radical to the heme-containing soluble guanylate cyclase of the smooth muscle cells (13,14).

Hyperpolarizing Factor

In several isolated blood vessels, acetylcholine and carbachol evoke endothelium-dependent relaxations which are associated with a transient hyperpolarization of the vascular smooth muscle cells (15-22). Similar endothelium-dependent hyperpolarizations of vascular smooth muscle are evoked also by kinins, purines, histamine, substance P, and the calcium ionophore A23187 (21-24).

Bioassay experiments suggest that the hyperpolarization of vascular smooth muscle is due to the release by the endothelial cells of a humoral factor called endothelium-derived hyperpolarizing factor (EDHF) (17,25). The chemical identity of EDHF is unknown. However, the endothelium-mediated hyperpolarizations are not affected by indomethacin or by inhibitors of nitric oxide, such as methylene blue (an inhibitor of soluble guanylate cyclase activation) or oxyhemoglobin (a scavenger of nitric oxide) (18-21). These observations suggest that EDHF is distinct from nitric oxide and from metabolites of arachidonic acid by the cyclooxygenase pathway. This view is supported by the fact that, in the rabbit, the production of EDHF by endothelial cells is due to the activation of M_1 muscarinic receptors, while that of nitric oxide results from stimulation of M_2 muscarinic receptors (16). It is supported further by the observation that authentic nitric oxide does not alter the membrane potential of the canine coronary artery (18). However, authentic nitric oxide hyperpolarizes, and an inhibitor of nitric oxide production inhibits the endothelium-dependent hyperpolarization of the guinea-pig uterine artery (26). Therefore, the production of nitric oxide by endothelial cells may contribute to the hyperpolarization of the vascular smooth muscle in certain species.

The endothelium-mediated hyperpolarizations are impaired by the absence of extracellular calcium in the rabbit carotid artery (22) and by calmodulin inhibitors in the canine coronary artery (27). These observations indicate that calcium, probably following its interaction with calmodulin, plays an important role in the production of EDHF by endothelial cells.

EDHF increases the potential of the vascular smooth muscle by either stimulating the Na^+-K^+ ATPase and/or increasing the K^+ conductance (17,19,25). The release of EDHF by endothelial cells is likely to participate, in coordination with endothelium-derived nitric oxide, in the regulation of vascular tone.

ENDOTHELIUM-DERIVED CONTRACTING FACTOR(S)

Shortly after the discovery that the endothelium mediates vascular relaxation, DeMey and Vanhoutte (28) reported that the endothelium can also induce contractions of isolated canine veins. The importance of the endothelium in mediating contractile responses has been stressed in several other isolated vascular beds (29-32). The contractions caused by the endothelium are due, in part, to the release by endothelial cells of contractile factors. The most important ones are likely to be metabolites of the metabolism of arachidonic acid by the cyclooxygenase pathway (thromboxane A_2 and/or PGH_2), superoxide anions, and the endothelins.

Thromboxane A_2/Endoperoxides H_2

In the aorta of the spontaneous hypertensive rat, high concentrations of acetylcholine evoke endothelium-dependent contractions which are sensitive to inhibitors of cyclooxygenase (29). The endothelium-generated contractions are unaffected by thromboxane and prostacyclin synthetase inhibitors but are impaired by thromboxane A_2/PGH_2 receptor antagonists (33,34). These observations indicate that the activation of the thromboxane A_2/PGH_2 receptor (probably by PGH_2) mediates the endothelium-dependent contractions in the aorta of the hypertensive rat. The production of a contractile factor by endothelial cells following their activation by acetylcholine may contribute to the blunted relaxation of isolated blood vessels from hypertensive and from old animals (35-37). The endothelium-dependent contractions evoked by acetylcholine in the spontaneously hypertensive rat aorta are prevented by the calcium antagonist diltiazem (38).

Superoxide Anions

In the canine basilar artery, arachidonic acid, acetylcholine, and stretch evoke endothelium-dependent contractions which are prevented by indomethacin (31,38). Since the calcium ionophore A23187 also causes endothelium-dependent contractions, increases in cytosolic free-calcium concentration in the endothelial cells must play a role in the response. The contractions caused by stretch are likely to involve the influx of extracellular calcium by voltage-activated calcium channels as they are sensitive to diltiazem and flunarizine (39). As it is unlikely that endothelial cells contain voltage-operated calcium channels, the effect of the calcium antagonists probably reflects inhibition of the response of the smooth muscle to the EDCF. The endothelium-dependent contractions evoked by A23187 are prevented by superoxide dismutase and in this tissue superoxide anions cause contractions (32). These observations support the hypothesis that superoxide anions, generated by the cyclooxygenase pathway, are an endothelium-derived contracting factor in the canine basilar artery.

Endothelin(s)

In 1988, Yanagisawa and collaborators (40) found that cultured, endothelial cells produce a 21-amino acid peptide termed endothelin-1 which is the most potent contractile agonist yet discovered. The production of the peptide by cultured as well as native endothelial cells, is stimulated by hormones, neurotransmitters, growth factors, and shear stress (41). Endothelin-1 is produced from a 39-amino acid precursor peptide by metallo-sensitive peptidase(s) termed endothelin converting enzyme (40). Endothelin-1 evokes slow and long-lasting contractions of isolated blood vessels which are partially inhibited by calcium channel antagonists (40).

CONCLUSIONS

The activation of the endothelial cells can produce both potent relaxing and contractile factors. The production of the relaxing factors is triggered by an increase in the free concentration of calcium in the endothelial cells while the tension evoked by the contractile factors are sensitive to calcium channel modulators.

REFERENCES

1. Furchgott R.F. and Zawadzki J.V. (1980): *Nature*, 288: 373-376.
2. Furchgott R.F. and Vanhoutte P.M. (1989): *FASEB J.*, 3: 2007-2018.
3. Moncada S., Palmer R.M.J. and Higgs E.A. (1991): *Pharmacol. Rev.*, 43: 109-142.
4. Singer H.A. and Peach M.J. (1982): *Hypertension*, 4: II19-II25.
5. Long C.J. and Stone T.W. (1985): *Blood Vessels*, 22: 205-208.
6. Palmer R.M.J., Ferrige A.G. and Moncada S. (1987): *Nature*, 327: 524-526.
7. Palmer R.M.J., Rees D.D., Ashton D.S. and Moncada S. (1988): *Biochem. Biophys. Res. Commun.*, 153: 1251-1256.
8. Förstermann U., Pollock J.S., Schmidt H.H.H.W., Heller M. and Murad F. (1991): *Proc. Natl. Acad. Sci. USA*, 88: 1788-1792.
9. Mayer B., Schmidt K., Humbert P. and Böhme E. (1989): *Biochem. Biphys. Res. Commun.*, 164: 678-685.
10. Busse R. and Mülsch A. (1990): *FEBS Lett.*, 265: 133-136.
11. Weinheimer G. and Osswald H. (1986): *Naunyn-Schmiedeberg's Arch. Pharmacol.*, 332: 391-397.
12. Spedding M., Schini V., Schoeffter P. and Miller R.C. (1986): *J. Cardiovasc. Pharmacol.*, 8: 1130-1137.
13. Holzmann S. (1982): *J. Cyclic Nucleotide Res.*, 8: 409-419.
14. Rapoport R.M. and Murad F. (1983): *Circ. Res.*, 52: 352-357.
15. Bolton T.B., Lang R.J. and Takewaki T. (1984): *J. Physiol. Lond.*, 351: 549-572.
16. Komori K. and Suzuki H. (1987): *Br. J. Pharmacol.*, 92: 657-664.
17. Feletou M. and Vanhoutte P.M. (1988): *Br. J. Pharmacol.*, 93: 515-524.
18. Komori K., Lorenz R.R. and Vanhoutte P.M. (1988): *Am. J. Physiol.*, 255: H207-H212.
19. Chen G., Suzuki H. and Weston A.H. (1988): *Br. J. Pharmacol.*, 95: 1165-1174.
20. Huang A.H., Busse R. and Basseng E. (1988): *Naunyn-Schmiedeberg's Arch. Pharmacol.*, 338: 438-442.
21. Chen G. and Suzuki H. (1989): *J. Physiol. Lond.*, 410: 91-106.

22.   Chen G. and Suzuki H. (1990): *J. Physiol. Lond.*, 421: 521-534.
23.   Chen G. and Suzuki H. (1991): *Am. J. Physiol.*, 260: H1037-H1042.
24.   Beny J.L., Brunet P. and Huggel H. (1987): *Regulatory Peptides*, 17: 181-190.
25.   Kauser K., Stekiel W.J., Rubanyi G. and Harder D.R. (1989): *Circ. Res.*, 65: 199-204.
26.   Tare M., Parkington H.C., Coleman H.A., Neild T.O. and Dusting G.J. (1990): *Nature*, 346: 69-71.
27.   Illiano S.C., Nagao T. and Vanhoutte P.M. (1991): *Circulation* (abstract), (in press.)
28.   DeMey J.G. and Vanhoutte P.M. (1982): *Circ. Res.*, 51: 439-447.
29.   Lüscher T.F. and Vanhoutte P.M. (1986): *Hypertension*, 8: 344-348.
30.   Harder D.R. (1987): *Circ. Res.*, 60: 102-107.
31.   Katusic Z.S., Shepherd J.T. and Vanhoutte P.M. (1988): *Stroke*, 19: 476-479.
32.   Katusic Z.S. and Vanhoutte P.M. (1989): *Am. J. Physiol.*, 257: H33-H37.
33.   Auch-Schwelk W., Katusic Z.S. and Vanhoutte P.M. (1990): *Hypertension*, 15: 699-703.
34.   Kato T., Iwama Y., Okumura K., Hashimoto H., Ito T. and Satake T. (1990): *Hypertension*, 15: 475-481.
35.   Konishi M. and Su C. (1983): *Hypertension*, 5: 881-886.
36.   Winquist R.J., Bunting P.B., Baskin E.P. and Wallace A.A. (1984): *J. Hypertension*, 2: 541-545.
37.   Koga T., Takata Y., Kobayashi K., Takishita S., Yamashita Y. and Fujishima M. (1989): *Hypertension*, 14: 542-548.
38.   Vanhoutte P.M. (1988): *J. Cardiovasc. Pharmacol.*, 12: S21-S28.
39.   Katusic Z.S., Shepherd J.T. and Vanhoutte P.M. (1987): *Am. J. Physiol.*, 252: H671-H673.
40.   Yanagisawa M., Kurihara H., Kimura S., Tomobe Y., Kobayashi M., Mitsui Y., Yazaki Y., Goto K. and Masaki T. (1988): *Nature*, 332: 411-415.
41.   Schini V.B. and Vanhoutte P.M. (1991): *Pharmacol. Toxicol.*, 69: (in press).

POSTNATAL DEVELOPMENT OF CALCIUM CHANNELS IN CARDIAC MUSCLE

M. Wibo and T. Godfraind

Laboratoire de Pharmacologie
Université Catholique de Louvain
UCL 73.50, Av. E. Mounier 73, B 1200 Brussels Belgium

During the postnatal period, excitation-contraction coupling changes dramatically in rat ventricular muscle, as evidenced by modifications in the sensitivity of systolic contraction to external Ca^{2+} (1) and to modulators of Ca^{2+} channels (2,3). The potency of L-channel blockers (nifedipine, verapamil) as inhibitors of systolic contraction is lower in adult than in neonatal ventricle and conversely, the negative inotropic effect of ryanodine, a selective modulator of sarcoplasmic reticulum (SR) Ca^{2+} release channels, is more pronounced in adult than in neonatal tissue. These findings suggest that, in neonatal ventricle, most of the Ca^{2+} that is needed to activate the myofilaments originates from the extracellular milieu, whereas, in adult tissue, the SR Ca^{2+} stores become progressively more important as a source of activator Ca^{2+}. These functional changes are parallelled by marked ultrastructural modifications. In particular, the volume percentage of SR increases 2-fold in the early postnatal period and the transverse tubular system, which is not detectable at birth, develops over the first postnatal weeks (4). In view of the different response of neonatal and adult heart to L-channel blockers and ryanodine, we decided to compare the properties of their respective receptors in cardiac membranes isolated at various ages, using $[^3H](+)$-PN200-110 ($[^3H]$-isradipine) and $[^3H]$ryanodine to label sarcolemmal dihydropyridine-sensitive Ca^{2+} channels and SR ryanodine-sensitive Ca^{2+} channels, respectively.

POSTNATAL CHANGES IN CHANNEL NUMBER

In agreement with previous reports (5,6), we found that binding affinities did not differ in neonatal and adult tissue. Therefore, an alternative explanation must be sought for the different sensitivities of neonatal and adult ventricle to Ca^{2+} channel modulators. The B_{max} $[^3H](+)$-PN200-110 increased by about 35% over the postnatal period, which was less than the increase in sedimentable protein (table 1). The B_{max} of $[^3H]$ryanodine increased more than that of $[^3H](+)$-PN200-110,

T. Godfraind et al. (eds.), Calcium Antagonists, 37–42.
© 1993 *Kluwer Academic Publishers and Fondazione Giovanni Lorenzini.*

since it was 2-fold higher in adult than in neonatal tissue. The activity of cytochrome c oxidase increased by 76% after birth. These increases are in line with the known postnatal proliferation of SR membranes and mitochondria, respectively. In contrast, [^{3}H]ouabain binding was reduced by more than half in adult tissue. Ouabain binding was measured at a low concentration of radioligand that labeled preferentially the high affinity isoforms of NA$^+$, K$^+$-ATPase (7).

TABLE 1.   Postnatal evolution of receptor numbers in rat heart[a]

	1-3 days	7-13 days	>30 days
	(pmol/g tissue wet wt.)		
[^{3}H](+)-PN200-110	10.0	12.1	13.5
[^{3}H]Ryanodine	44.0	76.1	92.7
[^{3}H]Ouabain	37.0	31.4	15.7
	(mg/g tissue wet wt.)		
Protein	83	107	126

[a]Results are total amounts sedimentable from homogenate (100,000 g, 35 min) related to 1 g tissue wet weight. The are B_{max} values for [^{3}H](+)PN200-110 and [^{3}H]ryanodine. For [^{3}H]ouabain, specific binding was measured at a free concentration of 25 nM. Values are means from 2-8 experiments. Data 1-3 days and >30 days are significantly different (P<0.02; t test). Adapted from Wibo et al. (3).

POSTNATAL CHANGES IN CHANNEL LOCALIZATION

Ryanodine receptors in skeletal and cardiac muscle have now been identified with the so-called feet, which are situated in junctional areas between the SR membrane and the sarcolemmal membrane (8). In adult rat ventricle, transverse (T) tubules account for about 75% of the junctional surface area of the sarcolemma (9). In skeletal muscle, T tubules are especially rich in dihydropyridine receptors (10), which probably serve as voltage sensors that trigger the opening of SR Ca^{2+} release channels. Since the development of T tubules was likely to play a particularly important role in the maturation of excitation-contraction coupling occurring after birth in rat ventricle, we decided to compare the subsarcolemmal distribution of dihydropyridine receptors in adult and neonatal ventricle. As first shown by Brandt (11), after grinding of cardiac tissue, junctional structures can be separated from "free" sarcolemma and SR elements by density gradient centrifugation. Sarcolemmal fragments that remain associated with SR by feet structures are characterized by high-equilibrium densities in sucrose gradient, while fragments from nonjunctional areas equilibrate at lower densities.

Fig. 1 shows density frequency distributions obtained after subfractionation of microsomal fractions from 3-days (thick-line histograms) and 30-days (thin-line histograms) ventricle by density gradient centrifugation. A major influence of age is obvious only in the case of $[^3H](+)$-PN200-110 binding.

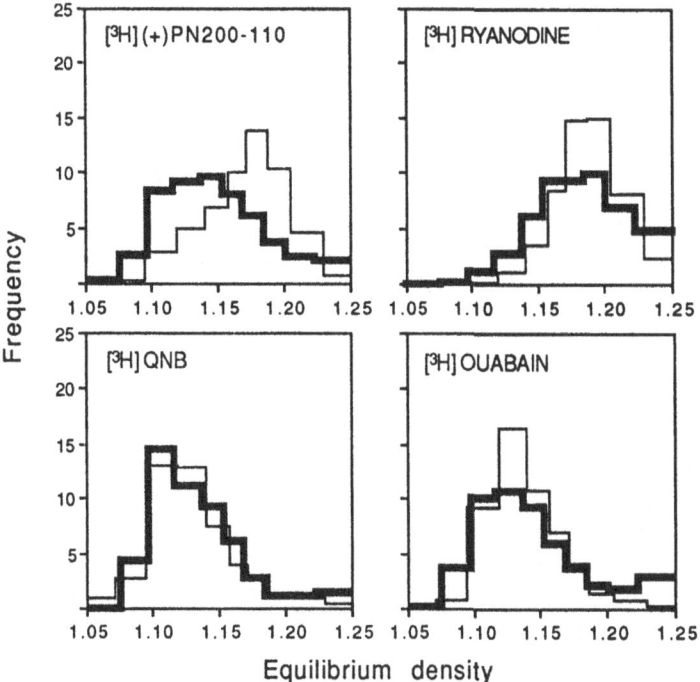

FIG. 1. Subfractionation of microsomes from neonatal and adult rat ventricle by density gradient centrifugation. Thick- and thin-line histograms correspond to 3-days and 30-days rats, respectively. Each subfraction is represented on the abscissa by its density boundaries. The frequency is the fractional amount of binding site recovered in a given subfraction divided by the density increment across that subfraction. QNB, quinuclidinyl benzilate. Adapted from Wibo, et al (3).

In neonatal heart, $[^3H](+)$-PN200-110 binding sites are mainly recovered at densities lower than 1.17, whereas in adult tissue, their distribution is characterized by a rather sharp peak centered around a density of 1.18 and a distinct shoulder at lower densities. In the neonate, the distribution pattern of dihydropyridine receptors is similar to those of the other sarcolemmal constituents (Na^+, K^+-ATPase labeled by [^3H]ouabain, muscarinic receptors labeled by [^3H]QNB), but dissociates markedly from that of ryanodine receptors, whereas the reverse is true for adult heart. Data from five density gradient experiments are summarized in fig. 2.

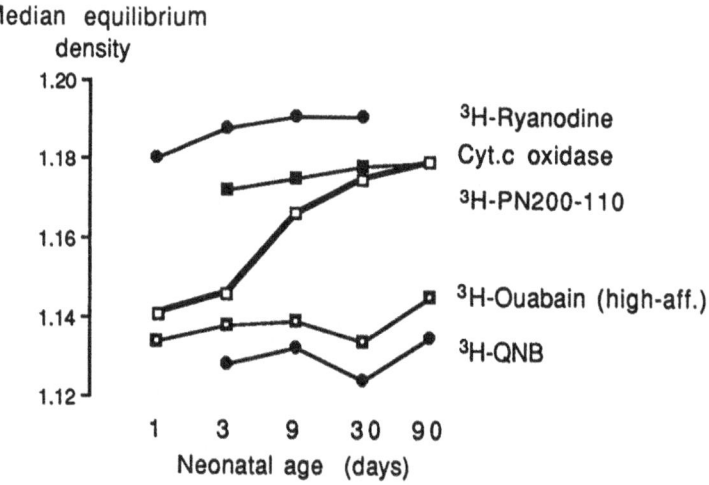

FIG. 2. Median equilibrium densities of receptors and cytochrome c oxidase in microsomal fractions from rat heart as a function of age. Median densities were estimated from density distribution histograms such as those shown in fig. 1. Adapted from Wibo et al. (3).

These results are in agreement with the view that most of the dihydropyridine receptors in microsomal fractions from adult ventricle are associated with junctional structures, and are thus predominantly located in T tubules, whereas early after birth they are mainly situated in nonjunctional areas at the cell surface together with the other sarcolemmal constituents.

SURFACE DENSITY OF CALCIUM CHANNELS IN ADULT VENTRICLE

A quantitative evaluation of density distribution data (fig. 1) suggests that some 75% of dihydropyridine receptors in microsomal fractions from adult rat ventricular muscle are associated with junctional structures. (We have recently obtained very comparable density distributions with microsomal fractions from adult guinea-pig ventricle, as illustrated by fig. 3. Extrapolating this percentage to the whole homogenate [see (3)] and taking into account the surface area of junctional and nonjunctional sarcolemma reported by Page and Surdyk-Droske (9), we estimated the surface density of dihydropyridine receptors in junctional sarcolemma to be 84 per μm^2 as compared to 7 per μm^2 in nonjunctional sarcolemma (3). A similar calculation for junctional SR yields a value of 765 per μm^2 for the surface density of ryanodine receptors. Therefore, the

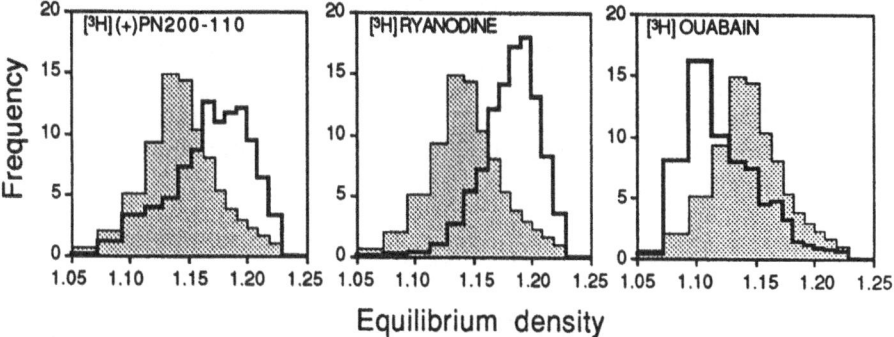

FIG. 3. Subfractionation of microsomal fractions from adult guinea-pig ventricle by density gradient centrifugation. Microsomes were subfractionated in a linear sucrose gradient as described previously (3), except that centrifugation was carried out, in a vertical rotor (Kontron TV-850) for 2 hours as 160,000 g. The shaded histogram reproduced on each panel is the distribution of the SR Ca-ATPase activity, which was measured as described by Simonides and van Hardeveld (12).

stoichiometric ratio dihydropyridine receptors:ryanodine receptors would be approximately 0.1 in junctional structures of adult rat ventricle. In contrast, the number of dihydropyridine receptor particles has been estimated to be twice that of feet (ryanodine receptors) in fish skeletal muscle (13).

FUNCTIONAL IMPLICATIONS

In summary, the number of ryanodine receptors (per gram tissue) doubles after birth in rat ventricle, in line with the proliferation of SR membranes, whereas the number of dihydropyridine receptors increases only moderately. Concomitantly with the postnatal development of T tubules, there is a progressive redistribution of dihydropyridine receptors, which concentrate in junctional areas. This predominant localization, in adult tissue, of the main Ca^{2+} entry pathway (dihydropyridine-sensitive channels) in close proximity to the Ca^{2+} release pathway of the SR (ryanodine-sensitive channels) appears especially appropriate in view of the presumed mechanism of activation of Ca^{2+} release in cardiac tissue, that is Ca^{2+}-induced Ca^{2+} release (14). Indeed, it ensures a rapid elevation of the Ca^{2+} concentration in the immediate vicinity of SR release channels, which should enhance the efficiency of the coupling between Ca^{2+} influx and Ca^{2+} release, and, therefore, the "gain" of this amplification mechanism. The development of this amplification system after birth accounts for the postnatal decrease in sensitivity of systolic contraction to inhibitors of Ca^{2+} entry, and the corresponding increase of the negative inotropic effect of ryanodine. Finally, the low stoichiometric ratio dihydropyridine-sensitive Ca^{2+} channels: ryanodine-sensitive Ca^{2+} channels in cardiac muscle is compatible with the amplification mechanism of Ca^{2+}-induced Ca^{2+} release, whereas the high stoichiometric ratio in skeletal muscle is in line with the proposed role of skeletal muscle dihydropyridine receptors as voltage-sensors that physically interact with SR Ca^{2+} release channel.

REFERENCES

1. Bers D.M., Philipson K.D. and Langer G.A. (1981): *Am. J. Physiol.*, 240: H576-H583.

2. Tanaka H. and Shigenobu K. (1989): *J. Mol. Cell. Cardiol.*, 21: 1305-1313.

3. Wibo M., Bravo G. and Godfraind T. (1991): *Circ. Res.*, 68: 662-673.

4. Page E., Earley J. and Power B. (1974): *Circ. Res.*, 35 (suppl.2): II12-II16.

5. Kazazoglou T., Schmid A., Renaud J.F. and Lazdunski M. (1983): *FEBS Lett.*, 164: 75-79.

6. Kojima M., Ishima T., Taniguchi N., Kimura K., Sada H. and Sperelakis N. (1990): *Br. J. Pharmacol.*, 99: 334-339.

7. Noël F., Wibo M. and Godfraind T. (1991): *Biochem. Pharamcol.*, 41: 313-315.

8. Lai F.A. and Meissner G. (1989): *J. Bioenerg. Biomembr.*, 21: 227-246.

9. Page E. and Surdyk-Droske M. (1979): *Circ. Res.*, 45: 260-267.

10. Fosset M., Jaimovich E., Delpont E. and Lazdunski M. (1983): *J. Biol. Chem.*, 258: 6086-6092.

11. Brandt N. (1985): *Arch. Biochem. Biophys.*, 242: 306-319.

12. Simonides W.S. and van Hardeveld C. (1990): *Anal. Biochem.*, 191: 321-331.

13. Block B.A., Imagawa T., Campbell K.P. and Franzini-Armstrong C. (1988): *J. Cell. Biol.*, 107: 2587-2600.

14. Fabiato A. (1989): *Mol. Cell. Biochem.*, 89: 134-140.

THE HEMODYNAMIC EFFECTS OF CALCIUM ANTAGONISTS UNDER PATHOLOGICAL CONDITIONS

P.A. van Zwieten

Departments of Pharmacotherapy and Cardiology
Academic Medical Center, Meibergdreef 15,
1105 AZ Amsterdam, The Netherlands.

The hemodynamic effects of calcium antagonists (CA) under normal conditions are well established. All current types of CA, that is the dihydropyridines (nifedipine and related drugs), phenylalkylamines (verapamil and related drugs), and benzothiazepines (diltiazem) are vasodilators which cause relaxation, in particular in the arterial vascular bed, with differential effects on specialized circulatory beds, like those of the coronary, renal, and cerebral systems (1,2). The dihydropyridines usually induce reflex tachycardia via the baroreflex/sympathetic nervous system. The reflex tachycardia is usually mild and transient. Verapamil, diltiazem, and related drugs do not cause reflex tachycardia but will rather reduce heart rate and impair A-V conduction (1,2). Flunarizine has a predominantly cerebral cytoprotective action with little or no influence on the cardiovascular system (3).

Pathological conditions may considerably influence and alter the hemodynamic pattern of CA. For this reason, it seems of interest to discuss the hemodynamic profile of the current types of CA when used in their major fields of application, such as hypertension, angina pectoris, and cerebral ischemia. In addition, a few remarks will be made on the hemodynamic effects of Ca in experimental diabetes.

ESSENTIAL HYPERTENSION

From a hemodynamic point of view, established hypertension is characterized by a rise in peripheral vascular resistance and slowly decreasing cardiac output in the course of several years (4). Since the elevated peripheral resistance is the most consistent hemodynamic change, vasodilatation at the level of the arterioles (resistance vessels) appears to be a logical approach in treatment. All three different categories of CA are indeed vasodilators which in hypertensives cause a reduction in peripheral resistance and a concomitant fall in blood pressure (2,5,6). For instance, diltiazem in hypertensives causes the circadian variations in pressure to be largely maintained, though at a

T. Godfraind et al. (eds.), Calcium Antagonists, 43–48.
© 1993 *Kluwer Academic Publishers and Fondazione Giovanni Lorenzini. Printed in the Netherlands.*

lower level (7). The degree of antihypertensive activity caused by various types of CA is very similar when given in appropriate doses, but the effect on heart rate is different among the various groups of CA (5,6). In hypertensives, nifedipine and other dihydropyridines will cause reflex tachycardia, triggered by the baroreceptors and the sympathetic nervous system. The reflex tachycardia is usually mild and transient. As in normotensives, verapamil, diltiazem, and related drugs do not cause reflex tachycardia because of their depressant effect on the SA- and A-V nodes. In the long term, the baroreflex adapts to the prevailing pressure level thus allowing heart rate and cardiac output to return to values close to (dihydropyridines) or just below (verapamil, diltiazem) control levels (3,4). Vasodilatation occurs especially in the vascular beds of skeletal muscle and the coronary system. Vasodilatation may also occur in the gastrointestinal, cerebral, and renal beds, although usually less clearly than in skeletal muscle and the coronary system. Skin vessels are usually hardly affected. Fluid retention, as frequently occurs with classical vasodilators, is prevented by a direct renal tubular effect of the CA, resulting in moderately enhanced natriuresis This effect will probably contribute to the antihypertensive activity of the CA (8). On longterm use, CA may induce the regression of myocardial and vascular hypertrophy (9,10). Techniques are now becoming available which allow the investigation of the influence of CA on the microcirculation (11). In spontaneously hypertensive rats, verapamil, nifedipine, and felodpine in antihypertensive doses caused a pronounced dilatation of the smallest precapillary arterioles, which appear to play an important role in the vasodilator action of CA in the hypertensive animals. Furthermore, the spontaneous rhythmic vasomotion which occurs in 60% of the small arterioles of the hypertensive rats was suppressed by the CA (11).

At the cellular level, the selective blockade of the influx of extracellular calcium ions through specific L-channels is the major mechanism explaining vasodilatation. In addition, the subtle interaction between CA and α-adrenoceptors and angiotensin II-receptors may play a role.

Several attempts have been made to establish a causal role of calcium metabolism and fluxes in the pathogenesis of hypertension, but such a relationship has not been demonstrated convincingly (12). Accordingly, CA, although effective and useful antihypertensives, cannot be considered as a causal treatment.

ANGINA PECTORIS

At the macroscopic level two major effects of CA may be expected: 1) coronary vasodilatation usually not associated with a steal effect, in particular the relief of spasm, causing an improved supply of oxygen and enhanced regional perfusion; 2) peripheral arteriolar vasodilatation, causing a reduction in peripheral vascular resistance and cardiac afterload. In addition, verapamil, diltiazem, and related compounds will cause slowing of the heart rate and hence a reduction in myocardial oxygen consumption. The complex interrelationship between these various effects is visualized in fig. 1. The outcome of the various effects is an improvement of the unfavorable balance between cardiac oxygen supply and consumption which is known to be the background of angina pectoris. The type of angina (stable, vasospastic, mixed) and the hemodynamic characteristics of the individual patient largely determine which hemodynamic changes caused by the CA will be the major background of its therapeutic efficacy (13). At the cellular level, the mechanisms involved are the same as discussed above for the vasodilator activity in antihypertensive treatment. The hemodynamic effects of CA are the major background of their therapeutic potency in angina. In addition, a cytoprotective mechanism may also play a role. A few cytoprotective agents are now available, for instance, ranolazine, R 56865, CERM 11956, and trimetazidine (14,15). Such compounds are effective anti-ischemic drugs with very little hemodynamic activity. They will counteract the deleterious rise in cellular calcium levels (calcium overload) as a result of ischemia.

CEREBRAL ISCHEMIA

Nimodipine and flunarizine are the two CA which have been investigated in some detail as potential therapeutics in cerebrovascular disease associated with ischemia. The effect of flunarizine is predominantly based upon cytoprotective activity, possibly associated with the blockade of a special type of T-channels in cerebral tissues (16). Nimodipine has been reported to improve the outcome of subarachnoidal hemorrhage (17), probably as a result of vasodilatation and the relief of spasm. Preferential cerebral vasodilatation by nimodipine has been claimed on the basis of *in vitro* experiments, but this is very difficult to demonstrate *in vivo*. The beneficial effect of nimodipine on the outcome of acute stroke is uncertain. Isradipine has been shown to offer cerebroprotective activity, but so far only in animal models of cerebral ischemia.

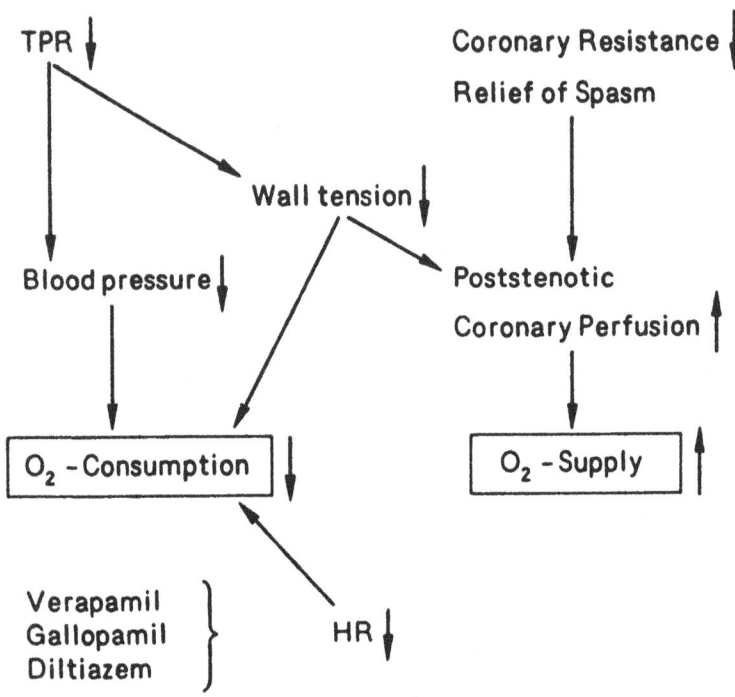

FIG. 1. Schematic presentation of the potential mechanism of CA with respect to their beneficial effect in angina pectoris. The final result is an improvement of the imbalance between myocardial oxygen consumption and supply. TPR = total peripheral resistance; HR = heart rate.

EXPERIMENTAL DIABETES

CA, although not antidiabetic agents, are frequently used by diabetic patients with concomitant cardiovascular diseases. For this reason it seems of interest to study the hemodynamic profile of CA in diabetic animals and patients. Apart from the well-known vascular changes associated with diabetes, attention should also be paid to diabetic cardiomyopathy. In recent experiments with diabetic rats (streptozocin-induced), we found an increased sensitivity of the (isolated) heart to various types of CA, both with respect to the negative inotropic and the coronary dilator activities of these drugs (18). This finding may support the hypothesis of different handling of calcium ions by the diabetic heart (19). We also found that nifedipine is more effective against myocardial ischemic injury in hearts from diabetic animals compared with appropriate controls (18). The investigations on the hemodynamic profile of CA in the diabetic state will be continued.

CONCLUSIONS

The macrohemodynamic effects of CA under pathological conditions are now reasonably well understood and most of the therapeutic activities and side-effects of these compounds can be explained satisfactorily on the basis of hemodynamic changes. There are considerable hemodynamic differences between the three major groups of CA. Most of the hemodynamic behavior of the CA is associated with peripheral arterial and coronary arterial dilatation, but attention should also be paid to the influence of verapamil, diltiazem, and related drugs on nodal tissues. Owing to recent technological developments investigations on the influence of CA on the microcirculation have now been initiated, also under pathological conditions. The hemodynamic background of the therapeutic activity of the CA in angina is particularly complex and very much dependent on the characteristics of the individual patient. The introduction of a new type of drug with a strong cytoprotective effect but little or no hemodynamic activity seems an interesting new development. The useful influence of a few CA in certain neurological disorders is partly based on cerebral vasodilatation and the relief of spasm, but other mechanisms probably play a role as well. It is uncertain whether calcium antagonistic activity is the only mechanism underlying the beneficial effects of CA in neurological disorders. Finally, interest has developed for the hemodynamic effects of CA in diabetic patients and in experimental diabetes.

REFERENCES

1.   Lehmann H.U., Hochrein H., Witt E. and Mies H.W. (1983): *Hypertension*, 5(suppl.2): 66-73.
2.   Struyker-Boudier H.A.J., De Mey J.G., Smits J.F.M. and Nievelstein H.M.N.W. (1989): In: *Clinical Aspects of Calcium Entry Blockers*, edited by P.A. van Zwieten pp. 21-66. Karger Verlag, Basel.
3.   Morel N. and Godfraind T. (1990): *Eur. Neurol.*, 30(suppl. 2): 10-15.
4.   Lund-Johansen P. (1983): In: *Clinical Aspects of Essential Hypertension. Handbook of Hypertension 1*, edited by J.I.S. Robertson pp. 151-173. Elsevier, Amsterdam.
5.   Lund-Johansen .P (1983): *Hypertension*, 5(suppl.3): 49-57.
6.   Muiesan G., Agabiti-Rosei E., Castellano M., Alicandri C.L., Corea L., Fariello R., Meschi M. and Romanelli G. (1982): *J. Cardiovasc. Pharmacol.*, 4(suppl.3): 325-329.
7.   Kiowiski W., Linder L. and Bühler F.R. (1990): *J. Cardiovas. Pharmacol.*, 16(suppl.6): S7-S10.

8.   Leonetti G., Cuspidi C., Sampieri L., Terzoli L. and Zanchetti A. (1982): *J. Cardiovasc. Pharamcol.*, 4: 319-324.
9.   Lundin S.A. and Hallbäck-Nordlander M.I.L. (1984): *J. Hypertens.*, 2: 11-18.
10.   Kobrin I., Sesoko S., Pegram B.L. and Frohlich E.D. (1984): *Cardiovas. Res.*, 18: 158-162.
11.   Messing M., van Essen H., Smith T.L., Smits J.F.M. and Struyker-Boúdier H.A.J. (1991): *Eur. J. Pharmacol.*, 198: 189-195.
12.   Hansson L. and Dahlöf B. (1989): In: *Clinical Aspects of Calcium Entry Blockers*, edited by P.A. van Zwieten pp. 103-124. Karger Verlag, Basel.
13.   Van Gilst W.H. and Lie K.I. (1989): In: *Clinical Aspects of Calcium Entry Blockers*, edited by P.A. van Zwieten pp. 88-102. Karger Verlag, Basel.
14.   Massingham R. and John G. (1990): *Progr. Pharmacol.*, 8: 109-138.
15.   Boddeke H.W.G.M., Hugtenburg J.G., Jap T.W., Heijnis J.B. and van Zwieten P.A. (1989): *Trends Pharmacol. Sci.*, 10: 397-400.
16.   Peters Th., Wilffert B., Vanhoutte P.M. and van Zwieten P.A. (1991): *J. Cardiovasc. Pharmacol.* (in press).
17.   Pickard J.D., Murray G.D. and Illingworth R. (1989): *Br. Med. J.*, 298: 636-642.
18.   Heijnis J.B., Mathy M.J. and van Zwieten P.A. (1991): *J. Cardiovasc. Pharmacol.*, 17: 983-989.
19.   Heyliger C.E., Prakash A. and McNeill J.H. (1987): *Am. J. Physiol.*, 252: H540-H544.

CALCIUM, ISCHEMIA, AND THE CALCIUM ANTAGONISTS

Winifred G. Nayler

Department of Medicine, University of Melbourne
Austin Hospital, Heidelberg 3084, Australia

Depending upon the duration and severity of an ischemic episode the heart either becomes "stunned" (1), "hibernates" (2), or infarcts (3). "Stunning" is best described as a postischemic condition, and is characterized by prolonged but not permanent depressed mechanical function of potentially jeopardized myocardium which has been salvaged by reperfusion after a period of ischemia of insufficient duration and intensity as to precipitate irreversible injury. "Stunning" therefore describes a state of reversible contractile dysfunction which occurs when coronary flow is reinstated after a short period of ischemia (4). The state of "hibernation" on the other hand refers to a state of sustained contractile dysfunction which occurs when coronary flow is impaired, but not to such an extent as to cause permanent loss of contractile function and tissue necrosis. The "hibernating" heart, therefore, is a chronically underperfused heart but, in contrast to the "stunned" or irreversibly injured heart, its normal contractile activity is rapidly restored once the defect in coronary flow is removed (2). "Stunning" and "hibernation" therefore are conditions which are quite dissimilar from that of infarction, since this latter condition is associated with an irreversible loss of structure and function which persists even if coronary perfusion is restored following a prolonged period of occlusion.

THE BIOCHEMISTRY OF THE ISCHEMIC HEART

The mammalian heart, even when quiescent, utilizes approximately 10 mol adenosine triphosphate (ATP)/g wet weight/minute in maintaining ionic homeostasis. A heart which is beating at the rate of 70 beats/minute requires the production of another 23 mol ATP/g wet weight/minute. Despite its absolute dependence on the availability of ATP, the cardiac reserves of ATP are relatively small, and even the maintenance of these reserves depends upon an uninterrupted coronary perfusion, with its attendant supply of O_2, glucose, and purine precursors. Early during an ischemic event, the heart utilizes its reserves of creatine phosphate (CP) (5), but these are exhausted within a few minutes, and the heart therefore switches to anaerobic metabolism. This provides a temporary respite in

49

terms of ATP production, but it creates additional problems because under these conditions the heart become acidotic, loses its osmotic equilibrium and generates free radicals. At the same time, the integrity of the sarcolemma is threatened, in part because the osmotic-induced swelling of the myocytes imposes an additional stress on an already fragile plasmalemma, in part because the free oxyradicals which are produced under these conditions favor lipid peroxidation, and in part because of phospholipase A_2 and protease activation.

Even this sequence of events fails to describe many aspects of the problem. For instance the mitochondria become swollen, vacuolated, and finally disrupt. Lysis of the myofibrils becomes evident, and the tissue loses its ionic homeostasis with respect to Na^+, and probably more importantly, to Ca^{2+}. It is this relatively early loss of homeostasis with respect to Ca^{2+} (6,7) which provides us with one of the vital clues as to the sequence of events which ultimately results in irreversible injury and cell death (6).

Calcium Antagonists (Calcium Channel Blockers) and the Ischemic Heart

The use of calcium antagonists as protective agents with respect to the ischemic heart involves their ability to act:

(a) as energy sparing agents (8,9);

(b) as vasodilators;

(c) as anti-atherosclerotic agents;

(d) as agents which, when used prophylactically, prevent the early ischemic-induced rise in cytosolic Ca^{2+} (7).

This paper describes the ability of one particular Ca^{2+} antagonist, verapamil, to slow the early ischemia-induced rise in cytosolic Ca^{2+}.

METHODS

Hearts excised from adult male Sprague Dawley rats anaesthetized with diethylether/O_2 mixture were used for these experiments. The isolated hearts were immersed in ice-cold Krebs-Henseleit (K-H) buffer gassed with 95% O_2 and 5% CO_2. The K-H buffer contained, in mmol/l: NaCl 119; $NaHCO_3$ 25.0; KCl 4.6; NaH_2PO_4 1.2; $MgSO_4$ 1.2; CaCl 1.3; and glucose 11.0. After contractions had ceased the hearts were mounted on a Langendorff perfusion apparatus for nonrecirculating perfusion at 37°C. The hearts beat spontaneously, and, unless otherwise stated, were perfused at a coronary flow rate of 10-12 ml/minute.

Perfusion Protocols.
Control hearts were continuously perfused with K-H buffer as already

described.

When globally ischemic hearts were required, hearts which had been initially perfused for 30 minutes with K-H buffer were made globally ischemic for 30 minutes by totally occluding coronary flow. Irrespective of coronary flow, the temperature of the hearts was maintained at 37°C, by means of a Bruker variable temperature unit, or temperature-controlled water jacket.

When hearts were reperfused, the perfusion buffer was K-H without verapamil.

Treatment Protocol.

In some of the experiments either (+), (-), or (±) verapamil was added to the perfusate 15-30 minutes before the hearts were made ischemic. Verapamil was added to provide a final concentration of 10^{-7} M.

Assays

Chemical determination of cellular ATP.

Hearts for chemical analysis of ATP and CP were freeze-clamped between large stainless steel tongs precooled in liquid N_2. ATP and CP were determined enzymatically, as described by Lamprecht and Trautschold (10) and Lamprecht et al. (11), respectively.

Tissue calcium content.

The method for determining total tissue Ca^{2+} has been described in detail elsewhere (12). In brief the coronary vasculature is flushed with 10 ml of ice-cold, Ca^{2+}-free solution containing 0.35 mM sucrose and 5 mM histidine at pH 7.4. Hearts are then blotted, dried at 100°C to constant weight, and digested in HNO_3. Samples of the HNO_3 extracts are then assayed for Ca^{2+} by atomic absorption spectrometry as previously described (12).

Measurement of $[Ca^{2+}]_i$ by NMR spectroscopy using 5F-BAPTA.

This technique has been described in detail by Marban and his colleagues (13). Hearts were loaded with the intracellular calcium indicator 5F-BAPTA (1,2-bis(e-amino-5-fluorophenoxy) ethan $N_1N_1N^1_1N^1_1$tetra-acetic acid) for up to 2 hours before the start of each experiment, and each heart was subjected to variable periods of global ischemia followed by aerobic reperfusion.

[19]F Spectra were acquired in 7-minute blocks (spectra accumulated over a 7-minute interval). The ratio of peak area of the Ca-5F-BAPTA complex and the free 5F-BAPTA was used to determine the average (average systolic/diastolic) intracellular Ca^{2+} concentration $[Ca^{2+}]_i$ for each 7-minute interval.

Statistical Analysis.

Results were analyzed for significance using Student's t test, and taking $p < 0.05$ as the limit of significance. Unless otherwise stated, results are expressed as mean $\pm$ SEM of 6 separate experiments.

Reagents.

$(+)$, $(-)$, and $(\pm)$ verapamil were obtained from Knoll AG, Ludwigshafen, Germany. All other reagents were of analytical grade, and were obtained from Sigma Chemical Co. (St Louis, Missouri, USA).

TABLE 1.   Effect of 10^{-7} M ($\pm$) Verapamil on the Ischemia-Reperfusion Induced Increase in Tissue Ca^{2+a}

Protocol	Tissue Ca^{2+} (mol/g dry wt)	Sig
Aerobic Perfusion		
30 min aerobic perfusion	4.1 ± 0.3	N.S.
10^{-7}M verapamil	4.0 ± 0.6	
Ischemia		
30 min global ischemia	4.5 ± 0.6	
30 min global ischemia + 10^{-7}M verapamil	4.6 ± 0.8	N.S.
Reperfusion after 30 min global ischemia		
5 min reperfusion	11.6 ± 0.9	
5 min reperfusion + 10^{-7}M verapamil	6.9 ± 1.1	$p < 0.01$
15 min reperfusion		
15 min reperfusion + 10^{-7}M verapamil	17.2 ± 0.6	$p < 0.01$
	8.3 ± 0.9	

[a]Results are mean $\pm$ of 6 separate experiments. Tests of significance calculated with respect to the difference caused by 10^{-7}M verapamil pretreatment, taking $P = 0.05$ as the limit.

Perfusion with 10^{-7}M verapamil for 30 minutes prior to the induction of ischemia failed to cause any change in total tissue Ca^{2+} (table 1). Upon reperfusion, however, the verapamil-pretreated hearts gained significantly less Ca^{2+} than the placebo-treated controls, irrespective of whether perfusion was for 5 or 15 minutes (table 1).

RESULTS

Effect of ($\pm$) Verapamil on Tissue Ca^{2+} During Ischemia and Post-ischemic Reperfusion

After 60 or 75 minutes aerobic perfusion the hearts contained 4.8 ± 0.3 and 4.6 ± 0.5 mmol Ca^{2+}/g dry weight. These levels are not significantly different from the levels obtained from freshly excised, nonperfused hearts (4.1 ± 0.3 mole Ca^{2+}/g dry weight). After 30 minutes K-H perfusion, 30 minutes global ischemia failed to cause any change in total tissue Ca^{2+} (table 1). However, reperfusion after 30 minutes global ischemia caused tissue Ca^{2+} to increase to 11.6 ± 0.9 mol/g dry weight during the initial 5 minutes of reperfusion. Fifteen minutes of reperfusion resulted in tissue Ca^{2+} levels of $17.2+0.6$ mol/g dry weight.

TABLE 2.   Effect of Verapamil Pretreatment on
the ATP and CP Content of Aerobically Perfused, Ischemic, and
Reperfused Hearts[a]

Protocol	ATP (moles/g dry wt)	CP
Aerobic Perfusion		
30 min aerobic perfusion	26.8 ± 1.6	35.3 ± 2.8
30 min aerobic perfusion + 10^{-7}M verapamil	27.5 ± 2.1	33.9 ± 1.7
Ischemia		
15 min ischemia	15.2 ± 2.1	5.0 ± 0.6
15 min ischemia + 10^{-7}M verapamil	20.6 ± 1.8	6.7 ± 1.2
sig		
30 min ischemia	$p < 0.05$	N.S.
30 min ischemia + 10^{-7}M verapamil	7.8 ± 0.4	3.5 ± 0.9
sig	11.9 ± 0.9	5.2 ± 1.1
	$p < 0.01$	N.S.
Reperfusion after 30 min ischemia		
15 min reperfusion		
15 min reperfusion + 10^{-7}M verapamil	14.2 ± 1.6	11.8 ± 2.1
sig	21.4 ± 2.1	16.6 ± 3.1
30 min reperfusion	$p < 0.01$	$p < 0.01$
30 min reperfusion + 10^{-7}M verapamil	16.0 ± 2.2	20.9 ± 3.1
sig	26.3 ± 1.8	25.6 ± 1.8
	$p < 0.01$	$p < 0.05$

[a]Results are mean ± SEM of 6 separate experiments. Tests of significance relate to the difference due to the presence of 10^{-7}M (±) verapamil. In each case, verapamil was added 15 minutes prior to making the hearts globally ischemic.

Adding (±) verapamil to the perfusion buffer failed to alter the tissue levels of ATP and CP in the aerobically perfused hearts but it slowed ($p < 0.01$) the loss of ATP during the ischemic episode, and hastened its recovery upon reperfusion (table 2).

Effect of Verapamil on Tissue Reserves of ATP and CP During Ischemia and Postischemic Reperfusion

Global ischemia causes a progressive reduction in the tissue reserves of ATP and CP. After 15 minutes of ischemia (table 2) tissue CP had fallen by 86%, and ATP by 57% (table 2). Even greater loss of ATP and CP occurred if the duration of the ischemic events was extended to 30 minutes (table 2). Reperfusion after 30 minutes of ischemia resulted in some recovery of the tissue reserves of ATP and CP (table 2) but even after 30 minutes reperfusion recovery was incomplete.

Effect of (+) and (-) Verapamil on Cytosolic Ca^{2+} $[Ca^{2+}]^i$ During Global Ischemia

During aerobic perfusion the mean (average between cytosolic and diastolic $[Ca^{2+}]_i$ levels remained constant (602 ± 20 nM). Adding 10^{-7}M (-) verapamil reduced this level to 501 ± 12 nM, but (+) verapamil had no effect. After 7 minutes of ischemia $[Ca^{2+}]_i$ in the placebo and (+) verapamil-treated hearts had increased significantly ($p < 0.001$), to 812 ± 16 and 812 ± 31 nM respectively. Hearts pretreated with (-) verapamil also showed an increase in cytosolic Ca^{2+}, with levels of 648 ± 12 nM being obtained, but the increase was significantly less ($p < 0.01$) than that exhibited by the placebo on (+) verapamil-treated hearts. This difference became more marked as the duration of the ischemic interval was prolonged.

DISCUSSION

These results show that in a model of global ischemia which shows the expected decline in the tissue reserves of ATP and CP and in which reperfusion is accompanied by substantial gains in tissue Ca^{2+}, cytosolic Ca^{2+} $[Ca^{2+}]^i$ increases early during the ischemic event, prior to reperfusion. These results agree well with those already described by other investigators (14). Other data presented here relates to the ability of verapamil-pretreatment to slow this ischemia-induced rise in $[Ca^{2+}]_i$ in an isomer-dependent manner. The results indicate that the (-) isomer, and hence the isomer which has the greatest calcium antagonist blocking activity, is more effective than the (+) isomer in attenuating the ischemia-induced increase in cytosolic Ca^{2+}. Presumably, therefore, the ability of verapamil to attenuate this early rise in cytosolic Ca^{2+} is a calcium antagonist-dependent property of the compound. Several questions now need to be answered. They include:
(a) what is the likely source of the Ca^{2+} that accumulates in cytosol;
(b) does the time course of the rise in $[Ca^{2+}]_i$ parallel that of the decline

in the tissue reserves of ATP and CP; and

(c) does the attenuation of this rise in $[Ca^{2+}]_i$ contribute to the cardioprotective effect of verapamil?

The Source of the Ca^{2+} Involved in the Early Rise in Cytosolic Ca^{2+}

The Ca^{2+} which accumulates in the cytosol early during the ischemic episode and which precedes any overall gain in tissue Ca^{2+} probably originates from several different sources. Possible sources include the penetration of extracellular Ca^{2+} across the sarcolemma by way of the voltage-sensitive Ca^{2+} channels or in exchange for Na^+ or because of H^+ ion-induced displacement of previously bound Ca^{2+} (H^+-ion accumulation being one of the earliest consequences of ischemia). Loss of Ca^{2+} from the sarcoplasmic reticulum either because of the failure of the C^{2+} accumulation process, or because of leakage of previously stored Ca^{2+} may also be involved. Feher and his colleagues (15) have recently shown that an enhanced rate of Ca^{2+} efflux by way of the Ca^{2+} release channels occurs after only ten minutes of ischemia. The cause of this enhanced rate of release is at present uncertain. Possibilities include increased release prompted by raised tissue levels of inositol triphosphate, and enhanced Ca^{2+}-induced Ca^{2+}-release.

Relationship Between the Time-course of the Ischemia-induced ATP and CP Depletion and the Accompanying Rise in Cytosolic Ca^{2+}

Although ATP (and CP) depletion is an early manifestation of ischemia-induced metabolic injury, the tissue reserves of ATP which exist after only ten minutes of ischemia should provide sufficient substrate for the various ATPase enzymes which are concerned with maintaining Ca^{2+} homeostasis, unless the tissue reserves of ATP are severely compartmentalized. Therefore, there seems to be little reason for anticipating a simple linear correlation between the rate of ATP depletion under these conditions and the appearance of a raised $[Ca^{2+}]_i$. A direct effect of ATP depletion on the Ca^{2+} release channels of the reticulum is probably not involved, because ATP holds these channels in their open state (16).

The Consequences of the Early Rise in $[Ca^{2+}]_i$

The consequences of the early rise in cytosolic Ca^{2+} are complex. They include:

(a) activation of the endogenous proteases and phospholipases;

(b) the mobilization of endogenous catecholamines;

(c) stimulation of O_2-derived free radical production;

(d) desensitization of the myofibrils to Ca^{2+};

(e) ATP wastage; and

(f) further inhibition of mitochondrial function.

Against this background of events it could be anticipated that any agent which attenuates the increase in cytosolic Ca^{2+} which occurs early during the ischemic event should be beneficial. Once again, however, this benefit will only be derived if treatment starts early, or preferably before, the onset of the ischemic episode. Clinical trials support this contention (17).

In conclusion, the present results show that pretreatment with verapamil attenuates the early rise in cytosolic Ca^{2+} which occurs during the first few minutes of ischemia. The effect shows stereospecificity, with (-) isomer being more effective than (+) isomer. Because the early rise in cytosolic Ca^{2+} precedes uncontrolled Ca^{2+} gain and severe ATP depletion, it may provide the primary focus for pharmacological interventions aimed at protecting the ischemic heart from the cascade of events which otherwise would culminate in cell death and tissue necrosis.

ACKNOWLEDGMENT

These investigations were supported by the National Health and Medical Research Council of Australia.

A similar paper on this topic was presented at a satellite symposium of the European Society for Hypertension meeting in 1991.

REFERENCES

1. Heyndrickx G.R., Millard R.W., McRitchie R.J., Maroko P.R. and Vatner S.F. (1975): *J. Clin. Invest.*, 56: 978-985.
2. Rahimtoola S.H. (1989): *Am. Heart J.*, 117: 211-221.
3. Jennings R.B. and Reimer K.A. (1982): *Am. J. Pathol.*, 102: 241-255.
4. Braunwald E. and Kloner R.A. (1982): *Circulation*, 66: 1146-1149.
5. Dunn R.B. (1984): *Circ. Res.*, 54: 405-413.
6. Nayler W.G. (1981): *Am. J. Pathol.*, 102: 262-270.
7. Steenbergen C., Levy L., Murphy E. and London R.E. (1987): *Circ. Res.*, 60: 700-710.
8. Nayler W.G. and Szeto J. (1972): *Cardiovasc. Res.*, 6: 120-128.
9. Nayler W.G., Ferrari R. and Williams A.J. (1980): *Am. J. Cardiol.*, 46: 242-248.
10. Lamprecht W. and Trautschold W. (1974): In: *Methods of Enzymatic Analysis* vol. 4, edited by H.U. Bergmeryer pp. 2101-2110. Academic Press, New York.
11. Lamprecht W., Stein P., Heinze F. and Weisser H. (1974): In: *Methods of Enzymatic Analysis* vol. 4, edited by H.U. Bergmeyer

pp. 1777-1781. Academic Press, New York.

12. Nayler W.G., Perry S.E., Elz J.S. and Daly M.J. (1984): *Circ. Res,* 55: 227-237.

13. Marban E., Kitzakaze M., Kusuoka H., Porterfield J.D., Yue D.T. and Chacko V.P. (1987): *Proc. Natl. Acad. Sci. USA,* 84: 6005-6009.

14. Jennings R.B., Hawkins H.K., Lowe J.E., Hill M.L. Klotman S. and Reimer K.A. (1978): *Am. J. Pathol.,* 92: 187-214.

15. Feher J.J., Le Bolt W.R. and Manson N.H. (1989) *Circ. Res.,* 65: 1400-1408.

16. Williams A.J. and Ashley R.H. (1989): *Ann. N.Y. Acad. Sci.,* 560: 163-173.

17. The Danish Study Group on Varapamil in Myocardial Infarction (1990): *Am. J. Cardiol.,* 66: 779-785.

MULTIPLE ORGAN PROTECTION THROUGH ENDOTHELIAL CELL STABILIZATION: A MECHANISM OF ACTION OF CALCIUM ANTAGONISTS?

Fred De Clerck, Walter Janssens, and Fons Verheyen

Department of Cardiovascular Pharmacology
Janssen Research Foundation
B-2340 Beerse, Belgium

Depending upon their pharmacological profile of activity (1,2), calcium antagonists differentially exert beneficial effects in experimental animals and in man. Such effects include alleviation of high blood pressure, myocardial ischemia and infarct size (3-6), reduction of cerebral infarct size (7), protection against experimental renal injury (8), atherosclerosis, and vascular hypertrophy (3,9,10).

Vascular smooth muscle (VSM) cell relaxation/anti-constriction in veins, resistance arterioles or coronary arteries, negative inotropic effects of heart cells via a blockade of voltage- or receptor-operated calcium channels (1-4), protection of myocardial and cerebral cells against exposure to anoxia via a blockade of calcium overload (5-7) are mechanisms mainly proposed to explain the beneficial effects of various calcium antagonists in hypertension, migration, proliferation, matrix protein synthesis, necrosis, and increase in number of low-density lipoprotein (LDL) receptors and increase in lysosomal cholesterol ester hydrolase activity are mechanisms that may underlie the protection provided by some calcium antagonists against vessel wall degeneration during atherogenesis or hypertension (9-16). However, several observations suggest that a functional and/or anatomical "stabilization" of endothelial cells (17) may contribute to the protection some calcium antagonists provide against vasculopathy in several vascular beds, for instance in hypertension or during a cerebral infarction (7,15-20).

Evidence for protection of endothelial cells by various calcium antagonists, including flunarizine, a type IV calcium antagonist (1) or a calcium overload blocker (2,21) is obtained in various experimental settings. Detachment of endothelial cells from the vessel wall (desquamation) as well as the subsequent adhesion of platelets to subendothelial tissue and the increase in vascular permeability elicited in rats by a transient imbalance of Ca^{2+}-homeostasis after intravenous injection of citrate is reduced by flunarizine, nifedipine, and verapamil (table 1).

T. Godfraind et al. (eds.), Calcium Antagonists, 59–63.

Stabilization of intercellular glycoproteins or the cellular glycocalix resulting in a reduction of cellular membrane or intercellular permeability for various molecules or, alternatively, inhibition of a transmembrane calcium influx attenuating endothelial cell contraction and gap formation, are possible mechanisms for such pharmacological effects (22,23).

TABLE 1.   Protection by calcium antagonists
                 against endothelial cell desquamation
                       induced[a] citrate IV in rats

Compound	Optimal protective dose (mg/kg orally)
Flunarizine	0.1
Nifedipine	0.09
Prenylamine	0.25
Verapamil	1

[a]From (22,23)

Supportive for the former possibility are the observations that flunarizine, at doses reducing VSM cell intimal proliferation and atheroma formation in electrically stimulated carotid arteries of cholesterol-fed rabbits, reduces the uptake of ruthenium red as well as of horseradish peroxidase into the cytoplasm of endothelial cells, indicating an action on the glycocalix and on the increased permeability of the endothelium for large molecules (24,25). The reduction by e.g. flunarizine, niludipine, nifedipine, and verapamil of edematous damage to pulmonary endothelial cells elicited by arachidonic acid IV in mice may involve such a mechanism as well (26).

Additionally, flunarizine and nimodipine curtail the deposition of horseradish peroxidase in the interendothelial spaces at sites of electrical stimulation of the carotid artery in rabbits (25); this effect may be the consequence of an inhibition of endothelial cell contraction, possibly via a reduction of membrane permeability for calcium or via an interaction with contractile filaments in these cells (27,28). A reduction by calcium antagonists such as flunarizine, nisoldipine, and verapamil of endothelial cell contraction is corroborated by the inhibition these compounds produce of gap formation between endothelial cells and of vascular permeability increases elicited by histamine, serotonin, bradykinin, and arachidonic acid in rat skin (29).

The molecular/biochemical mechanisms by which calcium antagonists act on endothelial cells to reduce contraction or maintain membrane integrity remain elusive.

A number of experiments probing into the role of calcium in the activation/release of endothelium-derived relaxing factors (EDRF) on the one hand suggest that endothelial cells may carry voltage-operated channels (VOC). Such VOC would differ from those in VSM cells; they are activated by the calcium channel agonist Bay K 8644 and by high K^+ depolarization and antagonized by nitrendipine. The observation that acetycholine, bradykinin, or adenine nucleotides can lead to calcium influx resulting in production/release of EDRF on the other hand suggest the presence of receptor-operated channels (ROC) in endothelial cells; however, such ROC, at least when linked to EDRF production, are not uniformly influenced by calcium antagonists such as flunarizine, verapamil, nifedipine, or diltiazem *in vitro* (30). Preservation of endothelium-dependent vascular relaxation by israpidine in aortas of cholesterol-fed rabbits (9,31) and by benidipine in mesenteric arteries of rats subjected to splanchnic ischemia/reperfusion (32) may occur indirectly via a reduction by these compounds of atherogenic vessel wall lesions and lipid accumulation of reperfusion damage respectively. Nevertheless, contraction of cultured endothelial cells, derived from bovine pulmonary microvessels, guinea pig inferior vena cava or canine aorta, elicited by agnonists such as thrombin, histamine, PAF, bradykinin, or leukotrienes C4, D4, E4, is inhibited by EGTA, TMB-8, trifluoperazine, and ML-7 (27,28). It is also reduced by verapamil and flunarizine (33). This indicates that the process requires extracellular calcium, involves intracellular calcium translocations, is dependent upon Ca^{2+}-calmodulin and myosin light-chain kinase activity, but may also involve transmembrane calcium fluxes attenuated by calcium antagonists.

Alternatively, an interference of some calcium antagonists with membrane components, unrelated to specific calcium channels, may be the key to explain their effects on endothelial cells. For instance, flunarizine, but not verapamil or diltiazem (0.5 mg/kg IV), reduces the thrombotic obstruction elicited by free radical-induced damage to the endothelial lining of microvessels in rats, possibly via a reduction of lipid peroxidation. These observations suggest that flunarizine protects plasma membrane structural phospholipids against various forms of insults.

The latter compound suppresses atherogenesis in aortae of cholesterol-fed rabbits (9) and prevents sclerosis, endothelial cell alterations, migration of VSM cells, infiltration of monocytes and subendothelial fibroid deposition in arteries of kidneys, heart, and brain of Skelton-hypertensive rats at blood pressure and lipid neutral doses (10,19,20,36). A stabilization of endothelial cells via one of the mechanisms discussed thus may contribute to multiple organ protection by some calcium antagonists.

REFERENCES

1. Vanhoutte P.M. and Paoletti R. (1987): *TIPS*, 8: 4-5.
2. Van Zweiten P. (1986): *Eur. Neurol.*, 25(suppl.1): 57-67.
3. Van Zwieten P. (1989): *Am. J. Cardiol.*, 64: 117I-121I.
4. Vanhoutte P.M. and Cohen R.A. (1983): *Am. J. Cardiol.*, 52: 99A-103A.
5. Nayler W.G., Panagiotopoulos S., Elz J.S. and Sturrock W.J. (1987): *Am. J. Cardiol.*, 59: 75B-83B.
6. Nayler W.G., Liu J. and Panagiotopoulos S. (1990): *Cardiovasc. Drugs Ther.*, 4: 879-886.
7. Van Reempts J., Van Deuren B., Van de Ven M., Cornelissen F. and Borgers M. (1987): *Stroke*, 18: 1113-1119.
8. Schrier R.W. (1991): *Am. J. Med.*, 90(S5A): 21S-26S.
9. Bond G., Purvis C. and Mercuri M. (1991): *J. Cardiovasc. Pharmacol.*, 17(S4): S87-S93.
10. Bernini F., Capatano A.L., Corsini A., Fumagalli R. and Paoletti R. (1989): *Am. J. Cardiol.*, 64: 129I-134I.
11. Henry P. (1988): *Ann. N.Y. Acad. Sci.* 522: 411-419.
12. Chobanian A. (1987): *J. Hypertens.*, 5: S43-S48.
13. Kjeldsen K. (1989): *Proc. Soc. Exp. Biol. Med.*, 190: 219-228.
14. Weinstein D.B. (1989): *Am. J. Med.*, 86(S4A): 27-32.
15. Jackson C.L., Bush R.C. and Bowyer D.E. (1988): *Atherosclerosis*, 69: 115-122.
16. Pelissou-Guyotat I., Guyotat J., Lievre M. and Chignier E. (1991): *J. Cardiovasc. Pharmacol.*, 17: 778-785.
17. Hugenholtz P.G. and Lichtlen P. (1986): *Eur. Heart J.*, 7: 546-559.
18. Knorr A., Garthoff B., Kazda S. and Stasch J.P. (1991): *J. Cardiovasc. Pharmacol.*, 17: S2, S94-S100.
19. Schwabedal P.E., Oestreich W. and Szathmary S. (1989): *Ather. Cardiovasc. Dis.*, 4: 829-834.
20. Verheyen A., Schwabedal P.E., Pulina M., Borgers M., Szathmary S. and Oestreich W. (1989): *Ather. Cardiovasc. Dis.*, 4: 813-818.
21. Borgers M., De Clerck F., Van Reempts J., Xhonneux R. and Van Nueten J. (1984): *Inter. Angiol.*, 3: 25-31.
22. De Clerck F. and Hladovec J. (1984): In: *Calcium Entry Blockers in Cardiovascular and Cerebral Dysfunctions*, edited by T. Godfraind, A.G. Herman and D. Wellens pp. 81-90. Martinus Nijhoff Publishers, The Netherlands.
23. Hladovec J. and De Clerck F. (1981): *Angiology*, 32: 448-462.
24. Betz E., Hammerle H. and Strohschneider T. (1985): *Res. Exp. Med.*, 185: 325-340.
25. Strohschneider T. and Betz E. (1989): *Atherosclerosis*, 75: 135-144.
26. De Clerck F., Loots W., Somers Y., Van Gorp L., Verheyen A.

and Wouters L. (1985): *Arch. Int. Pharmacodyn.,* 274: 4-23.

27. Northover A.M. (1989): *Agents Actions,* 26: 367-371.

28. Morel N.M., Petruzzo P.O., Hechtman H.B. and Shepro D. (1990): *Inflammation,* 14: 571-583.

29. De Clerck F., Van Gorp L. and Verheyen F. (1986): *Blood Vessels,* 23: 50-52.

30. Rubanyi G. and Vanhoutte P.M. (1988): *Ann. N.Y. Acad. Sci.,* 522: 226-232.

31. Habib J.B., Bossaler C., Wells S., Williams C., Morrisett J.D. and Henry P. (1986): *Circ. Res.,* 58: 305-309.

32. Karasawa A., Rochester J.A., Ma X. and Lefer A.M. (1991): *Circ. Shock,* 33: 135-141.

33. De Clerck F., De Brabander M., Neels H. and Van De Velde V. (1981): *Thromb. Res.,* 23: 505-520.

34. De Clerck F., Beerens M., Thone, F. and Borgers M. (1981): *Thromb. Res.,* 24: 1-12.

35. Thomas P.G. and Verkleij A. (1987): *Biochem. Soc. Trans.,* 15: 1062.

36. Schwabedal P.E., Oestreich W. and Szathmary S. (1989): *Z. Kardiol.,* 78(S5): 108-111.

NEW INSIGHT INTO THE VASCULAR PROPERTIES OF LACIDIPINE

G. Gaviraghi and T. Godfraind*

Glaxo Research Laboratories, 37135 Verona, Italy
*Laboratoire de Pharmacologie, UCL 7350,
Université Catholique de Louvain, 1200 Bruxelles, Belgium

The concept that calcium ions can have pathogenetic potential when they accumulate excessively into the cytoplasm is particularly true for the cardiovascular system. Hypertension has been described as a suitable model to observe the functional and morphological consequences of calcium overload into the arterial wall (1). In this context, salt-loaded Dahl-S rats and Stroke-Prone spontaneously hypertensive rats (SHR-SP) have been described as useful animal models for the study of hypertension and of associated tissue damage, such as necrotizing vasculopathy (2-3). In fact, their common anomaly is a marked inability to keep the intracellular calcium concentration of vascular smooth muscle at a normal level. The functional consequence is an increased peripheral vascular resistance and the structural consequence is a vascular damage similar to certain forms of arteriosclerosis in humans.

Calcium antagonists have proven their therapeutic value as antihypertensive agents by preventing excessive calcium influx in arteries (4), protecting animals from functional and structural alterations (3). However, it is noteworthy that very high dosages of first generation agents, e.g., nifedipine or verapamil, have had to be used because of their short duration of action (5). Lacidipine is a second-generation 1,4-dihydropyridine derivative endowed with potent and long-lasting antihypertensive properties. In SHRs lacidipine induces a prolonged reduction in blood pressure at doses lower than 1 mg/kg (a dose which reduces systolic blood pressure by 25% is about 0.3 mg/kg p.o.) (6-7).

The specific pharmacological profile of lacidipine prompted us to investigate whether once-a-day prophylactic oral administration to Dahl-S rats prevented salt-induced hypertension other than development of vasculopathy and accelerated mortality. To this end, lacidipine has been studied in a wide range of doses from 0.1 and 0.3 mg/kg (roughly equivalent to the human therapeutic dose) up to 10 mg/kg.

As recently published, a 100% survival rate was observed with all doses of lacidipine, in spite of the fact that only the highest doses tested

T. Godfraind et al. (eds.), Calcium Antagonists, 65–69.
© 1993 *Kluwer Academic Publishers and Fondazione Giovanni Lorenzini.*

prevented increased blood pressure, when measured at 24 hours after treatment (8). Most important, a selective prevention of brain lesions and a dose-related vascular protection in other organs were observed. These results raised the question as to whether lacidipine could still protect salt-loaded Dahl-S rats when administered therapeutically to animals which had already developed sustained hypertension. Based on our previous studies, treatment was begun in week 4, when generally hypertension is fully developed and survival rate is still close to 100%. Once again, although all doses of lacidipine (0.3, 1, and 3 mg/kg) did not significantly decrease the elevated values of blood pressure, an almost complete protection from mortality was observed (9).

Confirming our previous observations, fulminant hypertension was associated with a diffuse and severe necrotizing vasculopathy. In particular, macro- and microscopic alterations were detected in the distal branches of the mesenteric arteries, where lacidipine dose

dependently inhibited the development of macroscopic alterations, not present when active treatment started. A clear regression of microscopic abnormalities was observed at the highest dose. Essentially, the same picture could be drawn for brain damage, which, also in this study, was the main cause for accelerated mortality.

These properties were further confirmed in SHR-SPs which develop accelerated mortality as a result of salt-induced cerebral apoplexy and renal lesions (2-3). In fact, all untreated animals died within 6 weeks of a salt-rich diet (1% NaCl in the drinking water), whereas all rats survived during the same period when treated with lacidipine at 0.3 mg/kg p.o. once a day. About 30% of these animals were still alive at 6 months, whereas at 1 mg/kg all salt-loaded SHR-SPs, were protected (fig. 1).

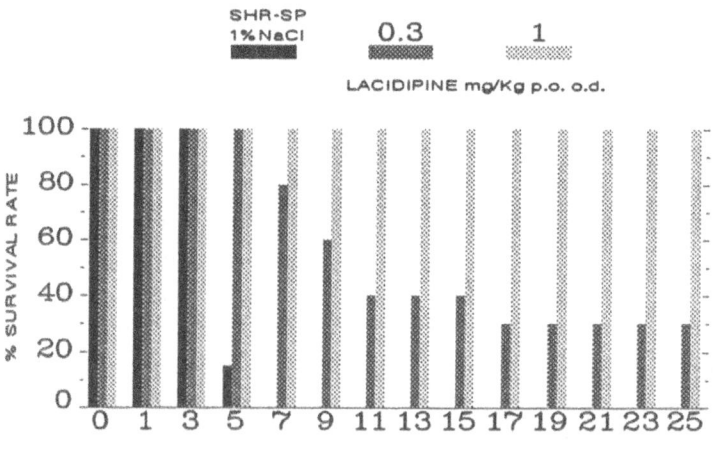

FIG. 1. Prevention of mortality in 1% NaCl-loaded Stroke-Prone SHRs by lacidipine after once-daily oral administration (n=20).

At 0.3 mg/kg lacidipine did not affect development of hypertension, whereas a slight reduction in systolic blood pressure, measured at 24 hours after daily treatment, was observed with 1 mg/kg (fig. 2).

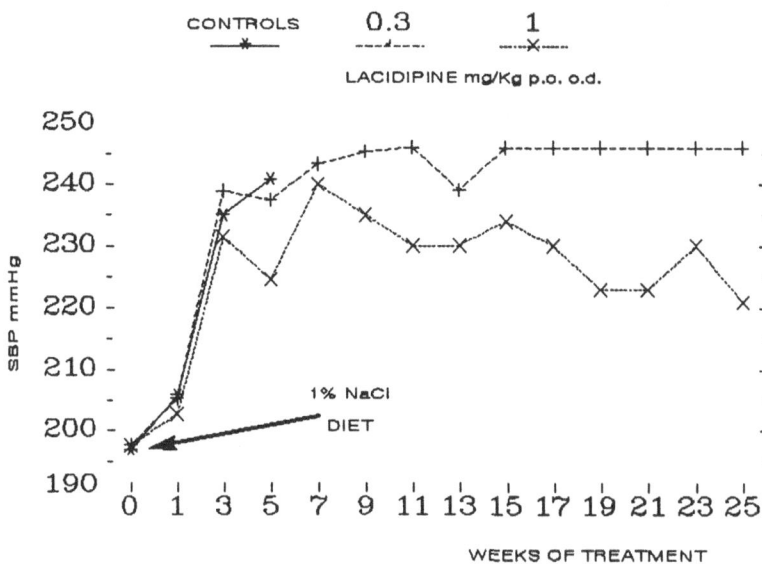

FIG. 2. Changes in systolic blood pressure (SBP) over 25 weeks in actively treated (0.3 and 1 mg/kg of lacidipine p.o.), compared with vehicle-treated (controls), salt-loaded Stroke-Prone SHRs.

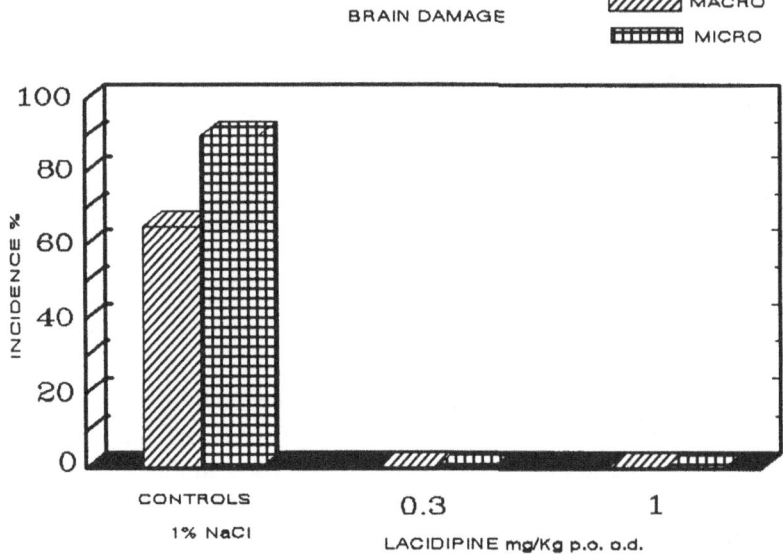

FIG. 3. Incidence of brain damage in actively treated (0.3 and 1 mg/kg of lacidipine p.o.) compared with vehicle-treated (controls), salt-loaded Stroke-Prone SHRs after 6 weeks of salt-rich diet.

Cerebral lesions were characterized by very marked brain edema and diffuse hemorrhages identified by macro- and microscopic examinations.

As shown in fig. 3, neither macro- nor microscopic abnormalities were observed in rats treated with lacidipine at 0.3 and 1 mg/kg, when all untreated animals had died.

As far as renal lesions are concerned, marked hypertrophy, necrosis and occlusions of renal arterioles, together with severe degenerative changes of tubules and glomeruli, were detected in almost all untreated animals. By contrast, normal histological appearance of renal structures was maintained in salt-loaded SHR-SPs treated for 6 months with lacidipine at 1 mg/kg.

The vasoprotective action of very low doses of lacidipine could be related to its potent and long-lasting calcium entry blocker activity. Indeed, washout studies in isolated vascular preparations have recently revealed a selective and sustained interaction of lacidipine with the inactivated state of voltage-operated calcium channels (10) (fig. 4), an observation consistent with the demonstration by Herbette et al. (11) that this drug has physicochemical properties which could be responsible for its peculiar interaction with DHP binding sites.

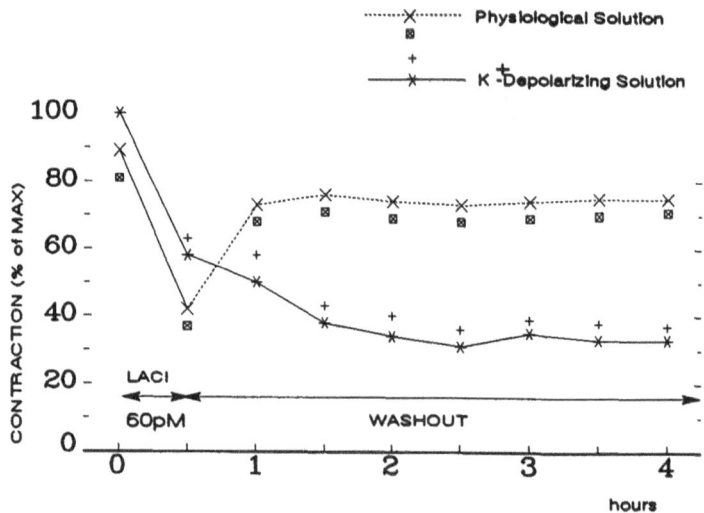

FIG. 4. Rat aorta: reversal of lacidipine inhibition in drug-free calcium-deprived physiological solution (-----) or in 40 mM K^+-depolarizing solution (——). Ordinate: response to 100 mM K^+-solution, pulse of 2 minutes expressed as % of controls; abscissa: time after preincubation in the presence of lacidipine (LACI) 60 pM.

Furthermore, it has been observed that accelerated mortality shown by salt-loaded SHR-SPs, is closely associated with a very pronounced increase in plasma lipid peroxides, which is completely inhibited by protective doses of lacidipine (F. Ursini, unpublished data); such an antioxidant activity might contribute to its tissue-protection action (7).

In conclusion, the results summarized above strongly suggest that vascular protection by lacidipine might be due to a combination of long-lasting calcium entry blocking activity and antioxidant properties.

REFERENCES

1. Fleckenstein A., Frey M. and Zorn J. (1987): *Trends Pharmacol. Sci.*, 8: 496-501.
2. Rapp J.P. and Dene H. (1985): *Hypertension*, 7: 340-349.
3. Luckhaus G., Nash G., Garthoff B., Kazda S. and Feller W. (1985): *Arzneimittelforschung (Drug Res.)*, 35: 115-121.
4. Godfraind T., Kazda S. and Wibo M. (1991): *Circ. Res.*, 68: 674-682.
5. Fleckenstein A., Frey M., Zorn J. and Fleckenstein-Grün, G. (1990): In: *Hypertension: Pathophysiology, Diagnosis and Management*, edited by H. Laragh and B.M. Brenner pp. 471-509. Raven Press, New York.
6. Micheli D., Collodel A., Semeraro C., Gaviraghi G. and Carpi C. (1990): *Cardiovasc. Pharmacol.*, 15: 666-675.
7. Micheli D., Ratti E., Toson G. and Gaviraghi G. (1991): *J. Cardiovasc. Pharmacol.*, 17(suppl.4): S75-S86.
8. Cristofori P., Terron A., Micheli D., Bertolini G., Gaviraghi G. and Carpi C. (1991): *J. Cardiovasc. Pharmacol.*, 17(suppl.4): S75-S86.
9. Gaviraghi G., Micheli D., Terron A. and Cristofori, P. (1991): *J. Cardiovasc. Pharmacol.* 18(suppl.11): S7-S12.
10. Godfraind T. and Salomone S. *J. Cardiovasc. Pharmacol*, 18(suppl.11): S1-S6.
11. Herbette L.G., Gaviraghi G. and Mason R.P. (1991): 5th International Symposium on Calcium Antagonists: Pharmacology and Chemical Research, Houston, Texas.

EFFECT OF FELODIPINE ON LEFT VENTRICULAR PERFORMANCE IN CONSCIOUS DOGS: ASSESSMENT BY LEFT VENTRICULAR PRESSURE-VOLUME ANALYSIS

W.C. Little and C.P. Cheng

Section of Cardiology, Department of Medicine
Bowman Gray School of Medicine
Wake Forest University
Medical Center Boulevard
Winston-Salem, North Carolina 27157-1045

Felodipine is a dihydropyridine calcium channel blocker with marked vascular selectivity. In isolated tissue preparations, felodipine is 100-fold more potent as a vasodilator than as a negative inotrope (1). At plasma concentrations relevant for treatment of hypertension in man, felodipine has been reported to have no direct cardiac effects (2) or to have a positive inotropic effect both in man (3) and *in vivo* (4). However, the direct effect of felodipine on left ventricular (LV) performance in intact, conscious animals is not known.

Conventional measures of LV performance (such as dP/dt_{max} and ejection fraction) are influenced not only by contractile state, but also by loading conditions. Thus, they can not be used to assess felodipine's direct effects on LV contractility because of felodipine's vasodilator effects. However, the influence of loading conditions can be minimized by evaluating LV performance in the pressure-volume plane (5,6,7). In addition, baroreceptor-reflex-mediated sympathetic cardiac stimulation can mask a vasodilating agent's direct negative inotropic actions (2). Thus, we recently used load-insensitive measures of LV performance, derived from variably loaded pressure-volume loops, to assess the direct inotropic effects of felodipine in normal conscious dogs with reflexes intact as well as after autonomic blockade (8).

Ten healthy, adult mongrel dogs were instrumented, as we have previously described (9) to measure three orthogonal LV diameters and LV pressure. Studies were performed after full recovery from instrumentation (7-10 days after the surgery).

Data were initially recorded with the animals lying quietly on their sides, without medication. Three sets of variably loaded pressure-volume loops were generated by sudden transient occlusions of the cavae. Immediately after the recording period, the caval occlusion was released and the

T. Godfraind et al. (eds.), Calcium Antagonists, 71–76.
© 1993 *Kluwer Academic Publishers and Fondazione Giovanni Lorenzini.*

animal was allowed to restabilize. After all parameters returned to their baseline level, felodipine (25 n moles/kg body weight) was injected intravenously over 2 minutes and followed by an infusion of felodipine (0.12 n moles/kg/min) for 15 minutes. Steady-state recording and three sets of variably loaded pressure-volume loops were then generated. To assess the interactions of felodipine with autonomic reflexes, the protocol was repeated on a subsequent day after administering hexamethonium (5 mg/kg, IV) and atropine (0.1 mg/kg, IV) to produce autonomic blockade.

LV volume was calculated as a modified general ellipsoid. This method of volume calculation gives a consistent measure of LV volume [$r > 0.97$, standard error of estimate (SEE) < 2 ml] despite changes in LV loading conditions, chamber configuration, and inotropic state. LV stroke work (SW) was calculated by point-by-point integration of the LV pressure-volume loop for each beat as described by Glower et al. (10). The effective arterial elastance (E_a) was calculated as the ratio of the LV end-systolic pressure (P_{ES}) to stroke volume (SV). The total systemic resistance (TSR) was estimated as P_{ES}/cardiac output.

Only caval occlusions that produced a fall in LV systolic pressure of at least 30 mm Hg were analyzed. From these beats, we determined linear approximations of the P_{ES}-V_{ES} relation, the SW-V_{ED} relation and the maximum rate of change of LV pressure (dP/dt_{max})-V_{ED} relations. The slopes and positions of these relations provide measure of LV contractile performance.

The plasma felodipine concentration was undetectable before and 16.1 ± 1.4 n moles/L at 15 minutes after felodipine infusion when the pressure-volume relations were assessed.

In the animals with intact autonomic nervous systems, there were significant decreases in LV P_{ES} (from 132 ± 13 to 109 ± 14 mm Hg, $p < 0.05$), total systemic resistance (from 6.06 ± 0.53 to 4.26 ± 0.58 mm Hg/ml/min, $p < 0.05$), and E_a (from 11.1 ± 1 to 8.7 ± 1.1 mm Hg/ml, $p < 0.01$). The heart rate was also increased (from 109 ± 15 to 118 ± 16 min^{-1}, $p < 0.05$). There were no significant changes in stroke volume (SV), V_{ED} and LV-end-ejection volume (V_{EE}). LV dP/dt_{max} tended to increase but this change did not reach statistical significance. In the four animals in which it did not increase, there was a fall in V_{ED}. After autonomic blockade, the heart rate remained unchanged. The LV end-systolic pressure and E_a were uniformly decreased. The dP/dt_{max} and the LV mechanical efficiency were also increased. There were tendencies for V_{EE} to decrease and for SV to rise, but the changes did not achieve statistic significance.

Variably loaded pressure-volume loops were generated from the data obtained following transient caval occlusion. From these loops were derived the P_{ES}-V_{ES}, dP/dt_{max}-V_{ED} and SW-V_{ED} relations.

With reflexes intact, felodipine increased the slopes of the LV P_{ES}-V_{ES}

relation (8.1 ± 0.6 vs 10.3 ± 1.0 mm Hg/ml, $p<0.05$), the dP/dt_{max}-V_{ED} relation (93.2 ± 7.2 vs 127.7 ± 9.9 mm/Hg/sec ml, $p<0.05$), and the SW-V_{ED} relations (82 ± 4.8 vs. 92.7 ± 6.4 mm Hg, <0.05). All three relations were shifted toward the left, manifested by a significant decrease in V_{ES} associated with P_{ES} of 100 mg Hg ($V_{100,ES}$) (26.1 ± 2.6 vs 24.6 ± 2.6 mm/Hg/ml, $p<0.05$), V_{ED} associated with dP/dt_{max} of 2000 mm Hg/sec ($V_{2000,-dP/dt}$) (42.9 ± 7.5 vs 35.4 ± 4.5 mm/Hg/sec/ml, $p<0.01$), V_{ED} associated with SW of 2000 mm Hg/ml ($V_{2000,SW}$) (52.1 ± 3.3 vs 45.1 ± 3.5 mm Hg, $p<0.01$). The increases in slopes and leftward shifts of all three relations with felodipine, indicating enhanced contractile effect, were also present after autonomic blockade.

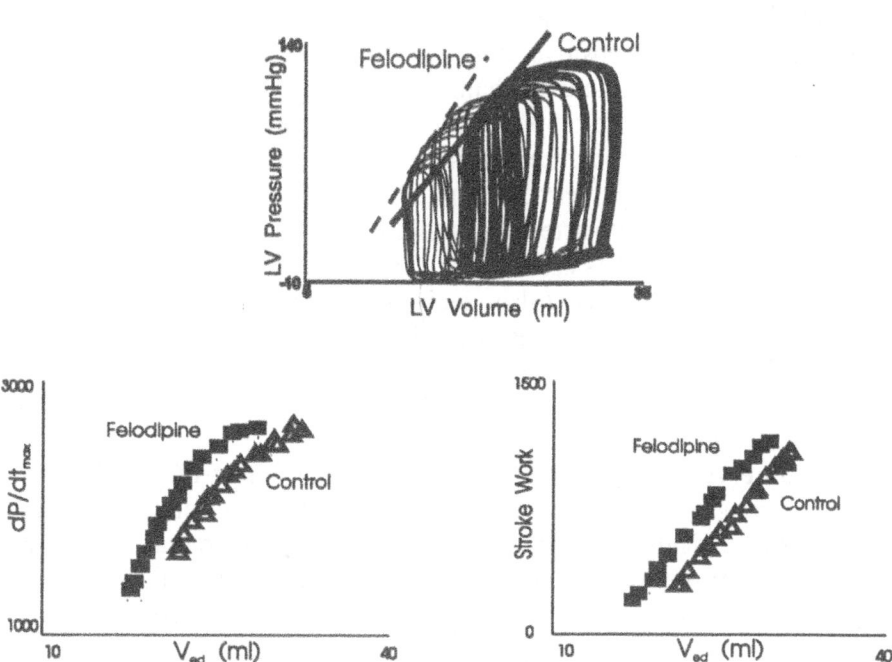

FIG. 1. Effect of felodipine after autonomic blockage on variably loaded pressure-volume loops and the three relations (P_{ES}-V_{ES}, dP/dt_{max}-V_{ED}, and SW-V_{ED}) indicating LV contractile function.

We investigated the acute effect of felodipine on LV contractile performance in conscious animals (fig. 1). The dose of felodipine used in this study produced a plasma felodipine concentration of 16.1 ± 1.4 n moles/L, which is in the range used for treatment of hypertension in man. Felodipine caused significant decreases in P_{ES}, E_a, and TSR, reflecting a direct vasodilating action of the drug on the arterial resistance vessels. These findings are in agreement with earlier studies in animals (11) and man (12). This vasodilatation produced a reflex activation of the sympathetic nervous system, increasing the heart rate. After autonomic blockade, felodipine produced a somewhat greater reduction in P_{ES} and E_a, suggesting that the baroreflex-induced sympathetic activation, induced by felodipine, partially offset its hypotensive action.

To avoid the potentially confounding effects of these changes in loading conditions on conventional measures of LV performance, we assessed LV contractive performance using variably loaded pressure-volume loops. All three measures of LV performance derived from the pressure-volume loops (the P_{ES}-V_{ES}, SW-V_{ED}, and dP/dt_{max}-V_{ED} relations) were shifted to the left with increases in slopes in response to felodipine. This indicates that in conscious animals, felodipine produced a mild enhancement of LV contractile performance. The enhancement of LV performance was apparent, both with intact reflexes and after autonomic blockade. Thus, it appears that the positive inotropic effect of therapeutic doses of felodipine is not dependent on changes in loading conditions or baroreflex activation induced by felodipine's arteriolar dilation.

It is important to recognize that the P_{ES}-V_{ES} relation may be shifted by changes in the arterial circulation (13,14). In contrast to the usual rightward shift seen with arterial vasodilation, felodipine produced a slight leftward shift of the P_{ES}-V_{ES} relation. It appears that felodipine's direct positive inotropic effect (also seen with the other two relations) overcomes the effect of arterial vasodilation to produce a leftward shift of the P_{ES}-V_{ES} relation. Thus, when evaluated by each of these complementary methods, felodipine's direct positive effect on LV performance is apparent.

The observation of a direct positive inotropic effect of felodipine in conscious dogs is consistent with the observations of Pettersson et al. (4) in the isovolumically contracting dog's heart. They found that felodipine, at plasma concentration from 7 to 20 n moles/L, produced a small, but significant, positive inotropic effect. However, in anesthetized dogs with intact circulation and arterial pressure kept constant, felodipine increased dP/dt_{max} only insignificantly (10%)(15). Similarly, Verdouw et al. (16) reported that intracoronary administered felodipine, in a dose that doubled coronary blood flow, did not significantly effect LV dP/dt_{max} in anesthetized pigs. The failure of felodipine to significantly increase dP/dt_{max} in these studies may have resulted from concomitant changes in V_{ED}.

The various calcium entry blockers affect LV performance differently in animals with intact circulation (17,18). For example, nisoldipine has little inotropic effect (19), verapamil has negative inotropic effect (20) and, in this study, felodipine has a mild positive inotropic effect. Therefore, calcium antagonists should be investigated for their myocardial/vascular properties, and the generalization that all vasodilating calcium antagonists have negative inotropic effects may not be correct.

It has previously been shown that other dihydropyridine-type drugs can be either C^{2+}-agonists or antagonists (21). Different vascular and cardiac binding sites for the agonist/antagonist effects (22) and/or different properties of the enantiomeres (23) have been suggested to explain these findings. Our study suggests that therapeutic doses of felodipine produce arterial vasodilation and mild positive myocardial inotropic effect.

In summary, LV pressure-volume analysis indicates that felodipine, at therapeutically relevant levels, directly increased LV contractile performance, independent of altered load and baroreflexes, and improved LV mechanical efficiency. Thus, a hemodynamically favorable effect of felodipine in the treatment of congestive heart failure (24) may result from not only reduction of peripheral resistance, but also from direct cardiac effects.

REFERENCES

1.   Ljung B. (1985): *Drug,* 29(suppl.2): 46-58.
2.   Culling W., Ruttley M.S.M. and Sheridan D.J. (1984): *Br. Heart J.,* 52: 431-434.
3.   Drake-Holland A.J., Pugh S., Mills C. and Noble M.I.M. (1987): *Drugs,* 34(suppl.3): 85-86.
4.   Pettersson K., Noble M.I.M., Bjorkman J.A., Hynd J. and Drake-Holland A.J. (1987): *J. Cardiovasc. Pharmacol.,* 10(suppl.1): S112-S118.
5.   Kass D.A., Maughan W.L., Guo Z.M., Kono A., Sunagawa K. and Sagawa K. (1987): *Circulation,* 76: 1422-1436.
6.   Kass D.A. and Maughan W.L. (1988): *Circulation,* 77: 1203-1212.
7.   Little W.C., Cheng C.P., Mumma M., Igarashi Y., Vinten-Johansen J. and Johnston W.E. (1989): *Circulation,* 80: 1378-1387.
8.   Cheng, C.P. (1991): *J. Pharmacol. Exp. Ther.,* 257: 163-69.
9.   Little W.C., Badke F.R. and O'Rourke R.A. (1984): *Circ. Res.,* 54: 718-730.
10.   Glower D.D., Spratt J.A., Snow N.D., Kabas J.S., Davis J.W., Olsen C.O., Tyson G.S., Sabiston, D.C. Jr and Rankin J.S. (1985): *Circulation,* 71: 994-1009.
11.   Norlander M. (1985): *Drugs,* 29(suppl.2): 90-101.
12.   Lund-Johansen P. (1990): *Cardiovasc. Pharmacol.,* 15(suppl.4): 534-539.
13.   Maughan W.L., Sunagawa K., Burkhoff D. and Sagawa K. (1984): *Circ. Res.,* 54: 595.
14.   Freeman G.l., Little W.C. and O'Rourke R.A. (1986): *Circulation,* 74: 1107-1113.
15.   Bjorkman J.A., Ek L., Gustafsson D., Ljung B. and Norlander M.

(1991): *Pharmacol. and Toxicol.*, 68: 310-315.

16.   Verdouw P.D., Wolffenbuttel B.H.R. and Scheffer M.G. (1983): *Naunyn-Schmiedeberg's Arch. Pharmacol.*, 323: 350-354.

17.   Nakaya H., Schwartz A. and Millard R.W. (1983): *Circ. Res.*, 52: 302-311.

18.   Millard R.W., Lathrop D.A., Grupp G., Ashraf M., Grupp I.L. and Schwartz A. (1982): *Am. J. Cardiology.*, 49: 499-505.

19.   Schipke J.D., Burkoff D., Alexander J. Jr, Schaefer J. and Sagawa K. (1988): *J. Pharmacol. Exp. Ther.*, 244(suppl.3): 2000-2004.

20.   Cheng C.P., Li K. and Little W.C. (1988): *FASEB J.*, 2: A382.

21.   Bechem M., Gross R., Hebisch S. and Schramm M. (1989): *Basic Res. Cardiol.*, 84(suppl.1): 105-116.

22.   Van Amsterdam F.T.M. and Zaagsma J. (1988): *Naunyn-Schmiedeberg's Arch. Pharm.*, 37: 213-219.

23.   Hof R.P., Ruegg U.T., Hof A. and Vogel A. (1985): *J. Cardiovasc. Pharmacol.*, 7: 689-693.

24.   Timmis A.D., Campbell S., Managham M.J., Walter L. and Jewitt D.E. (1984): *Br. Heart J.*, 51: 445-451.

LACIDIPINE IN THE TREATMENT OF HYPERTENSION: CLINICAL DATA

Giuseppe Mancia

Cattedra di Medicina Interna
Iª Divisione di Medicina
Ospedale S. Gerardo dei Tintori
Monza, Italy

My presentation will first review the evidence that lacidipine lowers blood pressure in hypertension. It will then focus on the hemodynamic mechanisms through which the antihypertensive effects of lacidipine is obtained. It will finally touch on the side effects of the drug and its potential advantages in some conditions included in the spectrum of the hypertensive population.

ANTIHYPERTENSIVE EFFECT OF LACIDIPINE

There is a large body of evidence that lacidipine is an effective antihypertensive drug [1,2]. An example is shown in fig. 1 which refers to a double-blind placebo-controlled study in essential hypertensive patients in whom lacidipine was administered once a day at the dose of 4 or 6 mg and blood pressure was measured by a sphygmomanometer 24 hours after dosing. Compared to placebo, diastolic blood pressure was reduced by both doses. The reduction (which involved also systolic blood pressure) became manifest following the first administration of lacidipine. It was even more evident, however, after one week administration, indicating no waning of the initial antihypertensive effect [3]. This has been observed also in other double-blind parallel-group studies which have documented that in the hypertensive group treated with a once-a-day lacidipine for four weeks [1] trough blood pressure is significantly reduced as compared to that of the reference group (fig. 2); 2) the reduction occurs not only at rest, but also during exercise [4]; and 3) the achieved blood pressure values are similar to those observed with traditional drugs such as hydrochlorothiazide, atenolol, and nifedipine [5,6,7]. Finally in a large number of patients followed in an open fashion, the antihypertensive effect of a once-a-day lacidipine has been found to remain unchanged during more than one-year treatment leaving no doubt as to the effectiveness of the drug in a chronic therapeutic regimen.

T. Godfraind et al. (eds.), Calcium Antagonists, 77–85.
© 1993 *Kluwer Academic Publishers and Fondazione Giovanni Lorenzini.*

G. MANCIA

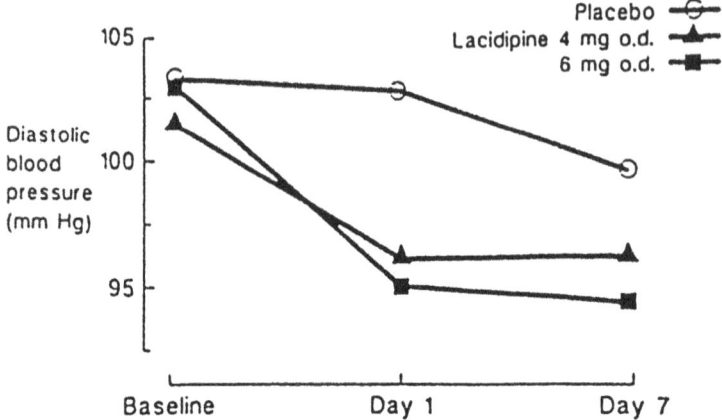

FIG. 1. Diastolic blood pressure at 24 hours after administration of lacidipine 4 mg, lacidipine 6 mg or placebo (from 3, by permission).

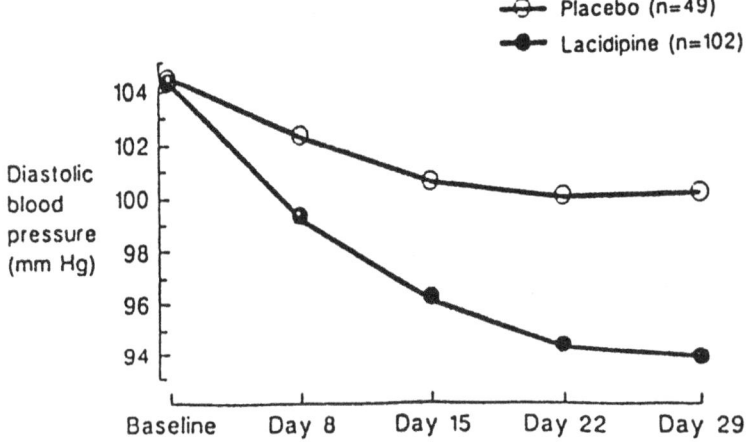

FIG. 2. Diastolic blood pressure at 24 hours after administration of lacidipine or placebo for a month. Lacidipine was given at 2 mg dose (n 26), 4 mg (42), 6 mg (15), and 8 mg dose (19) (from 3, by permission).

There is also a large body of evidence that lacidipine lowers blood pressure in daily life conditions and that this effect is consistent throughout the 24 hours (4,8,9,10). This is shown in fig. 3 which refers to a study on hypertensive patients in whom blood pressure was monitored over the day and night by an automatic noninvasive device. Clearly 24-hour mean blood pressure was significantly less following 6 weeks of treatment with once-a-day 4-6 mg lacidipine than following a similar period of placebo (8), with a ratio between the trough and the peak effect greater than 60%. Even more uniform blood pressure reductions have been described in a study measuring 24-hour blood pressure in elderly hypertensive patients treated with once-a-day lacidipine (9). The drug has been shown to lower an elevated day- and night-time blood pressure also by two intra-arterial ambulatory blood pressure monitoring studies (fig. 4) (4,10), i.e., a procedure far more accurate than the noninvasive approach (11).

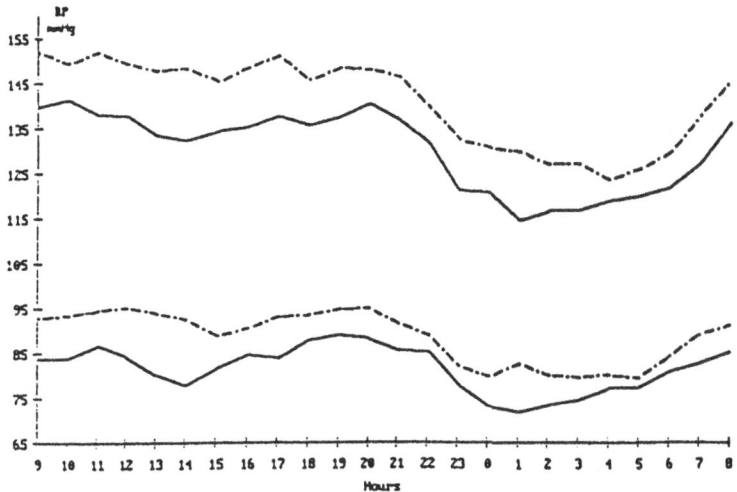

FIG. 3. Mean hourly blood pressure (BP) recorded by ambulatory monitoring at the end of a 3-week placebo period(-o-o-) and after 6 weeks of lacidipine treatment (-•-•-), 4-6 mg (from 8, by permission).

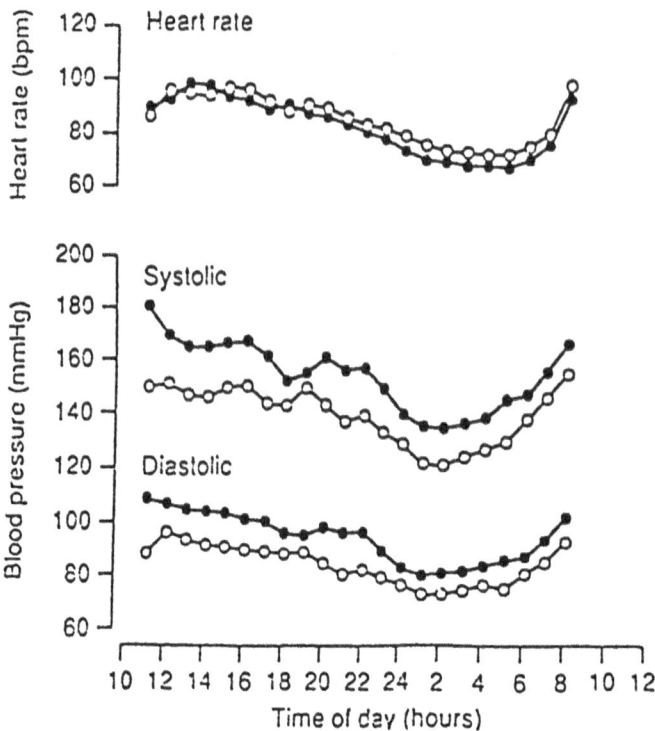

FIG. 4: Circadian curves of blood pressure and heart rate before (•-•-) and after (∘-∘-) treatment with 4-8 mg of lacidipine once daily. Data from 12 hypertensive patients treated for two weeks (from 10, by permission).

HEMODYNAMIC EFFECTS OF LACIDIPINE

Twenty-four hour intra-arterial blood pressure monitoring also shows that following prolonged treatment, the antihypertensive effect of lacidipine is not accompanied by any substantial tachycardia (fig. 4). This feature (which is in line with the observations made for most calcium antagonists of the dihydropyridine series) is associated with other favorable hemodynamic effects, i.e., a marked reduction in vascular resistance in the renal and coronary circulations leading to an increased perfusion of these vital organs (fig. 5) (12). Whether an increased renal blood flow delays the progressive deterioration of renal function occurring in hypertension is debated, although the potential adverse consequence of this change, namely glomerular hypertension (13), is prevented by the effective reduction in systolic blood pressure induced by lacidipine.

On the other hand, there is no question that an increased coronary blood flow represents an advantage considering that the reduced after-load couples the increased myocardial perfusion with a reduced cardiac work.

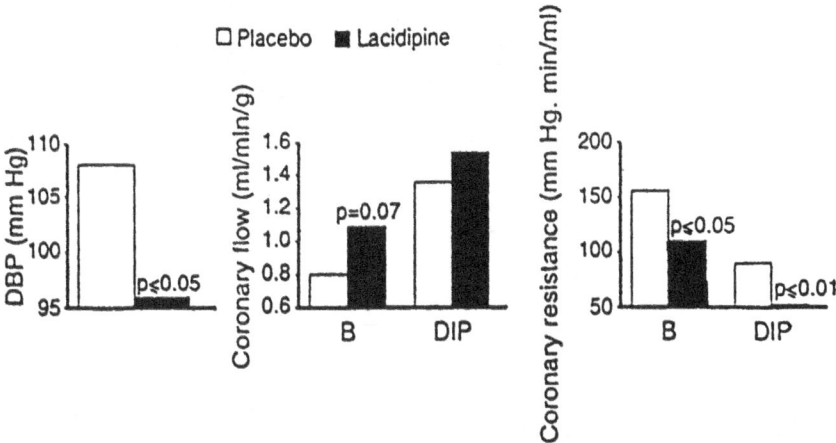

FIG. 5. Effects of lacidipine on coronary resistance (evaluated by PET); measurements in basal conditions (B) and after dipyridamole infusion (DIP).

SIDE EFFECTS OF LACIDIPINE

The side-effect profile of a new antihypertensive drug is more difficult to be precisely assessed because 1) its quantification cannot rely on criteria as much as those employed for drug efficacy and 2) the initial studies involved expert clinicians and checking systems that may make it more difficult for drug inconveniences to be fully disclosed. From the evidence obtained in a large number of subjects, however, it can be concluded that the side effects of lacidipine are those typical of the dihydropyridine substances, that is to say they mainly consist of headache, ankle edema, flushing, and palpitations (table 1). Interestingly, these effects were found to be less frequent after 14 months than after 5 months of administration of lacidipine, a feature of antihypertensive drug administration that has been reported in other instances and that imply that early appearance of nuisance problems should if possible not lead to treatment discontinuation. Furthermore, the side effects of lacidipine were comparable to those of nifedipine, except for a lower rate of pretibial edema (table 2). This may result from the more steady concentrations attained by administration of Lacidipine as compared to nifedipine but further evidence on a larger number of patients is obviously needed.

At any rate, there is little doubt that the overall tolerance profile of lacidipine is favorable and for the dose of 4 mg daily, the number of side effects appears to be not significantly different from placebo (table 1).

TABLE 1. Percentage of most commonly reported adverse
events in lacidipine (LAC)-treated patients:
comparison with Placebo (PL)

Event	Single dose		Repeat dose	
	LAC (n=95)	PL (n=30)	LAC (n=146)	PL (n=60)
Headache	18.9%	26.7%	0.7%	0
Flushing	4.2%	3.3%	1.4%	0
Palpitation	5.3%	6.7%	0.7%	0
Nausea	4.2%	6.7%	6.7%	0

TABLE 2.   Incidence of Side Effects with
Lacidipine and Nifedipine

Item	Lacidipine	Nifedipine SR	p Value
Number of patients	220	215	--
Headache	18.2%	19.1%	NS
Ankle edema	9.1%	17.7%	≤0.01
Flushing	8.6%	8.8%	NS
Palpitation	4.5%	1.9%	NS
Dizziness	4.5%	3.2%	NS
Chest pain	0.5%	0.5%	NS

OTHER EFFECTS OF LACIDIPINE

Lacidipine has no dislypidemic and disglycemic effects (14) indicating that, like other dihydropyridines, it can be appropriately administered to hypertensive patients with concomitant hypercholesterolemia and/or diabetes mellitus.

Like other dihydropyridines, its antihypertensive effect is accompanied by diuresis and natriuresis (15) which may account, at least in part, for the fact that lacidipine-dependent systemic vasodilatation is not accompanied by a fluid retention that may in the long-term attenuate its antihypertensive effect.

Two further reported features of lacidipine, again similar to those found for other calcium antagonists, are worthy of mention. One, in hypertensive patients aged more than 70 years, a once-a-day lacidipine regimen effectively reduced the elevated blood pressure throughout the day and night with no modification of heart rate values and no disturbance of the circadian blood pressure profile typical of aged individuals (fig. 6) (9). Thus, the drug can treat a common condition such as hypertension of the elderly.

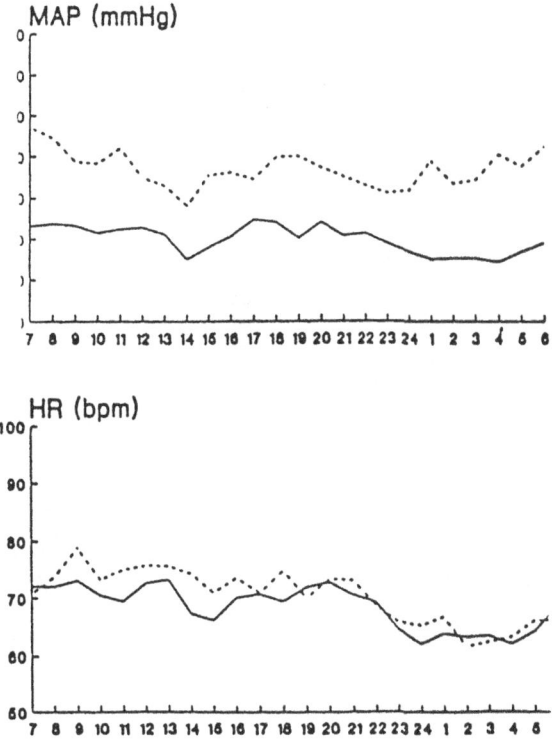

FIG. 6. Mean blood pressure (MAP) and heart rate (HR) profiles evaluated with 24-hour ambulatory monitoring after 4 weeks of treatment with lacidipine (---) or placebo (···) in elderly. (from 9, by permission).

Two, regression of left ventricular hypertrophy has been obtained in hypertensive patients in whom lacidipine administration for 6 months resulted in a reduction of septal and left posterior wall thickness similar to that observed with nifedipine (fig. 7) with no impairment of systolic and diastolic functions (16). The drug has also been observed to increase brachial artery compliance measured by an echodoppler device (17). Finally, recent evidence obtained by means of positron emission tomography indicates that in hypertensive subjects with no angiographic evidence of coronary heart disease lacidipine administration is accompanied by a reduction in arterial vascular resistance, i.e., the vasodilating effect of a full dose of dypiridamole became greater as compared to before lacidipine (12). Thus, the drug can regress left ventricular hypertrophy, which is a structural alteration of hypertension both common and risky. This is probably the case also for the hypertrophy of the arteriolar walls that develop as a consequence of an increased pressure load with favorable effects for long-term blood pressure reduction and tissue perfusion. It is likely that also the increased arterial stiffness accompanying hypertension is favorably affected by the drug which may thus operate towards normalization of the entire spectrum of structural alterations of the arterial system accompanying a chronic blood pressure elevation.

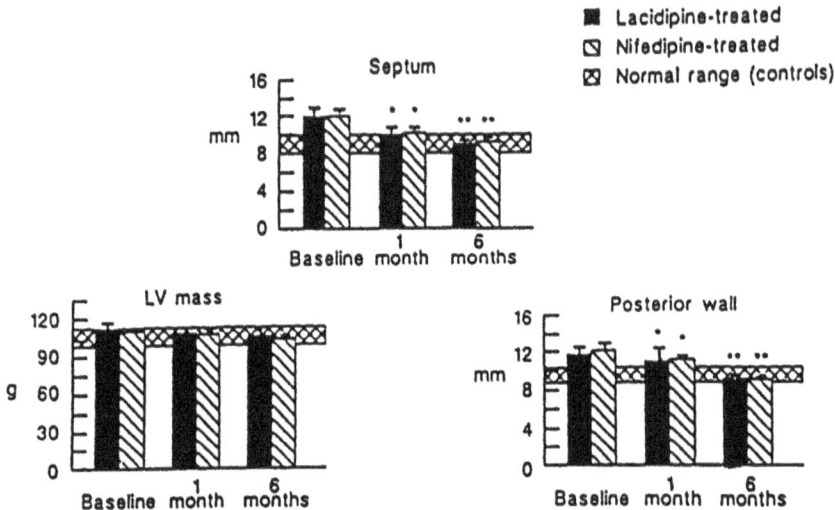

FIG. 7. Left ventricular wall thickness and mass in hypertensive patients before and after 1 and 6 months of therapy with lacidipine and nifedipine SR (*p < 0.05, **p < 0.01); LV mass = left ventricular mass (from 16, by permission).

REFERENCES

1. Harding S.M. and Boyd A.K. (1988): *J. Cardiovasc. Pharmacol.*, 12, 65: 155.
2. Marone C., Weldmann P., Kroutins D., Marcolli M. and Rizzini P. (1991): *Curr. Ther. Res.*, 50(6): 842-856.
3. Perelman M. (1991): *J. Cardiovasc. Pharmacol.*, 17(suppl.4): S14-S19.
4. Fariello R., Boni E., Corda L., Muiesan H.L. and Apobiti-Rosei E. (1991): *J. Hypertens.*, 9(suppl.3): S67-S72.
5. The United Kingdom Lacidipine Study Group (1991): *J.*

Cardiovasc. Pharmacol., 17(suppl.4): S27-S30.

6. Leonetti G. (1991): *J. Cardiovasc. Pharmacol.*, 17 (suppl.4): S31-S34.

7. Chiariello M. (1991): *J. Cardiovasc. Pharmacol.*, 17 (suppl.4): S35-S37.

8. Palatini P., Penzo M., Guzzardi G., Anaclesio M. and Pessina A. (1991): *J. Hypertens.*, 9(suppl.3): S61-S66.

9. Zito M., Abate G., Cervone C., Squassante L. and Calabrese G. (1991): *J. Hypertens.*, 9(suppl.3): S79-S84.

10. Raftery E.B. (1991): *J. Cardiovasc. Pharmacol.*, 17(suppl.4): S20-S26.

11. Omboni S., Ravogli A., Parati G., Zanchetti A. and Manci G. (1991): *J. Hypertens.*, 9(suppl.3): S25-S28.

12. Camici P., Palombo C. and L'Abbate A. (1991): *Am. J. Hypertens.*, 4(no.5, part2): 57A.

13. Brenner B.M., Troy J.L. and Dangharthy T.M. (1972): *Am. J. Physiol.*, 223: 1184.

14. Mancini M., Ferrara A.L., Strazzullo P. and Marotta T. (1991): *J. Hypertens.*, 9(suppl.3): S47-S50.

15. Rappelli A., Baldinelli A., Zingaretti O., Espinosa E., Salvi S. and Dessi-Fulgheri P. (1991): *J. Hypertens.*, 9(suppl.3): S37-S40.

16. Sheiban I., Arosio E., Montesi G., Tonni S., Montresor G. and Lechi A. (1991): *J. Cardiovasc. Pharmacol.*, 17(suppl.4): S68-S74.

17. Safar M.E. (1991): *J. Cardiovasc. Pharmacol.*, 17(suppl.4): S51-S54.

INSIGHTS IN THE CARDIOPROTECTIVE MECHANISM
OF THE NEW DRUG R56865

F. Verdonck, F.V. Bielen, L. Ver Donck*, and M. Borgers*

University of Leuven Campus Kortrijk,
B-8500 Kortrijk, Belgium

*Life Sciences, Janssen Research Foundation,
B-2340 Beerse, Belgium

The compound R56865, a benzothiazolamine derivative, is the prototype of a new class of substances that protect the heart in conditions of $[Ca^{2+}]_i$ overload. The substance improves postischemic functional recovery and significantly reduces the infarct size in dogs subjected to coronary artery occlusion [for review, (see 1)]. The drug is especially active against arrhythmias and mechanical deterioration induced by cardiac glycoside intoxication (2,3) and postischemic reperfusion (4). In both conditions Na^+/Ca^{2+} exchange is assumed to be involved in the rise of $[Ca^{2+}]_i$, most likely by a primary increase in $[Na^+]_i$.

Cardioprotection by R56865 may be related either to a reduction of $[Ca^{2+}]_i$ overload or to prevention of its detrimental effects on cellular function. $[Ca^{2+}]_i$ overload can be limited by inhibition of the Ca^{2+} channel or indirectly by reducing the $[Na^+]_i$ load. The potent action against electrical and mechanical signs of cardiac glycoside intoxication might indicate that the drug affects Na^+/K^+ pumping or interferes with the inhibition of the Na^+/K^+ pump by these substances. Alternatively, the drug might limit the Na^+ influx, thereby reducing secondary $[Ca^{2+}]_i$ loading. The drug might leave the $[Ca^{2+}]_i$ overload unchanged but might limit the deteriorating consequences of it: either by reducing the oscillatory Ca^{2+} release from the sarcoplasmic reticulum found in Ca^{2+}-overloaded cells or by blocking the effect of a raised $[Ca^{2+}]_i$ on membrane conductances.

INTERACTION WITH THE NA$^+$/K$^+$ PUMP

R56865 may antagonize cardiac glycoside toxicity by affecting the Na^+/K^+ pump directly. In view of this hypothesis, we investigated the effect of R56865 (2.10^{-6} M) on the Na^+/K^+ pump current (I_p) in single cardiac Purkinje cells and ventricular myocytes. At this concentration, afterdepolarizations are largely abolished (2,5). I_p was measured by

T. Godfraind et al. (eds.), Calcium Antagonists, 87–92.

superfusing the cell with a solution containing 5.4 mM $[K^+]_o$, interrupted with short superfusions containing 5.4 mM extracellular $K^+[K^+]_o$ [for details, see (6)]. Passive K^+ conductances were blocked by external Ba^{2+} (2 mM) and Cs^+ in the pipette solution. We could not find a significant effect of R56865 on the Na^+/K^+ pump current.

To investigate an eventual direct interaction with cardiac glycoside binding on the Na^+/K^+ ATPase molecule, we measured binding and unbinding of dihydroouabain (DHO) in the presence and absence of R56865 (2.10^{-6} M). Because of the well-known effect of $[Na^+]_i$ and cycling rate on DHO binding, comparison was made at similar pump rates. Neither the association rate constant nor the dissociation rate constant, and thus the dissociation constant, were affected by R56865.

Hence, our results, supported by findings reported in the literature (1,7), show that the anti-digitalis action of R56865 cannot be explained by an effect on the Na^+/K^+ pump.

Effect on the transient inward current

R56865 inhibits afterdepolarizations in guinea pig papillary muscle at concentrations which effectively block arrhythmias in heart-lung preparations (2). The transient inward current elicited in single ventricular myocytes of the guinea pig is highly sensitive to R56865 (5). I_{ti} oscillations are half maximally blocked at $7.5x10^{-8}$ M. The transient inward current consists of two components: the Na^+/Ca^{2+} exchange current, activated as an inward current during diastolic release of Ca^{2+}, and, to a smaller extent, a current carried through the nonspecific cation channel (8). Leyssens and Carmeliet (5) could exclude a direct effect on both currents. In accordance, Ver Donck and Borgers (9) found no direct effect of R56865 on Na^+/Ca^{2+} exchange; the veratridine-induced hypercontracture in ventricular myocytes could only be protected by the drug if added during the period of Na^+ loading. The absence of a direct effect on both components of the transient inward current suggests that the underlying $[Ca^{2+}]_i$ oscillations are affected by R56865. This is supported by the observation that the transient inward current oscillations and the aftercontractions disappear with similar time constants (5). Therefore, R56865 might diminish the magnitude of the Ca^{2+} overload or affect the threshold for Ca^{2+} release from the sarcoplasmic reticulum in a Ca^{2+}-overloaded cell.

Effect of R56865 on Na^+ currents

Because the Na^+/K^+ pump is not affected by R56865, $[Ca^{2+}]_i$ loading may be reduced by decreasing Na^+ influx. In the literature, controversy exists on the eventual role of a reduced $[Na^+]_i$. The reported action of the drug on dV/dt_{max} of the action potential and Na^+ current highly depends on the experimental conditions and on the cell type. No influence (3,10), antagonistic (11,12), as well as agonistic effects (12) on the Na^+ channel have been described [for review, (see 13)]. A tonic agonistic effect becomes evident when the action potential is elicited at low frequencies from a holding potential negative to -75 mV. Antagonistic effects become prominent in depolarized cells, at high stimulation frequency and at high concentrations. Although R56865 may reduce Na^+ loading in a beating preparation, the concentration affecting dV/dt_{max} or peak Na^+ current is still 10 times higher than the concentration needed to block ouabain-induced arrhythmias or the transient inward current (1,5). This finding strengthens the hypothesis that R56865 might interfere with Ca^{2+} release in conditions of Ca^{2+} overload as suggested by Leyssens and Carmeliet (5) and Daniels and ter Keurs (14). However, in some conditions, Na^+ influx through tetrodotoxin (TTX)-sensitive channels becomes highly sensitive to blockage by R56865. We could show (13) that the noninactivating Na^+ current, enhanced in the presence of veratridine, is much more sensitive to R56865. Half maximal block occurs at 2×10^{-7} M, a concentration which is nearly ineffective on dV/dt_{max}. Similar findings were reported for the DPI-modified Na^+ channel (5). Although the normal action potential is not influenced by R56865, prolongation by DPI was effectively blocked. Fig. 1 illustrates the time course of block of the TTX-sensitive current by 10^{-6} M R56865. The noninactivating Na^+ current was induced by veratridine (15×10^{-6} M). Blocking rate and removal of block are slow compared to TTX. Also in these experiments block was enhanced by depolarization.

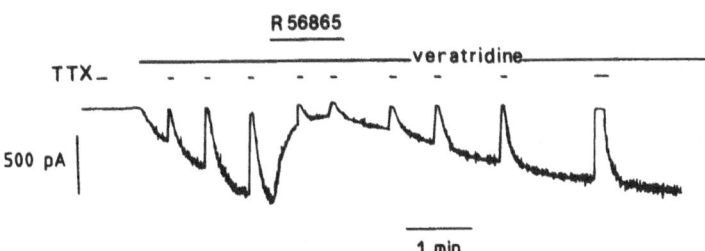

FIG. 1 Effect of R56865 (10^{-6} M) on the tetrodotoxin (TTX, 30×10^{-6} M)-sensitive, veratridine (15×10^{-6}M)-induced current in an internally dialyzed, single cardiac Purkinje cell. Holding potential -20 mV. R56865 is a potent blocker of the veratridine-induced, non-inactivating Na^+ current.

It has been shown that during ischemia and reperfusion, factors may be liberated which increase the noninactivating Na^+ current as veratridine and DPI do. Reactive oxygen species generated during reperfusion may induce slow inactivation (15). Lysophosphatidyl-choline accumulating in ischemic myocardium induces long openings of the Na^+ channel (16). Although the mechanism and extent of rise in $[Na^+]_i$ in ischemia (17) is still debated, part of it may be related to Na^+ influx via the Na^+ channel (18). The increases in $[Na^+]_i$ and $[Ca^{2+}]_i$ in hypoxic myocytes are blunted by R56865 in association with an improvement in cell survival (19).

Summarized, the effects of submicromolar concentrations of R56865 on the Na^+ channel in depolarized cells, especially in those cells with an increased noninactivating Na^+ current, may contribute to reduction of $[Ca^{2+}]_i$ overload in ischemia and reperfusion.

Effect on Ca^{2+} currents

In guinea pig ventricular myocytes only negligible effects on the L-type Ca^{2+} current have been reported at high concentrations (10^{-5} M) of R56865 (5,11). This explains the absence of an effect on slow Ca^{2+}-mediated action potentials and on force of contraction in papillary muscle (1). The T-type Ca^{2+} channel is only suppressed at concentrations higher than 10^{-6} M (5).

Therefore, the suppression of $[Ca^{2+}]_i$ overload by R56865 is not caused by a reduction of the Ca^{2+} current.

Effect on K^+ currents

R56865 has no effect on the inward rectifier in guinea pig ventricular myocytes (5,11). Experiments on inside-out patches and on whole cells have shown that the Na^+-activated current induced by addition of cardiac glycosides can be suppressed by R56865 at low concentrations [$< 10^{-7}$ M; (20)]. This effect of R56865 counteracts the shortening of the action potential by R56865 in the presence of cardiac glycosides (1). Part of the effect may be related to suppression of the delayed outward K^+ current (5). However, the expected increase in action potential duration has not been observed in normal conditions. The influence on the delayed rectifier may become apparent when the current is enhanced at increased $[Ca^{2+}]_i$ and $[Na^+]_i$ (21) as occurs in ischemia and cardiac glycoside intoxication.

CONCLUSION

The potent action of R56865 on cellular dysfunction caused by $[Ca^{2+}]_i$ overload may be related to a combination of effects on different ion channels. It suppresses Na^+ influx in depolarized cells; reduces oscillatory Ca^{2+} release; and blocks shortening of the action potential in Na^+- and

$[Ca^{2+}]_i$-overloaded cells. Although experimental evidence for a direct effect on Ca^{2+} release from the sarcoplasmic reticulum is lacking, the difference in sensitivity of the Na^+ channel and the transient inward current for R56865 supports that the drug might interfere with the Ca^{2+} release mechanism itself. Inhibition of the Na^+-dependent K^+ current counteracts K^+ loss and inhomogeneous repolarization. In the concentration range affecting these processes, the drug does not influence the normal action potential and force of contraction.

REFERENCES

1. Koch P., Wilffert B. and Peters T. (1990): *Cardiovasc. Drug Rev.*, 8: 238-254.
2. Vollmer B., Meuter C. and Janssen P.A.J. (1987): *Eur. J. Pharmacol.*, 142: 137-140.
3. Schneider J., Beck E. and Borgers M (1988): *Naunyn-Schmiedeberg's Arch. Pharmacol.*, 337: R65.
4. Garner J.A., Hearse D.J. and Bernier M (1990): *J. Cardiovasc. Pharmacol.*, 16: 468-479.
5. Leyssens A. and Carmeliet E. (1991): *Eur. J. Pharmacol.*, 196: 43-51.
6. Bielen F.V., Glitsch H.G. and Verdonck, F. (1991): *J. Physiol.*, 442: 169-189.
7. Heers C., Scheuffler E., Wilhelm D., Wermselkirchen D., Wilffert B. and Peters T. (1988): *Br. J. Pharmacol.*, 93: 273P.
8. Kimura J. (1988): *J. Physiol.*, 407: 79P.
9. Ver Donck L. and Borgers M. (1991): *Am. J. Physiol. (Heart Circ. Physiol.)*, (in press).
10. Leyssens A., Vereecke J. and Carmeliet E. (1988): *Pflugers Arch.*, 412: S53.
11. Himmel H.M., Wilhelm, D. and Ravens, U. (1990): *Eur. J. Pharmacol.*, 187: 235-240.
12. Carmeliet E. and Tytgat J (1991): *Eur. J. Pharmacol.*, 196: 53-60.
13. Verdonck F., Bielen F.V. and Ver Donck L. (1991): *Eur. J. Pharmacol.*, (in press).
14. Daniels M.C.G. and ter Keurs H.E.D.J. (1990): *Circulation*, 81: 143.
15. Bhatnagar A., Srivastava S.K. and Szabo G. (1990): *Circ. Res.*, 67: 535-549.
16. Burnashev N.A., Undrovinas A.I., Fleidervish I.A. and Rosenshtraukh (1989): *Pflugers Arch.*, 415: 124-126.
17. Pike M.M., Kitakaze M. and Marban E. (1990): *Am. J. Physiol.*, 259: H1767-H1773.

18.   Iida H., Teragaki M., Toda I., Yasuda M., Akioka K., Takeuchi K., Takeda T., Ishikawa M. and Miura I. (1991): *J. Am. Coll. Cardiol.*, 17: 132A.

19.   Haigney M.C.P., Ver Donck L., Stern M.D. and Silverman H.S. (1991): *Circulation*, (suppl.), (in press).

20.   Luk H.-N. and Carmeliet E. (1990): *Pflugers Arch.*, 416: 766-768.

21.   Scamps F. and Carmeliet E. (1989): *Am. J. Physiol.*, 257: C1086-C1092.

DO THE DIFFERENCES BETWEEN CALCIUM ANTAGONISTS MATTER WHEN TREATING HYPERTENSION?

T. Morgan and A. Anderson*

Department of Physiology, University of Melbourne,
Parkville, 3052, Australia

*Hypertension Clinic, Repatriation General Hospital
Heidelberg West, 2081, Australia

Slow-calcium-channel blocking drugs lower blood pressure but there are differences in their pharmacokinetics, effectiveness, and side effects. Certain differences are observed because classes of slow-calcium-channel blocking drugs have different affinities for receptors associated with voltage-activated and other calcium channels (1). Falls in blood pressure achieved with these agents are due to relaxation of arteriolar smooth muscle. Effects on other muscles (e.g., gastrointestinal, cardiac, and skeletal) are, in general, unwanted. Slow calcium channels exist in most cells of the body and slow-calcium-channel blocking drugs may have unwanted or beneficial effects related to these sites of action. This paper reviews the efficacy of the various drug classes, the duration of action of the dihydropyridine drugs, the side-effect profile of the drugs and possible beneficial effects based on theoretical concepts and animal studies.

EFFICACY

There have been few studies comparing the efficacy of different classes of slow-calcium-channel blocking drugs. Theoretically, it would be predicted that vascular selective agents would be more effective as the dose can be increased without concern for negative inotropic effects on the heart. This appears to be confirmed by our observational studies (table 1). Patients had been treated with verapamil or diltiazem to the usual considered maximal dose and had not achieved satisfactory control at trough or peak level of the drug. All other medication was maintained the same and verapamil or diltiazem was replaced with nifedipine or felodipine which was titrated to achieve control at trough level of the drug. Blood pressure supine and erect was also measured between 2 and 4 hours after taking the drug (peak level). The results indicated that better control was achieved with nifedipine and felodipine than with the other agents (table 1).

T. Godfraind et al. (eds.), Calcium Antagonists, 93–98.
© 1993 *Kluwer Academic Publishers and Fondazione Giovanni Lorenzini.*

TABLE 1.   Response to different calcium blockers

	n	Dose mg/d	Trough BP Sys	mmHG Dias	Peak BP Sys	mmHG Dias
Study 1						
Verapamil	4	280	177	101	168	97
Nifedipine	4	45	167*	97	160	90*
Study 2						
Verapamil	5	320	172	104	165	99
Felodipine	5	12	159*	92	157	89*
Study 3						
Diltiazem	6	180	175	102	164	95
Felodipine	6	13	162*	92*	155*	88*

*$p < 0.05$

Control was better at trough and peak level of the drug inferring that this was intrinsic to the drug and not due to its pharmacokinetics. These results indicated that some patients resistant to verapamil and diltiazem respond to the dihydropyridine group of drugs. What is unknown is whether patients nonresponsive to felodipine or nifedipine may respond to verapamil or diltiazem.

Differences Between Dihydropyridine Group of Drugs

Studies have compared felodipine with nifedipine (2,3) and nitrendipine (4). The blood pressure control achieved with felodipine assessed either as the mean fall in blood pressure or the percentage achieving control was greater than with the other agent. This control was greater during both the peak action of the drug and immediately predose.

Slow-calcium-channel blocking drugs have a plasma concentration-response relationship and from their pharmacokinetic profile the result presented was predicted. The ability of felodipine to give adequate 24-hour control has been confirmed in ambulatory monitoring studies. These studies (2,5,6) have been relatively short term and with long-term management this difference may become insignificant. However, it would appear preferable to use a slow-calcium-channel blocking drug that has an antihypertensive profile with a duration of 24 hours.

Renal effects

The dihydropyridine class of calcium antagonists have natriuretic and diuretic effects. DiBona and Sawin (7) showed a direct effect on distal tubule transport and more recently Thomas and Morgan (8) have shown that felodipine also inhibits directly sodium reabsorption by the proximal tubule. Significant natriuretic and diuretic activity has usually not been demonstrated with verapamil or diltiazem, and when felodipine and diltiazem were compared in a group of 12 healthy subjects (9), the natriuretic and diuretic effect of felodipine could be clearly demonstrated while the response to diltiazem was the same as to placebo. A number of studies have shown that addition of a diuretic does not have an additional hypotensive effect when added to nifedipine (10). This may be due to its intrinsic natriuretic activity or due to its cellular mechanism of action. The natriuretic activity could be one explanation of the superior antihypertensive effect of the dihydropyridine calcium blocking drugs.

Felodipine when used in people with hypertension associated with renal disease and/or renal failure has been remarkably effective, allowing control to be obtained with a smaller number of drugs. These are not direct comparative studies, but in patients with renal impairment whose blood pressure has been controlled with felodipine there has been stabilization or improvement of renal function over a 5- to 8-year interval (11). In patients with intrinsic renal disease the rate of deterioration of renal function has been markedly reduced. Whether this is an intrinsic property of the dihydropyridine group or whether it reflects the response to good hypertensive control is unknown.

Cardiac effects

Verapamil and diltiazem have negative inotropic effects and decrease cardiac conduction. Verapamil is very useful in supraventricular cardiac arrhythmias but this limits its usefulness in the management of high blood pressure. However, the absence of acute vasodilatory symptomatology have made these the preferred drug used by cardiologists in the treatment of angina. There is no objective evidence that these drugs are superior to the dihydropyridine drugs in the management of angina. If a person has hypertension and coronary artery disease and is treated using a drug with negative inotropic effects this would cause a reduction in the basic contractility of cardiac muscle. This is usually not evident because the reduction in cardiac contractility is recognized and increased release of catecholamines return contractility to what is required. However, work performed under catecholamine influence is more oxygen expensive and thus if blood supply is limited, angina may be precipitated. This is rarely a clinical problem as the reduction in blood pressure reduces dramatically cardiac work. However, if a person is on a beta-blocking drug and verapamil or diltiazem is added, this may precipitate cardiac failure or atrioventricular block.

Among the dihydropyridine group of drugs there are differences. Nifedipine does have negative inotropic effects compared to felodipine, particularly if used in association with beta-blocking drugs (12). This negative inotropic effect is usually not clinically important because the reduction in afterload causes a major reduction in cardiac work. However, if nifedipine and felodipine are injected into the coronary artery this difference can be readily illustrated (13).

Felodipine was compared with tiapamil, another slow-calcium-channel blocking drug, in 16 previously untreated hypertensive patients (14). Felodipine lowers blood pressure somewhat more than tiapamil and the fall was due solely to a fall in peripheral resistance while the fall with tiapamil was due to a reduction in both peripheral resistance and cardiac index.

The effects of different drugs on left ventricular hypertrophy (LVH) have been poorly studied. However, the dihydropyridine group has reduced LVH in animals and man (15). LVH is primarily due to increased pressure and if pressure is reduced, this will cause LVH regression. Nevertheless, there are also trophic effects of growth factors, including angiotensin II. These growth factors work through alterations in cytoplasmic calcium concentrations and the slow-calcium-channel blocking drugs would prevent and reverse these effects.

Patients with hypertension usually have increased peripheral resistance and it is logical to use a drug that alters this basic defect rather than reducing blood pressure by using drugs that alter cardiac function.

Atherosclerosis

Felodipine and nifedipine reduced atherosclerosis in animals (16) and nifedipine in man has prevented progression of lesions (17). Whether these effects are class specific is unknown. This may become an important reason for their use, particularly as most people with high blood pressure die of atherosclerotic complications. Felodipine reduced the symptoms of Raynaud's disease (18,19) and the slow-calcium-channel blocking drugs with primarily vasodilating actions would be expected to be more useful in people with peripheral vascular disease.

Side-Effect Profile

The major side effects with dihydropyridine are flushing, headaches, and palpitations (20). These result from the acute vasodilating effect of the drug and are prominent with the first dose. If the drug is continued, these side effects tend to resolve themselves, particularly if the drug has a relatively long-half life so that extreme trough and peak blood levels are reduced. The side effect that causes the dihydropyridine group to be stopped is swollen ankles. This results from precapillary relaxation of blood vessels and transmission of pressure to the capillary bed. The vascular side effects and swollen ankles do not appear to be as common with verapamil and diltiazem, probably reflecting their lesser effect on blood pressure. Verapamil and to a lesser extent diltiazem cause constipation. This may be moderately severe particularly in elderly patients and cause the drug to be stopped. If blood pressure had responded, felodipine or nifedipine can be substituted without this problem. Verapamil and diltiazem may cause worsening of cardiac failure.

CONCLUSION

The dihydropyridine group of drugs are more effective at lowering blood pressure than verapamil or diltiazem. There are theoretical considerations that the more vascular selective drugs will have less adverse cardiac effects and this may be a specific advantage. There are suggestions that various drugs affect different vascular beds to a varying extent. Whether this is of definitive importance in the treatment of hypertension is unknown. At present it would seem preferable to use a vascular selective, slow channel, calcium blocking drug that has a pharmacokinetic profile that allows it to be used once a day.

REFERENCES

1.   International Society and Federation of Cardiology Working Group (1987): *Am. J. Cardiol.*, 60: 630-632.
2.   Littler W.A. (1990): *J. Clin. Pharmacol.*, 30: 871-878.
3.   Goudie A.W., Gupta O.M., Gray P.L., Ross J.R.M., McKenna B.J., Blomfield I.A., Bramley, K.W., Sharp R.W., Jones D.F., Hadley R.J., Blychert E., Richardson P.D.I., Timerick S.J.B. and Watts D.A. (1991): *Br. J. Clin. Pharmacol.*, 31: 580P-581P.
4.   Lassen E., Ahlstrom F., Bogeskov-Jensen I., Frimodt-Moller J., Rishoj Nielsen M., Wickers-Nielsen N. and Winkel O. (1990): *J. Cardiovasc. Pharmacol.*, 15(suppl.4): 94.

5.   McGrath B.P., Langton D., Matthews P.G., Syme S., Treloar K. and McNeil J.J. (1989): *J. Hypertension.*, 7: 645-651.
6.   Fariello R., Boni E., Corda L., Cantalamessa A., Zaninelli A.,

Pollavini G., Alicandri C. and Muiesan G. (1991): *Am. J. Hypertens.*, 4: 21-33.

7. DiBona G.F. and Sawin L.L. (1984): *J. Pharmacol. Exp. Ther.*, 228: 420-424.

8. Thomas D. and Morgan T.O. (1991): *Clin. Exp. Pharmacol. Physiol.*, (submitted).

9. Johansson P., Edgar B. and Bergstrand R. (1990): *Eur. Heart J.*, 11(abstr. suppl.): 69.

10. Morgan T.O. and Anderson A. (1988): *J. Hypertens.*, 6(suppl.4): S652-S654.

11. Herlitz H. (1991): *Kidney Int.*, (in press).

12. Culling W., Ruttley M.S.M. and Sheridan D.J. (1984): *Br. Heart J.*, 52: 431-434.

13. Koolen J.J., Piek J., Hoedemaker G., van Wezel H., David G.K. and Visser C.A. (1991): *Eur. Heart J.*, 12(abstr. suppl.): 164.

14. Lund-Johansen P. and Omvik P. (1990): *J. Cardiovasc. Pharmacol.*, 15(suppl.4): 42-47.

15. Cerasola G., Cottone S., Nardi E., Fulantelli M.A., Carone M.B. and D'Ignoto G. (1990): *J. Hum. Hypertens,* 4: 703-708.

16. Pettersson K., Bjork H. and Nordlander M. (1991): *9th International Symposium on Atherosclerosis.* Chicago, October 6-11 (accepted for presentation).

17. Lichtlen P.R., Hugenholtz P.G., Rafflenbeul W., Hecker H., Jost S. and Deckers J.W. (1990): *Lancet,* 335: 1109-1113.

18. Nilsson H., Blychert E., Jonasson T., Leppert J. and Rinqvist I. (1990): *J. Cardiovasc. Pharmacol.,* 15(suppl.4): 108-110.

19. Schimdt J.F., Valentin N. and Levin-Nielsen S. (1989): *Eur. J. Clin. Pharmacol.,* 37: 191-192.

20. Saltiel E., Ellrodt A.G., Monk J.P. and Langley M.S. (1988): *Drugs,* 36: 387-428.

EFFECTS OF LACIDIPINE ON ISCHEMIC AND REPERFUSED MYOCARDIUM

*A. Boraso, *A. Cargnoni, A. Maggi, R. Ferrari, and O. Visioli

Cattedra di Cardiologia, Università degli Studi di Brescia
and *Fondazione Clinica del Lavoro,
Centro di Gussago, Brescia, Italy

As a general rule, calcium antagonist protection has usually been obtained in the experimental setting when the drugs were added before the onset of ischemia (1-15). This ability to protect applies to verapamil, nifedipine, diltiazem, as well as to their derivatives such as gallopamil (16) and anipamil (16-19).

The drugs' success depends on the experimental model employed, and protection has been proved with many different indices, either biochemical, electrophysiological, istological or ultrastructural in nature. Conversely, protection has not been obtained when calcium antagonists have been added after the onset of the ischemic episode or during reperfusion (6-7,12,15-16,20-22).

The protection exerted by calcium antagonists has usually been explained in terms of the diminished oxygen requirement associated with reduced heart rate, force of contraction and afterload, as well as to an improved oxygen supply to the ischemic zone.

MECHANISM OF PROTECTION

At the cellular level, maintenance of calcium homeostasis seemed for many years to be the most important effect of calcium antagonists. This goal is achieved not by a reduction of the activity of the calcium channels, but indirectly, by way of the energy-sparing properties. Thus it has been argued that calcium antagonists are protective because of their negative inotropic and chronotropic effects (which, in turn, depend on inactivation of calcium channels), indirectly resulting in preservation of the endogenous myocardial stores of high-energy phosphates, such as adenosine triphosphate (ATP) and creatin phosphate (CP) in the myocardium; the rate of fall in ATP and CP during ischemia is directly dependent on the physical activity and on the rate of ATP hydrolysis just before the onset of ischemia (23).

If sufficient ATP remains available during ischemia to maintain membrane ultrastructure and calcium homeostasis, myocardial overloading with calcium on reperfusion will not occur and the

T. Godfraind et al. (eds.), Calcium Antagonists, 99–106.

mitochondria would not accumulate calcium, insuring functional survival and utilization of the restored oxygen for ATP production. Late administration of calcium antagonists, after the onset of ischemia, when the calcium channels are already closed and the ischemic cells have lost their contractile activity, does not result in any ATP preservation. Thus, this important molecular mechanism of protection is no longer operating. Occasionally, small degrees of protection have been obtained even when calcium antagonists have been added after the onset of the ischemic episode (5,16). Interestingly, these results have been obtained with diltiazem and gallopamil, the only two drugs able to reduce mitochondrial calcium accumulation (24). It is likely that these compounds are able to prevent the reperfusion-induced accumulation of mitochondrial calcium directly, thus preserving ATP production on reperfusion. These effects, however, are obtained with extremely high concentrations, unlikely to be used clinically.

Apart from these molecular events linked with an amelioration of oxygen balance for the ischemic myocardium, calcium antagonists can be advantageous in the reduction of ischemia-reperfusion-induced injury for other reasons. They slow the loss of adenine precursors (8,25), exert a direct protective effect on the endothelial cells of the vasculature (26), inhibit platelet aggregation (27-28), and directly protect the sarcolemma, particularly the more lypophilic derivates (29). A typical example is lacidipine, a dihydropyridine derivative which is able to protect the myocardium in the absence of a negative inotropic effect.

PROTECTIVE EFFECT OF LACIDIPINE

Fig. 1 describes the effects of lacidipine, a new 1,4-dihydropyridine derivative (30), with effective oral antihypertensive action (31) on the mechanical performance of the isolated rabbit heart perfused under ischemic and reperfusion conditions. Lacidipine was administered directly in the perfusate 60 minutes before ischemia at three different concentrations: $10^{-10}, 10^{-9}$, and 10^{-8} M, respectively. In control hearts, reduction of coronary flow to 1 ml/min induced a rapid decline of developed pressure (fig. 1). Diastolic pressure began to rise progressively 10 minutes after the onset of ischemia and by the end of 60 minutes it had increased to 33 ± 6.8 mm Hg. Readmission of coronary flow resulted in a rapid, further increase of diastolic pressure, reaching a peak 5 minutes after reperfusion with only a 22% recovery of developed pressure. Fig. 1 shows that one hour of aerobic perfusion with lacidipine at the concentration of 10^{-10} M and 10^{-9} M did not produce any alteration of developed pressure, while lacidipine at 10^{-8} M induced a progressive decline of developed pressure with a 36% reduction the rise of diastolic pressure that occurs during ischemia and reperfusion. The recovery of developed pressure was 34%.

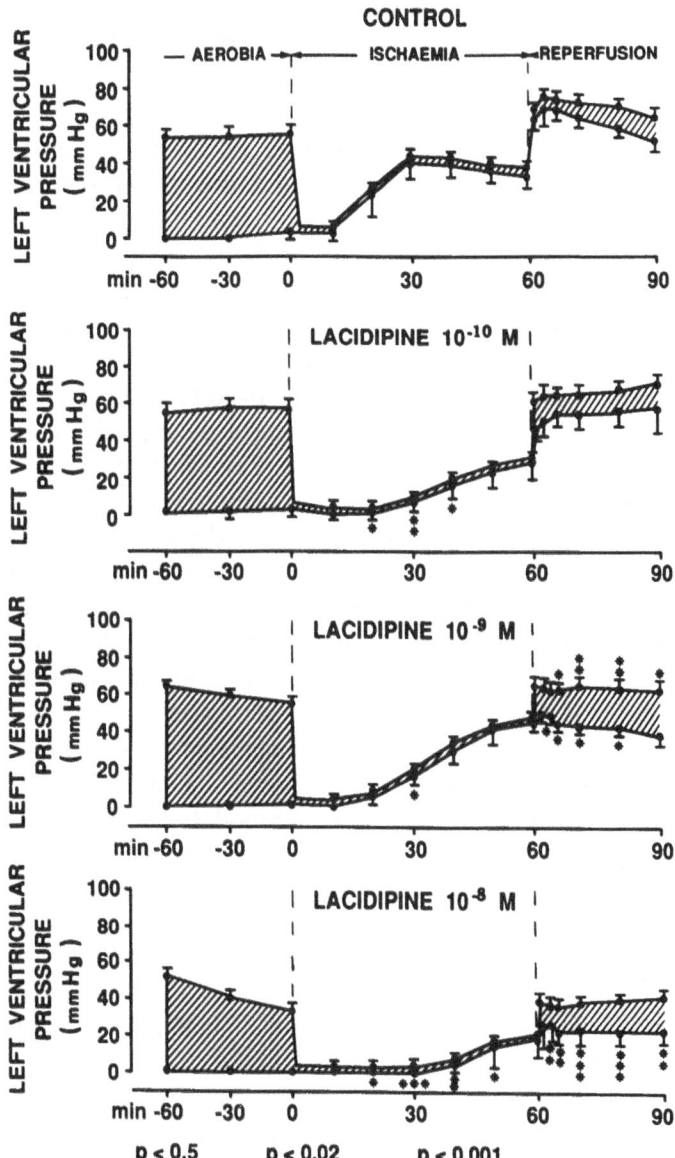

FIG. 1. Effect of different dosages of lacidipine on the developed and diastolic pressure of isolated rabbit hearts. Lacidipine was perfused 60 minutes before ischemia and during the ischemic period. Data are expressed as mean of 6 experiments, bars indicate SEM.

This later parameter was not significantly different from the 22% recovery of the control hearts, but it should be considered that the negative inotropic effect of lacidipine in aerobically perfused hearts persisted for as long as four hours. It is therefore likely that the negative inotropic effect was present also in our experiments during reperfusion. When the percentage of recovery was calculated with respect to the pressure developed by the hearts after 60 minutes of perfusion with lacidipine, it was of 54%. Administration of lacidipine at 10^{-9} M only delayed the ischemic-induced increase of diastolic pressure (fig. 1). However, it significantly reduced contracture during reperfusion and it increased the percentage of recovery of developed pressure to 39%.

Lacidipine at concentration of 10^{-10} M failed to significantly modify the increase of diastolic pressure (during ischemia and reperfusion), or to improve the recovery of developed pressure on reperfusion (fig. 1).

The intimate mechanism by which lacidipine protects the ischemic heart is unknown. When employed at 10^{-8} M, lacidipine probably acts through the ATP-sparing effect as a consequence of the negative inotropic effect present before the onset of ischemia (23). This mechanism, however, cannot be operating when lacidipine was employed at 10^{-8} M. Under these circumstances, developed pressure was not diminished by lacidipine and heart rate was kept constant in all groups, suggesting that the two major determinants of oxygen consumption were not affected. We have previously shown that, in the same experimental model, lacidipine at 10^{-9} M reduced CPK, noradrenaline, and oxidized glutathione release suggesting that the ischemia- and reperfusion-induced occurrence of oxidative stress and membrane disarrangement were, at least, prevented (32). As a consequence, intracellular calcium homeostasis and mitochondrial function were also maintained. It is known that liposoluble long-lasting dihydropyridines exert a membrane effect, by initially dissolving into the phospholipid bilayer and subsequently migrating towards the slow channels (33). Lacidipine would be a perfect candidate for such a mode of action because of its highly lipophilic properties.

PROTECTIVE EFFECT OF CALCIUM ANTAGONISTS: CLINICAL STUDIES

As a general assumption, when used clinically, calcium antagonists do not afford protection and there is great discrepancy between experimental and clinical data. This assumption, however, in its simplest form is unlikely to be correct. Recent publications have shown that calcium antagonists, when administered after the signs of human myocardial infarction, cannot reduce infarct size or the risk of initial or recurrent infarction or death. But this is in perfect agreement with what experimental studies have shown. It has never been demonstrated by any author that calcium antagonists have a significant contribution to make once the cell membrane is disrupted (as in the case of infarction when

CPK release is an important diagnostic criteria) and the cardiac cell has died (34).

However, many physicians continue to think that, because calcium antagonists have been so effective in improving coronary blood flow and because they have been shown experimentally to have direct cardioprotective effects, they should be used even in patients with proven acute myocardial infarction (35).

We believe that this way of thinking is too simple and too optimistic. The ischemic condition is extremely complex even in the experimental setting when several parameters can be precisely controlled and determined. In patients, myocardial ischemia might result in a series of different conditions including angina, unstable angina, myocardial infarction, post-infarction ischemia, each of which has several different facets.

In addition, recent therapeutic approaches against ischemia, such as cardiac surgery, thrombolysis, and percutaneous transluminal coronary angioplasty, have given rise to a further aspect of the problem, including the possible contribution of reperfusion to ischemic damage.

Another complication arises from the meaning of the words cardiac protection. In strict terms cardiac protection should mean the ability for intervention to make cardiac cells less vulnerable or more resistant to ischemic insult by acting directly on the myocyte.

The possibility for myocyte protection against ischemic insult can only be established by reperfusion of the myocardium and later quantification of the amount of necrotic tissue. However, reperfusion itself may cause further deterioration of the ischemic changes, depending on duration and severity of the previous ischemic period and, as already discussed by Hearse (36), several uncertainties exist regarding reduction of infarct size.

Furthermore, calcium antagonists could be useful for the ischemic myocardium, not necessarily because of their direct action on myocytes. Admittedly, the peripheral vasodilating action of these drugs may lower afterload and reduce the work of the heart as a whole and the coronary vasodilating action may lead to an increase of coronary flow. Obviously, even these positive effects should be considered to be of minimal or no efficacy once the ischemic cells have died.

However, danger from excessively reduced arterial perfusion pressure should be considered, particularly when combined with tachycardia which is usual with dihydropyridines. Similarly, there is the coronary steel syndrome reported to occur under some circumstances.

These later observations introduce another aspect of the problem when considering the cardioprotective effects of calcium antagonists in man: although there are substantial similarities in their mechanism of action, these drugs differ somewhat in their ancillary properties (37). Some agents, such as dihydropyridines can cause tachycardia and increase myocardial contractility (38), whereas phenylalkylamines and diltiazem reduce heart rate and contractility, actions that may influence their clinical effects.

Availability of new compounds, such as lacidipine which are cardioprotective even in the absence of a negative inotropic effect, may be important in the clinical setting and appropriate trials should be conducted to establish the role of this new calcium antagonist in myocardial protection.

SUMMARY

There are several reasons for using calcium antagonists to protect ischemic and reperfused myocytes. At the molecular level, calcium antagonists induce coronary dilation and improvement of the supply of oxygen and substrate to the area at risk; energy sparing secondary to dilatation of the peripheral vasculature and, at least in the case of phenylalkylamines and benzothiazepines, a reduction in contractility and heart rate; attenuation of reperfusion-induced arrhythmias; inhibition of platelet aggregation (only at high concentrations); reduction of loss of adenosine precursor; reduction of the release of lysosomal enzymes; a direct protective effect on the sarcolemma; a protective effect of the mitochondria and specific reduction of mitochondrial calcium transport (for benzothiazepines and some phenylalkylamines); and attenuation of the ischemia-reperfusion-induced displacement of endogenous noradrenaline.

Evidence of a protective role has been obtained in a variety of experiments, provided the calcium antagonists were introduced prior to ischemia. It is believed that this protective effect is an indirect consequence of the ability to modulate the function of the calcium channels and the energy-sparing effect is of major importance. Lacidipine, a new 1,4-dihydropyridine derivative is able to cardioprotect even at a concentration (10^{-9} M) which lacks negative inotropism and ATP sparing. The likely mechanism of protection under these conditions relates to the ability of lacidipine to maintain sarcolemmal integrity as a result of its high lipophilicity. Conversely, protection has not been obtained when calcium antagonists are added after ischemia or during reperfusion. Thus it is not surprising that when administered after the signs of human myocardial infraction, calcium antagonists do not reduce infarct size or avoid subsequent complications or decrease mortality.

ACKNOWLEDGEMENT

This work was performed under a research agreement between Glaxo Italia, the Department of Cardiovascular Physiopathology, University of Brescia, Italy, and the National Research Council (CNR) targeted project "Prevention and Control Disease Factors" n.91.00156PF41.

REFERENCES

1.	Nagao T., Matlib M.A., Franklin D., Millard R.W. and Schwartz A. (1980): *J. Mol. Cell Cardiol.,* 12: 29-43.
2.	Nayler W.G., Ferrari R. and Williams A. (1980): *Am. J. Cardiol.,* 46: 242-248.
3.	Perez J.E., Sobel B.E. and Henry P.D. (1980): *Am. J. Physiol.,* 239: H658-H663.
4.	Watts J.A., Koch C.D. and LaNoue K.F. (1980): *Am. J. Physiol.,* 238: H909-H916.
5.	Weishaar R.E. and Bing R.J. (1980): *J. Mol. Cell Cardiol.,* 12: 993-1009.
6.	Fagbemi O. and Paratt J.R. (1981): *Br. J. Pharmacol.,* 74: 12-14.
7.	Bourdillon P.D.V. and Poole-Wilson P.A. (1982): *Circ. Res.,* 50: 360-368.
8.	De Jong J.W., Harmsen E., De Tombe P.P. and Keyzer E. (1981): *Eur. J. Pharmacol.,* 81: 89-96.
9.	Bersohn M.M. and Shine K. (1983): *J. Mol. Cell Cardiol.,* 15: 659-671.
10.	Lange R., Ingwall J., Hale S.L., Alker K.J., Braunwald E. and Kloner R.A. (1984): *Circulation,* 70: 734-741.
11.	Klein H.H., Schubothe M., Nebendal K. and Kreuzer H. (1984): *Circulation,* 69: 1000-1005.
12.	Reimer K.A. and Jennings R.B. (1984): *Lab Inv.,* 51: 655-667.
13.	Fujibayashi Y., Yamazaki S., Chang B., Rajagopalan R.E., Meerbaum S. and Corday E. (1985): *J. Am. Coll. Cardiol.,* 6: 1289-1298.
14.	Yoon S.B., McMillin-Wood J.B., Michael L.H., Lewis R.M. and Entman M.L. (1985): *Circ. Res.,* 56: 704-708.
15.	Ferrari R., Curello S., Ceconi C., Di Lisa F., Raddino R. and Visioli O. (1986): *J. Mol. Cell Cardiol.,* 18: 487-488.
16.	Ferrari R., Boffa G.M., Ceconi C., Curello S., Boraso A., Ghielmi S. and Cargnoni A. (1989): *Basic Res. Cardiol.,* (in press).
17.	Curtis M.J., Walker M.J.A. and Yaswack T. (1986): *Br. J. Pharmacol.,* 88: 355-361.
18.	Kirkels J.H., Ruigrok T.J., VanEchteld C.J. and Meijler F.L. (1988): *J. Am. Cardiol.,* 11: 1087-1093.

19. Kirchengast M. and Raschack M. (1989): *Cardiovasc. Pharmacol.*, 13(suppl.4): 73-75.

20. Karlsberg R.P., Henry P.D. and Ahmend S.A. (1977): *Eur. J. Pharmacol.*, 42: 339-346.

21. Lefer A.M., Polansky E.W., Bianchi C.P. and Narayan S. (1979): *Basic Res. Cardiol.*, 74: 555-567.

22. Foster E., de Jong D., Connelly C. and Apstein C.S. (1984): *Circulation*, 70: 506-512.

23. Kubler W. and Spieckerman P.C. (1970): *J. Mol. Cell Cardiol.*, 1: 351-371.

24. Ferrari R., Boraso A., Cargnoni A., Pasini E., Raddino R. and Albertini A. (1990): *Eur. J. Pharmacol.*, 1-13.

25. Nigidikar S.V., Bowditch J. and Down J.W. (1986): *Cardiovasc. Res.*, 20: 604-608.

26. McDonagh P.F. and Roberts D.J. (1986): *Circ. Res.*, 58: 127-136.

27. Ikeda Y., Kikuchi M., Toyama K., Watanabe K. and Ando Y. (1981): *Thromb. Haemo.*, 45: 158-161.

28. Ware J.A., Johnson P.C., Smith M. and Salzman E.W. (1986): *Circ. Res.*, 59: 39-42.

29. Daly M.J., Elz J.S. and Nayler W.G. (1985): *J. Mol. Cell Cardiol.*, 17: 667-674.

30. Carpi C., Gaviraghi G. and Semeraro C. (1986): *Br. J. Pharmacol.*, (proc. suppl.) 89: 758.

31. Heber M.E., Broadhurst P.A., Biragen G.S. and Raftery E.B. (1990): *Am. J. Cardiol.*, 66: 1228-1232.

32. Boraso A., Cargnoni A., Gaia G., Bernocchi P. and Ferrari R. (1992) *J. Mol. Cell Biol.*, (in press).

33. Ferrari R. (1991): In: *Proceedings: Symposium held during the VII° Congress of the European Society of Cardiology*, Amsterdam, pp. 41-55.

34. Ferrari R. and Visioli O. (1991): *Eur. Heart J.*, 12: 18-24.

35. Hlatky M.A., Cotugno H.E., Mark D.B., O'Connor C., Califf R.M. and Pryor D.B. (1988): *Am. J. Cardiol.*, 61: 515-518.

36. Hearse D.J. (1977): *J. Mol. Cell Cardiol.*, 9: 605-616.

37. Opie L.E., Buhller F.R. and Fleckenstein A. (1987): *Am. J. Cardiol.*, 60: 630-632.

38. Visioli O., Bolognesi R., Cucchini F. and Ferrari R. (1987): *Cardiovasc. Drugs Ther.*, 1: 353-358.

ATHEROGENIC ACTIVITY CAUSED BY EXCESS MEMBRANE CHOLESTEROL IN ARTERIAL SMOOTH MUSCLE: ROLE OF CALCIUM CHANNELS

Thomas N. Tulenko, R. Preston Mason*, Meng Chen,
Hiromi Tasaki, Daniel Rock, and David Stepp

Department of Physiology and Biochemistry
Medical College of Pennsylvania
Philadelphia, Pennsylvania 19129

*Department of Radiology and Psychiatry
Biomolecular Structural Analysis Center
University of Connecticut Health Center
Farmington, Connecticut 06030

The concept of elevated serum cholesterol level as a principal risk factor in the etiology of atherosclerotic lesions has evolved from several large-scale epidemiological studies (1,2,3). These studies clearly link increased serum low-density lipoprotein (LDL) levels to mortality due to myocardial infarction. LDL is a cholesterol-rich, apo B_{100} containing particle which has been implicated in the forward transport of cholesterol to peripheral cells, including the smooth muscle cells of the arterial wall, by way of the classical LDL receptor (4). While the role of this particle, and its ligand (apo B_{100})-receptor complex, in the central hepatic metabolism of cholesterol has been well studied, little is known of the mechanism by which elevated LDL levels initiate and maintain lesion development and progression in the arterial wall.

As the cholesterol-rich particle, LDL contains nearly 5 times more cholesteryl esters (37% of the total mass) than unesterified cholesterol (8%). However, in the unesterified form, cholesterol [free-cholesterol (FC)] may play an important role in atherogenesis. Because FC is highly insoluble in aqueous media, in cells it partitions readily, and predominantly, into cell membranes. In fact, it has been suggested that over 90% of the cellular FC is contained into the plasma membrane (5) where its presence is thought to be important to proper membrane function (6,7). Recently, we began publishing a series of studies designed to elucidate the effects of cholesterol enrichment on arterial smooth muscle (ASM) plasma membranes. In these studies, we modified ASM plasma membrane cholesterol content as a single and isolated variable, and then examined alterations in membrane cation movements,

T. Godfraind et al. (eds.), Calcium Antagonists, 107–115.

membrane structure, and ASM cell function.

EXCESS MEMBRANE CHOLESTEROL ALTERS
CALCIUM CHANNELS

When isolated arterial segments are perfused with a medium containing FC-rich phospholipid (PL) liposomes, cholesterol readily exchanges into the ASM cells, without measurable changes in ASM or liposome PL content (8). Following enrichment of these cells with FC, the cellular FC content increases as does the FC/PL molar ratio, a dependable index of membrane FC content (9). Interestingly, the biologic effect of excess membrane cholesterol emerged as an increase in arterial contractility to the arterial spasmogens norepinephrine (8,10) and thromboxane (R.A. Bialecki and T.N. Tulenko, unpublished observation), but not to depolarization, histamine, or serotonin. Contractility experiments demonstrated between a 5- and 10-fold amplification in sensitivity to these agents. To determine the source of calcium mediating this increase in vasoconstrictor responses, $^{45}Ca^{2+}$ flux studies were performed. As illustrated in fig. 1, we observed a marked increase in basal (unstimulated) calcium influx in this procedure, and this FC-induced increase in calcium influx was abolished with the calcium channel blocker diltiazem. Equally noteworthy was the observation that diltiazem abolished norepinephrine-induced calcium influx as well. This was an unexpected finding as the calcium channel blockers are, for the most part, selective for voltage-operated (L-type) calcium channels (VOC), and rather ineffective on receptor-operated calcium channels (ROC) in ASM, as indicated by their lack of action in control arterial segments.

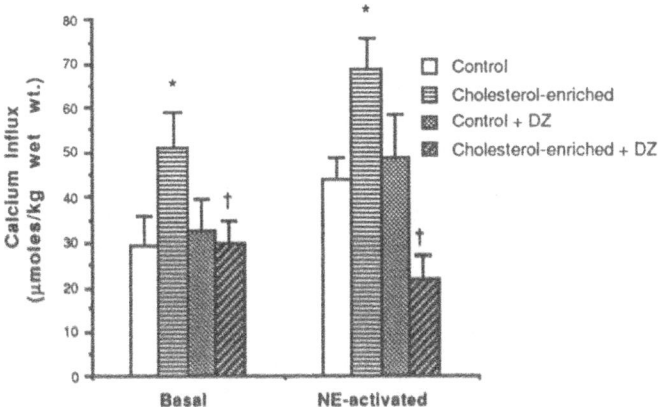

FIG. 1. The effects of cholesterol-enrichment on basal and norepinephrine-stimulated calcium influx in intact arterial segments. Control (open bars) and cholesterol-enriched arterial segments were obtained from carotid arteries of normal New Zealand White rabbits. Cholesterol enrichment was achieved by perfusion with cholesterol-donor liposomes and control arteries were perfused with cholesterol-nondonor arteries as previously described (8,10). Reproduced by permission of the authors and publishers. *$p < .05$ cholesterol-enriched vs control; DZ-diltiazem.

We interpret this alteration in calcium influx during stimulation with norepinephrine as indicative of a fundamental alteration in the ROC calcium channel following cholesterol enrichment of the cell membrane, since excess membrane cholesterol converted the ROC to a VOC-like channel, i.e., it became susceptible to the organic calcium channel blockers. Fig. 2 demonstrates the effects of cholesterol enrichment on ASM cells in culture. As in intact arterial segments, exposure to cholesterol-donor liposomes markedly increases calcium influx. These data also demonstrate that the calcium influx pathway activated by cholesterol enrichment is susceptible to all three classes of the calcium channel blockers. Thus, cholesterol enrichment of ASM activates what appears to be an L-type calcium channel in the membrane. Not only did cholesterol enrichment increase basal calcium influx but it also increased calcium efflux and cytosolic calcium levels as well (9). However, it is important to note that cholesterol enrichment did not have vasoconstrictor effects, even though cytosolic calcium levels appear to increase.

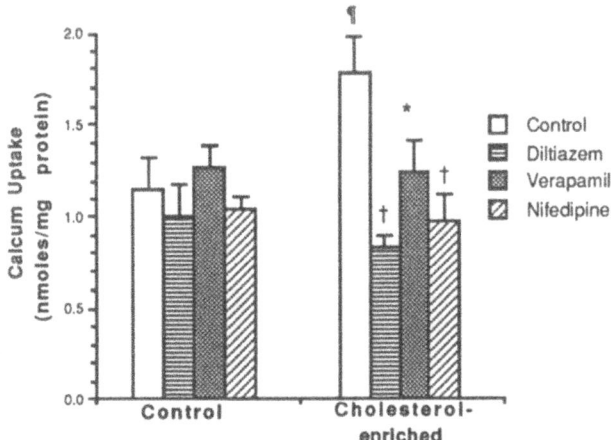

FIG. 2. The effects of calcium channel blockers on cholesterol-induced calcium influx in cultured ASM cells. Control (left panel) and cholesterol-enriched (right panel) cells were obtained from normal New Zealand White rabbit aorta. Cholesterol enrichment was achieved by incubation of cells with cholesterol-donor liposomes and LDL and control cells were incubated with cholesterol-nondonor liposomes as previously described (9). Reproduced by permission of the authors and publishers. ¶p < 0.05 control vs cholesterol enriched; ˙p < 0.05 control enriched vs cholesterol enriched + verapamil (VP); †p < 0.05 control-enriched vs cholesterol-enriched + diltiazem (DZ) or + nifedipine (NF)

In these studies, cholesterol enrichment was achieved using liposomes to promote the transfer of cholesterol to the ASM membranes. We have also observed similar effects using human LDL (9), indicating that this biological cholesterol-rich lipoprotein can serve to enrich ASM membranes with cholesterol as well. That LDL and liposomes can enrich the plasma membrane of ASM with cholesterol is indicated by the observation of increased cholesterol content and increased FC/PL molar ratio of a microsomal membrane fraction isolated from ASM which in itself is highly enriched with the plasma membrane markers 5'-nucleotidase (9) and Na/K-ATPase (11).

Clues as to the biological or functional significance of keeping cell membrane cholesterol levels constant in ASM can be obtained from the data illustrated in fig. 3. In this study (9), the steady-state calcium efflux rate constant was measured in cultured ASM using a standard efflux protocol (12). By definition, under steady-state conditions, calcium efflux and influx rates must be equal, permitting this expression of calcium movements to report on overall membrane calcium permeability. As can be seen, calcium permeability in ASM is directly and linearly related to cell (membrane) cholesterol content. The mean ($\pm$SEM) of cell membrane cholesterol content in control cells (middle point) reflects the "normal" range of membrane cholesterol. Likewise, the mean (±SEM) of cell membrane calcium permeability in control cells (middle point) reflects the "normal" range of membrane permeability to calcium. With cholesterol enrichment (using human LDL) of the ASM cell membrane, calcium permeability is increased (upper point), and with cholesterol depletion (using a cholesterol-acceptor liposome) calcium permeability is decreased (lower point). In other words, only when "normal" membrane cholesterol levels are maintained is calcium permeability also "normal." Increases or decreases in membrane cholesterol levels markedly disturb the kinetics of transmembrane calcium movements. The corollary to this is that ASM, if not all mammalian cells, must know exactly how much of this membrane sterol should be present in its plasma membrane, and furthermore, its level there should be tightly regulated by some physiological mechanism.

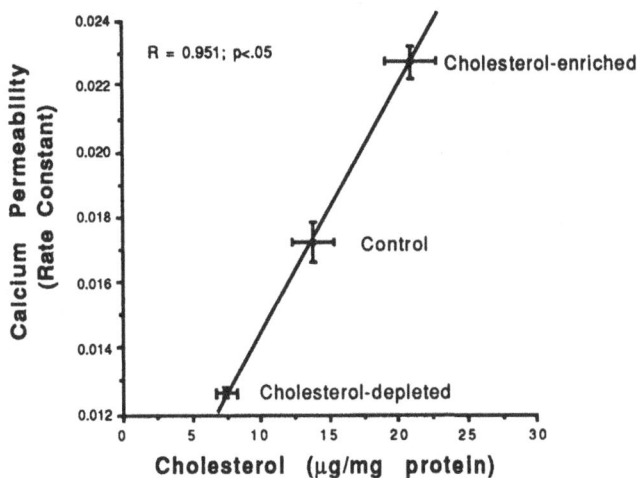

FIG. 3. The effects of cholesterol enrichment and depletion on calcium permeability in cultured ASM. Calcium permeability was measured using a steady-state $^{45}Ca^{2+}$ efflux permeability described in detail previously (12). Reproduced by permission of the authors and publishers.

Perhaps the most interesting alteration in ASM function as a result of cholesterol enrichment is it mitogenic effect. Preliminary data demonstrate that cholesterol enrichment initiates ASM cell proliferation in culture as measured by both cell counting and ^{3}H-thymidine uptake techniques (12). Furthermore, this proliferative effect of excess membrane cholesterol is very sensitive to the calcium channel blockers diltiazem, nifedipine, verapamil, and amlodipine. Of these, however, amlodipine is by far the most potent inhibitor, demonstrating up to 10^3 to 10^6 times greater potency than the others. Amlodipine's antiproliferative effects at low aqueous concentrations ($< 10^{-12}$ M) appear to be independent of its calcium channel blocking activity, and we propose that this dihydropyridine has membrane effects (14) separate and distinct from its calcium antagonist activity, and it is this membrane effect that accounts for its antiproliferative action following cholesterol enrichment.

Membrane alterations with cholesterol enrichment

As has been demonstrated in several other cell lines (15), cholesterol enrichment of the ASM membrane reduces membrane fluidity as indicated by an increase in fluorescence anisotropy when examined with the fluorophore diphenlyhexatriene (9). It is tempting to speculate that this membrane ordering effect causes the alterations in calcium influx and calcium ion channel proteins described above. Just how changes in membrane fluidity translate into alterations in calcium channels and transmembrane calcium movements is hard to envision, but has recently been invoked to explain this phenomenon (16). However, we have recently observed another physical parameter affected by cholesterol enrichment on the ASM membrane. Using small angle x-ray diffraction, preliminary data indicate that the membrane bilayer thickness increases with cholesterol enrichment (14) (table 1) which can be substantial (up 20%). It is well known that the conformation of amphipathic, membrane-spanning proteins (including ion channels proteins) is affected by their lipid environment. Conceivably, swelling of the membrane bilayer would result in distortion of the ion channel protein's tertiary structure and thereby alter its pore permeation characteristics, and/or ligand binding site availability and topography.

TABLE 1. Summary of alterations in cultured and intact
ASM function and calcium permeability characteristics
with enrichment of the plasma membrane with cholesterol

Arterial Smooth Muscle	Altered Vaso-constriction	Membrane cholesterol content†	Calcium Permeability	Effective Calcium Antagonists	Membrane Width¶
Cultured					
Control	-	0.31	1X	-	
Cholesterol-enriched	-	0.69	2X	DZ/NF/VP/AML*	↑
Intact					
Control		0.31	1X	-	
Cholesterol-enriched	NE-TXA2* (5-10X)	0.65	2X	DZ/AML*	↑

*DZ-diltiazem, NF-nifedipine, VP-verapamil, AML-amlodipine, NE-norepinephrine; TXA$_2$-thromboxane
† expressed as the cholesterol/phospholipid molar ratio
¶ based on d-space determinations using small-angle x-ray diffraction

RELEVANCE TO ATHEROSCLEROSIS

Alterations in membrane cholesterol clearly result in important changes in ASM behavior. Whether these changes in cell function following enrichment of membrane cholesterol using experimental techniques *in vitro* (liposomes or LDL) are relevant to dietary or genetic atherosclerosis is a compelling question. We have recently gathered data on membrane alterations in ASM isolated from rabbits fed a 2% cholesterol diet for 1-12 weeks. We observed a marked (2-fold) increase in membrane cholesterol content in ASM cells obtained from the atherosclerotic group compared to age-matched normal diet control animals (11,13). Furthermore, this change in membrane cholesterol content correlated with an increase in basal (unstimulated) calcium influx (17). These preliminary results clearly support the hypothesis that increases in membrane cholesterol content occur in dietary atherosclerosis, and that this alteration in the ASM membrane lipid composition mediates the changes in vasoconstrictor sensitivity of atherosclerotic arteries (18-21) as well as the increased proliferative activity that have been reported by other investigators (22-26). This hypothesis is further supported by the observations of decreased membrane fluidity of ASM from atherosclerotic rabbits (27,28), the prediction of altered calcium permeability based on modelling studies using $^{45}Ca^{2+}$ efflux kinetics in atherosclerotic rabbit aorta (29) and by the reports of the relative effectiveness of the calcium channel blockers and anti-atherogenic agents in experimental animal models (30,31) as well as human subjects (32).

ACKNOWLEDGEMENT

Supported in part, by NIH grant HL-30497 and Grants-in-Aid from the Delaware Chapter and Southeastern Pennsylvania Affiliate of the American Heart Association (TNT,HT,DR). DR and DS were supported by NIH training grant HL-07443, RPM by a Fellowship from the Connecticut Affiliate of the American Heart Association and HT by a Fellowship from the Uehara Memorial Foundation.

REFERENCES

1.   Kannel W.B., Castelli W.P. and Gordon T. (1979): *Ann. Intern. Med.*, 90: 85-91.
2.   Kannel W.B., Neaton J.D., Wentworth D., Thomas H.E., Stamler J., Hulley S.B. and Kjelsberg M.O. (1986): *Am. Heart J.*, 112: 825-836.
3.   Solberg L.A., Enger S.C. and Hjermann I. (1980): In: *Atherosclerosis V*, edited by A.M. Gotto, L.C. Smith and B. Alaaen pp. 57-62. Springer-Verlag, Inc., New York.
4.   Brown M.S. and Goldstein J.L. (1986): *Science*, 232: 34-37.
5.   Lange Y., Swaisgood M.H., Ramos B.V. and Steck T.L. (1989): *J. Biol. Chem.*, 264: 3786-3793.
6.   Yeagle P. (1985): *Biochem. Biophys. Acta.*, 822: 267-287.
7.   Tulenko T.N., Bialecki R.A., Gleason M. and D'Angelo G. (1990): In: *Potassium Channels: Basic Function and Therapeutic Aspects*, edited by T.J. Colatsky pp. 187-203. A.R. Liss, New York.
8.   Bialecki R.A. and Tulenko T.N., (1989): *Am. J. Physiol.*, 257: C306-C314.
9.   Gleason M., Medow M. and Tulenko T.N. (1991): *Circ. Res.*, 69: 216-227.
10.   Broderick R., Bialecki R.A. and Tulenko T.N. (1989): *Am. J. Physiol.*, 257: H170-H178.
11.   Chen M., Mason R.P. and Tulenko T.N. (1991): *FASEB J.*, 5: A531.
12.   Tulenko T.N. and Cox R.H. (1991): *Cardiovasc. Drug Rev.*, 9: 59-77.
13.   Tulenko T. (1991): *J. Am. Coll. Cardiol*, 17: 24A.
14.   Mason P., Herbette L.G. and Tulenko T.N. (1991): In: *Proceedings of the 5th International Symposium on Calcium Antagonists: Pharmacology and Clinical Research*, edited by T. Godfraind, R. Paoletti and P.M. Vanhoutte. This monograph.

15. Barenholz Y. (1984): In: *Physiology of Membrane Fluidity* vol. 1, edited by M. Shinitzky. CRC Press, Boca Raton.

16. Zhou Q., Jimi S., Smith T.L. and Kummerow F.A. (1991): *Biochem. Biophys. Acta.*, 1085: 1-6.

17. Stepp D. and Tulenko T.N., (1991): *FASEB J.*, 5: A532.

18. Shimokawa H., Tomoike H., Nabeyama H., Yamamoto H., Araki H., Nakamura M., Ishil Y. and Tanaka K. (1983): *Science*, 221: 560-561.

19. Heistad D.D., Armstrong M.L., Marcus J.M., Piegors D.J. and Mark A.L. (1984): *Circ. Res.*, 54: 711-718.

20. Verbeuren T.J., Jordaens F.H., Zonnenkeyn L.L., Van Hove C.E., Coene, M.-C. and Herman, A.G. (1986): *Circ. Res.*, 58: 552-564.

21. Heistad D.D., Mark A.L., Marcus M.L., Piegors D.J., Armstrong M.L. (1987): *Circ. Res.*, 61: 346-351.

22. Gordon D., Schwartz S.M., Benditt E.P. and Wilcox J.N. (1989): *Transplant Proc.*, 21: 3692-3694.

23. Gordon D., Reidy M.A., Benditt E.P. and Schwartz S.M. (1990): *Proc. Natl. Acad. Sci. USA*, 87: 4600-4604.

24. Gordon D. and Schwartz S.M. (1991): *Trends in Cardiovasc. Med.*, 1: 24-28.

25. Ross R. and Glomset J.A. (1976): *N. Eng. J. Med.*, 295: 369-377, 420-425.

26. Schwartz S.M. and Ross R. (1984): *Progr. Cardiovasc. Dis.*, 25: 355.

27. Gillies P. and Robinson C. (1988): *Atherosclerosis*, 70: 161-164.

28. Gillies P., Robinson C. and Chapple R. (1987): *Exptl. Molec. Pathol.*, 47: 90-97.

29. Strickberger S.A., Russek L.N. and Phair R.D. (1988): *Circ. Res.*, 62: 75-80.

30. Henry P.D. (1985): *Circulation,* 72: 456-459.

31. Henry P.D. (1988): *Ann. N.Y. Acad. Sci.*, 522: 411-419.

32. Lichtlen P.R., Hugenholtz P.G., Rafflenbeul W., Hecker H., Jost S. and Deckers J.W. (1990): *Lancet,* 335: 1109-1113.

UNDERSTANDING THE INITIATING FACTORS IN ATHEROSCLEROSIS: IMPLICATIONS FOR INTERVENTION

Winifred G. Nayler

Department of Medicine, University of Melbourne
Austin Hospital, Heidelberg 3084, Australia

Although atherosclerosis is a primary cause of death in many "western-type" civilizations (1), the disease is not new. Thus, arteries taken from ancient Egyptian mummies have been found to be heavily encrusted with atherogenic plaque, and the medical literature of the eighteenth and nineteenth centuries contains detailed descriptions not only of the morphology of atherosclerotic lesions but also of the relationship between their occurrence in the coronary vasculature and the onset of angina pectoris (1). Only in the past few years, however, has the complexity of the pathogenic mechanisms which are involved in the development of these lesions been recognized, and effective treatment regimes developed. These regimes involve dietary control, and the use of either lipid-lowering (2), anti-oxidant (3), or calcium channel blocking (calcium antagonist) drugs (4-7). An explanation of why such a diversity of treatment regimes provides effective therapy for this condition requires an understanding of the mechanisms which are responsible for transforming normal arterial tissue into "an acellular, fibrotic section of tissue containing the cellular debris of generations of dead and decaying cells and a variety of substances trapped amidst this debris" (8).

The steps involved in the growth of an atherosclerotic plaque can be conveniently divided into four stages: an initiating stage, a fatty streak stage, a proliferative stage, and a "mature" or "advanced" stage. Since the advanced plaque stage concerns the presence of acellular and fibrotic structures it is unlikely to provide a useful target for therapeutic interventions. By contrast, elucidating the mechanisms responsible for initiating the fatty streak stage and its progression to the proliferative and advanced stages may uncover target sites for effective interventions.

INITIAL EVENT

Although endothelial damage contributes markedly to the events involved in lesion formation, it can no longer be regarded as the initial event or trigger for atherogenesis. Instead, provided that plasma cholesterol levels are high, the earliest detectable event which can be

T. Godfraind et al. (eds.), Calcium Antagonists, 117–124.

related to the subsequent development of an atherogenic lesion appears to be the focal accumulation of low-density lipoprotein (LDL) in the subintimal space of an apparently otherwise normal artery (9). This focal accumulation of LDL does not occur haphazardly. Instead it occurs in lesion-prone areas, located mainly at sites of arterial bifurcations and branching, and hence at areas where blood flow patterns are complex, i.e., with vortex formations and adjacent areas of low and high flow.

FATTY STREAK STAGE

Soon after focal accumulations of LDL appear, previously circulating monocytes begin adhering to the luminal surface of the overlying endothelium (8). Reasons for this surface accumulation of monocytes include the generation of chemotactic factors, possibly generated by oxidized LDL (10), and the appearance of surface ligands which bind the monocytes (11). The monocytes, presumably attracted by chemoattractants, now begin migrating into the subintimal space below the endothelium, even though the endothelium is still intact. Presumably, as Witztum (8) has already suggested, the monocytes penetrate the gap junctions separating adjacent endothelial cells. Once resident in the subintimal space, the monocytes convert to macrophages and in so doing become avid accumulators of lipid. The end-result is the appearance of lipid-laden "foam" cells. As these foam cells accumulate excess cholesterol, they increase in volume, with the inevitable result of stretching and weakening the overlying endothelium. The end-result is the exposure of the macrophages to circulating blood elements, including platelets.

Once this stage has been reached, platelet aggregation takes place with the subsequent release of some of the platelet-derived growth factors which mediate smooth muscle cell migration and proliferation. The key to the sequence of events which has been described so far, involves the conversion of circulating monocytes into macrophages which, once resident in the arterial wall, become capable of accumulating large amounts of lipoprotein. Since circulating monocytes do not become "fat-laden," even when plasma cholesterol levels are relatively high, the conversion of the monocytes to macrophages must be accompanied by a modification of either the macrophages or the "resident" LDL to trigger a "scavenger"-like activity in macrophages. Current evidence indicates that it is the peroxidation of polyunsaturated fatty acids in the LDL which provides the vital key. Thus:

(a)   lipid peroxidation enhances the capacity of macrophages to recognize and accumulate LDL (8);

(b)   oxidized LDL serves as a chemoattractant for monocytes, attracting them into the subintimal space where they convert to macrophages (11);

(c) oxidized LDL inhibits the egress of macrophages out of the lesions (11); and

(d) peroxidized lipids are cytotoxic (12).

Several lines of investigation [including direct chemical analysis (13), the use of immunological probes (14), and the interaction of oxidized LDL with cultured macrophages under *in vitro* conditions (13)] lend support to the idea that oxidized LDL is a key link in the cascade of events that results in the avid accumulation of lipids by macrophages, and hence, in the generation of the "fatty-streak" stage.

THE PROGRESSION OF THE FATTY STREAK STAGE TO THE ADVANCED PLAQUE

The avid accumulation of lipid by macrophages and their accumulation in the subintimal space occurs long before there is any symptomatic evidence of atherosclerosis. The progression of this stage to the advanced, calcified, and fibrotic plaque, which is symptomatically evident and detectable at angiography, involves the migration of smooth muscle cells into the arterial intima (15). Here they proliferate, and some undergo phenotypic changes to become more like macrophages than smooth muscle cells. There are at least two reasons why these smooth muscle cells are an important component of the developing lesion. Firstly, they contribute substantially to its bulk. Secondly, they secrete the connective tissue matrix which holds the lesion together. A variety of growth factors, including platelet-derived growth factor (PDGF), angiotensin II, endothelin-1, interleukin-1 (IL-1), ß-transforming growth factor (TGFß), and tissue necrosis factor (TNF) (16), stimulate the proliferation of these smooth muscle cells.

In summary, the events responsible for atherosclerotic lesion formation are complex. The initiating factor appears to be lipid infiltration followed closely by lipid peroxidation and the adhesion of previously circulating monocytes to the endothelium, followed by their migration to the subendothelium where they convert to macrophages. Late events include the migration and proliferation of smooth muscle cells from the arterial wall. Ultimately, the mass of the lesion is such that it disrupts the endothelium, thrombi form, and the smooth muscle cells begin to die, necrose, and calcify. At this stage, the lesion is angiographically detectable.

THE INVOLVEMENT OF CALCIUM IN THE FORMATION OF ATHEROSCLEROTIC LESIONS

Many of the events which contribute to lesion formation are Ca^{2+} dependent. The most obvious but relatively late event is the deposition of Ca^{2+}, as calcium apatite, in the advanced lesion [17]. However, as table 1 shows, Ca^{2+} is crucial to many of the earlier stages of lesion formation (for references, see ref 18). Thus, monocyte infiltration, as well as smooth muscle cell proliferation and migration, are Ca^{2+}-dependent events, as are platelet aggregation, the liberation of growth factors, and the secretion of matrix material. In addition, hypercholesterolemia enhances membrane Ca^{2+} permeability [19] and hypertension, one of the main risk factors for atherosclerosis, is Ca^{2+} dependent [17].

TABLE 1.   Calcium-dependent Events Involved in
the Development of Atherosclerotic Lesions

Monocyte infiltration
Smooth muscle cell migration
Smooth muscle cell proliferation
Liberation of growth factors
Secretion of matrix materials
Endothelial injury
LDL metabolism

POSSIBLE THERAPEUTIC INTERVENTIONS

Against this background of events which contribute to lesion formation, it is now possible to identify possible target sites for successful therapeutic interventions. For example, as described in table 2, agents which prevent the adhesion of monocytes to the endothelium should be protective, as should agents which prevent lipid peroxidation and the uptake of oxidized lipids by macrophages. Likewise, agents which prevent smooth muscle cell migration, proliferation and release of matrix material, or which stimulate reverse cholesterol transport from the arterial wall to the blood stream, should be protective.

TABLE 2.   Potential Targets for Therapeutic
              Interventions Relating to the Control of
                        Atherosclerosis

(a)	Agents which prevent monocyte adhesion to the endothelium.
(b)	Substances which inhibit lipid uptake by macrophages.
(c)	Agents which prevent lipid peroxidation and metabolism.
(d)	Agents which slow smooth muscle cell proliferation and migration.
(e)	Regimes for lowering plasma cholesterol.
(f)	Agents which restrict Ca^{2+} availability.
(g)	Antichemotactic agents.
(h)	Risk-reducing drugs, e.g., antihypertensives.

CALCIUM ANTAGONISTS AS ANTI-ATHEROGENIC AGENTS

Recognition of the importance of Ca^{2+} in the etiology of atherosclerotic plaque formation has resulted in investigation aimed at determining whether calcium antagonists slow the atherogenic process. Initially, these studies were undertaken on cholesterol-fed animals (table 3) but more recently the trials have extended to include humans. As table 3 shows the results have been positive - provided that the test subjects possessed normal LDL receptors.

First- and second-generation calcium antagonists appear to be equi-effective, although the human studies completed so far have been restricted to first-generation drugs. The question which concerns us here is why these drugs are effective? Their effect on arterial resistance can

TABLE 3.   Studies Relating to the
                  Anti-atherogenic Effects of Calcium Antagonists

Agents	Test Animal	Response	Ref
Animal Studies			
Diltiazem	Rabbit	+	20
Nifedipine	Rabbit	+	21
Nicardipine	Rabbit	+	22
Verapamil	Rabbit	+	23
Anipamil	Rabbit	+	24
Nilvadipine	Rabbit	+	25
Amlodipine	Rabbit	+	26
Isradipine	Rabbit	+	27
Felodipine	Rabbit	+	28
Human Studies			
Nifedipine		+	4
Nifedipine		+	5
Nifedipine		+	6
Nicardipine		+	7

Where + denotes a slowed rate of atherogenesis.

be discounted as a mechanism (21), and since they do not lower plasma cholesterol (6) this cannot be a contributory factor. Probably a multiplicity of effects is involved. They include lowering cytosolic Ca^{2+} as a direct response to their inhibitory effect on Ca^{2+} influx through the L-type Ca^{2+} channels (17), inhibition of monocyte adhesion (29), inhibition of smooth muscle cell proliferation and migration (29,30), inhibition of cholesterol ester deposition (31), promotion of cholesterol efflux (32), inhibition of platelet aggregation (33), slowed release of growth factors (18), inhibition of superoxide production (34), protection against the consequences of lipid peroxidation (35), and inhibition of collagen synthesis (27,36).

IMPLICATIONS FOR INTERVENTION

Based on an understanding of the mechanisms which contribute to the formation of an atherosclerotic plaque, and of the crucial role played by Ca^{2+} in the early stages of plaque formation, it is now relatively easy to understand why calcium antagonists slow the development of new lesions (6,7). Their efficacy is not dependent on any blood-pressure-lowering effect these drugs may exert nor does it depend upon a lipid-lowering activity. Instead, these compounds interfere directly with the biochemical events responsible for lesion formation. Their use as anti-atherogenic agents, therefore, is not aimed at the removal of existing adult plaques but rather at preventing their development. Prophylactic therapy is an obvious requirement, as is the use of either a long-acting drug or a slow-release formulation of a short-acting drug to ensure protection on a twenty-four-hour basis.

REFERENCES

1. Born G.V.R. (1991): In: *Calcium Antagonism and Atherosclerosis,* edited by G.V.R. Born, D.J. Triggle and P.A. Poole-Wilson pp. 1-25. Science Press, London.
2. Grown G. (1990): *N. Engl. J. Med.,* 323: 1289-1298.
3. Daugherty A., Zweifel B.S. and Schonfeld G. (1991): *Br. J. Pharmacol.,* 103: 1013-1018.
4. Gottlieb S.O., Brinker J.A., Mellit E.D., Achuff S.C., Baughman K.L., Traill T.A., Weiss J.L., Reitz B.A., Weisfeldt M.L. and Gerstenblith G. (1989): *Circulation,* 80(suppl.2): 228 (abstract).
5. Loaldi A., Polese A., Montorosi P., De Cesare N., Fabbiocchi F., Ravagnani P. and Guazzi M. (1989): *Am. J. Cardiol.,* 64: 433-439.
6. Lichtlen P.R., Hugenholtz P.G., Rafflenbeul W., Hecker H., Jost S. and Deckers J.W. (1990): *Lancet,* 1: 1109-1113.
7. Waters D., Lespérance J., Fracetich M., Causey D., Théroux P., Chiang Y.-K., Hudon G., Lemarbre L., Reitman M., Joyal M., Grosselin G., Dyrda I., Macer J. and Havel R.J. (1990): *Circulation,* 82: 1940-1953.
8. Witztum J. (1990): *Atherosclerosis Rev.,* 21: 59-69.
9. Schwenke D.C. and Carew T.E. (1989): *Arteriosclerosis,* 9: 908-918.
10. Bevilacqua M.P., Pober J.S., Wheeler M.E., Cottran R.S. and Gimbrone M.A., Jr (1985): *J. Clin. Invest.,* 76: 2003-2011.
11. Morel D.W., Di Corleto P.E. and Chisolm G.M. (1989): *Arteriosclerosis,* 4: 357-364.
12. Cathcart M.K., Morel D.W. and Chisolm G.M. (1985):*J. Leukocyte Biol.,* 38: 341-350.
13. Palinski W., Rosenfeld M.E., Ylä-Herttuala S., Gurtner G.C., Socher S.S., Butler S.W., Parthasarathy S., Curtiss L.K. and Witzum J. (1989): *Proc. Natl. Acad. Sci. USA,* 86: 1372-1376.
14. Palinski W., Ylä-Herttuala S., Rosenfeld M., Butler S.W., Socher S.A., Parthasarathy S., Curtiss K.K. and Witzum J. (1990): *Arteriosclerosis,* 10: 325-335.
15. Ross R. (1986): *N. Engl. J. Med.,* 314: 488-500.
16. Nayler W.G. (1991): *The Second Generation of Calcium Antagonists,* pp. 139-151. Springer Verlag, Berlin.
17. Fleckenstein A., Frey M., Zorn J. and Fleckenstein-Grün G.

(1990): In: *Hypertension, Physiology, Diagnosis and Management,* edited by J.H. Larach and B.M. Brenner pp. 471-509. Raven Press, New York.

18. Nayler W.G. (1988): *Calcium Antagonists,* pp. 235-247. Academic Press, London.

19. Strickberger S.A., Russek L.N. and Phair R.D. (1988): *Circ. Res.,* 62: 75-80.

20. Ginsburg R., Davis K., Bristow M.R., McKennet K., Kodsi S.R., Billingham E.M. and Schroeder J.S. (1983): *Lab. Invest.,* 49: 154-158.

21. Henry P.D. and Bentley K.L. (1981): *J. Clin. Invest.,* 68: 1366-1369.

22. Willis A.L., Nagel B., Churchill V., Whyte M.A., Smith D.L., Mahmud I. and Puppione D.L. (1985): *Arteriosclerosis,* 5: 250-255.

23. Rouleau J.L., Parmley W.W., Stevens J., Wikman-Coffelt J., Sievers R., Mahley R.W. and Havel R.J. (1983): *J. Am. Coll. Cardiol.,* 1: 1453-1460.

24. Catapano A.L., Maggi F.M. and Cicerano U. (1988): *Ann. N.Y. Acad. Sci.,* 522: 519-522.

25. Nomota A., Hirosumi J., Sekiguchi C., Mutoh S., Yamaguchi I. and Aoki H. (1987): *Atherosclerosis,* 64: 255-261.

26. Nayler W.G. (1990): *Am. J. Cardiol.,* 66: 23H-27H.

27. Heider J.G., Weinstein D.G., Pickens C.E., Lan S. and Su C.-M. (1977): *Transplant Proc.,* 19(suppl.5): 96-101.

28. Staudt H., Nordlander M. and Kling E. (1991): *Proceedings of 2nd International Symposium on Calcium Antagonists in Cardiovascular Care,* Basel, Abstract 120.

29. Neuser D. and Rosen B. (1990): *Proceedings of Satellite Symposium on Anti-atherosclerotic Potentials of Calcium Antagonists,* 13th International Society of Hypertension meeting. Montreal, pp. 9-10.

30. Nilsson I., Sjolund M., Palmberg L., Von Euler A.M., Jonzon B. and Thyborg I. (1985): *Atherosclerosis,* 58: 109-112.

31. Daugherty A., Rateri D.L., Schonfeld G. and Sobel B.E. (1987): *Br. J. Pharmacol.,* 91: 113-118.

32. Schmitz G., Robernek H., Beuck M., Krause R., Schurek A. and Niemann R. (1988): *Arteriosclerosis,* 8: 46-56.

33. Ware J.A., Jonson P.C., Smith M. and Salzman E.W. (1986): *Circ. Res.,* 59: 39-42.

34. Irita K., Fujita I., Takeshige K., Minakami S. and Yoshitake J. (1986): *Biochem. Pharmacol.,* 35: 3465-3471.

35. Shridi F. and Robak J. (1988): *Pharmacol. Res. Commun.,* 20: 13-21.

36. Orekhov A.N., Tertov V.U., Khashimov K.A., Kudryashov S.A. and Smirnov V.N. (1986): *J. Hypertens.,* 4(suppl.6): S153-S155.

CALCIUM ANTAGONISTS AND LIPID-LOWERING AGENTS IN THE PROTECTION OF THE ARTERIAL WALL

R. Paoletti, M. Raiteri, M. Soma, and F. Bernini

Institute of Pharmacological Sciences, University of Milan
Via Balzaretti 9, 21033 Milan, Italy

INTRODUCTION

Calcium antagonists (CAs) may affect major processes involved in the atheroma formation. Among the possible mechanisms CAs modulate cholesteryl ester (CE) metabolism. We compared the effect of verapamil and nifedipine on cholesterol esterification in mouse peritoneal macrophages (MPM) in different conditions of Acyl-CoA: Cholesterol Acyltransferase (ACAT) activation. CE is formed in cells via ACAT, which senses free cholesterol (C) supplied by lysosomal hydrolysis of CE carried by lipoproteins. As expected, verapamil completely inhibited the ability of acetyl low-density lipoproteins (AcLDL) to stimulate cholesterol esterification when simultaneously incubated with these lipoproteins, but was less effective in cells previously loaded with cholesteryl esters or stimulated by 25hydroxycholesterol (25-OH). In the same experimental conditions, the results obtained with nifedipine indicate that this dihydropyridine has no major influence on cellular cholesterol esterification. However, the new dihydropyridine derivatives, lacidipine and elgodipine, very efficiently inhibited the ability of AcLDL and 25-OH to stimulate cholesterol esterification in MPM. It appears that CAs of different structure, even within the same class, may have various effects on esterification of cellular cholesterol.

Some lipid lowering agents, designed to reduce a risk factor of atherosclerosis (i.e., hyperlipidemia), may directly affect, like CAs, some processes involved in atheroma formation. We reported that lipophilic simvastatin, but not hydrophilic pravastatin, inhibits the proliferation of rat aorta myocytes at concentrations ranging between 0.1 and 10 μM. This inhibitory effect was prevented by addition of mevalonate. Preliminary data indicates that this effect is also present with the new lipophilic inhibitors HMG-CoA reductase fluvastatin.

CAs and some lipid-lowering drugs may directly inhibit atheroma formation. The identification of the mechanisms involved in their action will contribute to understanding atherogenesis and promote its pharmacological control.

T. Godfraind et al. (eds.), Calcium Antagonists, 125–132.
© 1993 *Kluwer Academic Publishers and Fondazione Giovanni Lorenzini.*

CALCIUM ANTAGONISTS

Calcium channel inhibitors (CAs) include a heterogeneous group of therapeutic agents. Based on their pharmacological profiles CAs can be classified into six categories (1). This classification recognizes three primary types: verapamil-like, nifedipine-like, and diltiazem-like compounds selective for calcium channels. The other three types, which include flunarizine-like, prenylamine-like, and other compounds, are not selective for calcium channels. The number of CAs, as well as the range of indications for their use, are being expanded (2). These indications include: cardiovascular and peripheral vascular diseases and cerebral circulatory diseases. New potential indications are central nervous system diseases, uterine and gastrointestinal disorders, morphine withdrawal and alcohol toxicity, enhancement of activity of antineoplastic agents, facilitation of immunosuppressing and antirejection therapy, and an increase in the effectiveness of various antimalarial regimes. They may also provide cytoprotection, not only within the myocardium, but also within the vasculature. Of particular interest is the potential anti-atherosclerotic activity of CAs (3).

CAs reduce the severity of experimentally induced atherosclerosis in cholesterol-fed animals. The reduction of aortic cholesterol is one of the most striking findings. This effect is achieved without decreasing plasma lipid, but may be related to an interference of CAs with lipid metabolism in the arterial wall (4).

We recently studied the effect of different CAs on receptor-mediated LDL catabolism by human fibroblasts in culture. Verapamil-like compounds and diltiazem are effective in stimulating receptor-mediated LDL uptake, while nifedipine-like compounds and flunarizine are inactive. However, we have shown that the dihydropyridine-derivative amlodipine, which has basic properties with a pKa of 8.6, similar to verapamil, effectively stimulates the LDL-receptor expression in human fibroblasts. This observation suggests that a basic group should be present on the CA molecule to modulate the LDL-receptor activity (5).

CAs not only may affect the cholesterol delivery to the cells by modulating lipoprotein metabolism, but may also interfere with cholesteryl ester hydrolysis and re-esterification.

In cytoplasm cholesterol undergoes a continuous cycle of esterification and de-esterification which implies a constant activation of ACAT (6). In the absence of a cholesterol acceptor in the extracellular space, this cycle remains activated even when simultaneously incubated with acLDL, but not when cholesterol esterification was induced by 25-OH (table 1). At the highest concentration tested (50 μM) an inhibitory effect was observed also in the latter condition (table 1).

TABLE 1.   Effect of Verapamil and Nifedipine
             on Cholesterol Esterification in Normal
             Mouse Peritoneal Macrophages

DRUGS (μM)	CONDITIONS OF ACAT STIMULATION	
	AcLDL	25-OH
	% of Control ($\pm$SD)	
VERAPAMIL 10	$21.1 \pm 0.7^*$	100.1 ± 3.7
VERAPAMIL 50	$1.1 \pm 0.3^*$	$37.9 \pm 0.5^*$
NIFEDIPINE 50	88.1 ± 5.4	118.5 ± 4.7

Values of control were (pmoles cholesteryl oleate formed/mg cell protein): acLDL 2988 ± 148; 25-OH cholesterol (25-OH) 1548 ± 58. $^*p < 0.001$; samples were run in triplicate.

Because ACAT activation by acLDL in the absence of a cholesterol acceptor in the medium is not reversible (6), we tested the effect of CAs on preloaded MPM after removal of acLDL. Our results indicate that, under these experimental conditions, verapamil (10 μM) only slightly reduced ACAT activity (table 2). Stein and Stein (8) observed a higher effect of verapamil in inhibiting ACAT activation induced by acLDL as compared to the enzyme stimulation obtained with cholesterol-rich liposomes. Moreover, these authors demonstrated that verapamil did not affect cholesterol esterification in cell-free homogenate obtained from macrophage pretreated with the drug. In our studies verapamil was completely inactive when added directly in the cell-free homogenate (data not shown). All together, our data supports the hypothesis suggested by Stein and Stein (8) that inhibition of cholesterol esterification determined by verapamil involves an interference of the drug with cholesterol transport to the ACAT esterification site.

The results obtained with nifedipine indicate that this dihydropyridine has no major influence on cholesterol esterification activity in MPM either stimulated by acLDL or by 25-OH (table 1). A slight inhibitory effect of the drug on this enzymatic system was observed in preloaded cells (table 2). This effect, however, may follow the stimulation of neutral cholesteryl ester hydrolase elicited by nifedipine (9). Daugherty et al. (10) demonstrated that nifedipine was able to inhibit cholesterol esterification induced by beta-very-low-density lipoproteins in rabbit alveolar macrophages. This discrepancy with our results may relate to a different sensitivity to the drugs of the two cellular models.

TABLE 2.   Effect of Verapamil and Nifedipine
        on Cholesterol Esterification in Cholesterol-Loaded
                Mouse Peritoneal Macrophages

DRUGS (μM)	
	% of Control ($\pm$SD)
VERAPAMIL 10	84.9 ± 4.3[*]
VERAPAMIL 50	28.2 ± 2.5[**]
NIFEDIPINE 50	79.1 ± 1.2[*]

Values of control were: 2468 ± 53 (pmoles cholesteryl oleate formed/mg cell protein). [*]$p < 0.01$; [**]$p < 0.001$; samples were run in triplicate.

Although our results suggest that CAs of different classes may have different effects on cholesterol esterification, data from our laboratory suggests that not all dihydropyridine derivatives behave like nifedipine on cholesterol esterification (11). Experiments with the new CA lacidipine suggest that this dihydropyridine derivative can completely inhibit cholesterol esterification in macrophages (11). These results suggest the view that inhibition of cholesterol esterification cannot be ascribed to specific classes of CAs. Whether these differences *in vitro* mediate different *in vivo* anti-atherosclerotic mechanisms or efficacy of CAs remains to be established.

LIPID-LOWERING AGENTS

A pivotal role in the formation of atherosclerotic lesions is also played by initiation of smooth muscle cells (SMC, myocyte) proliferation in the medial layer, their migration and further proliferation in the intima of the arterial wall (12). Factors controlling vascular SMC proliferation are thought to be important in the development of atherosclerotic disease (12). Several data indicate the role in cell growth played by the cholesterol synthetic pathway. While cholesterol seems to be required early in the cell cycle of arterial SMC (G1 phase), mevalonate itself and some of its nonsteroidal derivative (isoprenoids) are determinant in cell division and growth regulation (13,14). Compactin, a specific competitive inhibitor of hydroxy methylglutaryl coenzyme-A (HMG-CoA) reductase, the rate-limiting enzyme in cholesterol biosynthesis, inhibits the replication of different cell types in culture unless large amounts of mevalonate are supplied (13,14). Mevalonate requirement is reduced but not eliminated by cholesterol supplied in low-density lipoprotein (13,14). In view of the important role of mevalonate and of cholesterol biosynthesis in stimulating the proliferation of arterial smooth muscle cells, it was of interest to study the *in vitro* effect of two potent HMG-CoA reductase inhibitors, simvastatin and pravastatin, on the growth of these cells. The results provide evidence that the lipophilic compound simvastatin, but not the hydrophilic compound pravastatin, decreases the rate of growth of rat vascular myocytes, in concentrations ranging between 0.01 and 10 μM (15). Similar results have been obtained using human femoral artery myocytes (unpublished results). The inhibitory effect of simvastatin, already detectable at the lowest concentration tested, is statistically significant at 0.1 μM (table 3), the reported therapeutic concentration. A similar concentration of lovastatin has been found by Falke et al. (16) to inhibit the replication of fibroblasts and smooth muscle cells. Fluvastatin, a new HMG-CoA reductase inhibitor, also possesses the propriety to inhibit SMC proliferation at similar concentrations (unpublished data).

TABLE 3.   Mean Doubling Time for Smooth Muscle Cells Cultured From Rat Aorta: Effect of Simvastatin and Pravastatin

DRUGS	μM	Doubling Time (h) Mean $\pm$ SD
CONTROL		44.8 ± 1.6
SIMVASTATIN	0.1	50.3 ± 0.5[*]
	1	54.5 ± 1.1[*]
	5	72.8 ± 0.6[*]
PRAVASTATIN	0.1	45.5 ± 1.2
	1	46.5 ± 2.0
	10	46.0 ± 0.5

The doubling time was measured after 72 hours of incubation; each point was run in triplicate. Drug versus control: [*]$p < 0.01$.

The inhibition, evaluated as cell number and nuclear incorporation of ^{3}H-thymidine, was dose-dependent with an IC_{50} of 4.2 μM, and was completely prevented by addition of mevalonate (100 μM), confirming the role of this metabolite or its products in regulating cell division and growth (13,14).

Similar results were obtained *in vivo* utilizing the rabbit model described by Booth et al. (17). Fluvastatin, simvastatin, and lovastatin were able to inhibit neointimal formation induced by perivascular manipulation of carotid artery. Pravastatin had little effect in this respect (M. Soma, et al., unpublished observations).

No explanation is available for the failure of pravastatin (0.01-500 μM) to inhibit the replication of vascular myocytes. It can only be speculated that the hydrophilicity of the drug could impair its diffusion through the plasma membrane. However, the ability of pravastatin, as well as of simvastatin, to reduce cholesterol biosynthesis in myocytes (-80%, unpublished results) clearly suggests the intracellular presence of pravastatin.

The ability of simvastatin, lovastatin, and fluvastatin to inhibit the proliferation of arterial myocytes may be potentially important for clinical applications. An accelerated proliferation of SMC appears to be the cause of premature coronary occlusion in patients undergoing heart transplantation, coronary artery bypass graft (CABG), and percutaneous transluminal coronary angioplasty (PTCA) (18). This process of accelerated atherosclerosis accounts for significant morbidity and mortality in these patients (18).

Pharmacological approaches, from heparin to angiotensin converting enzyme inhibitors, to inhibit SMC proliferation have been tried with variable success (18). Lipophilic statins appear to be potentially of great benefit because of their ability to strongly reduce both myocyte proliferation and hypercholesterolemia, that has been shown to promote intimal proliferations after endothelial injury in normal artery. Aggressive lipid-lowering treatment should be undertaken in all patients undergoing PTCA and CABG because of the proven effects of delayed progression and induced regression of atherosclerotic lesion, as well as reduction of cardiovascular events (19). On the other hand, the lack of effect of pravastatin on smooth muscle cell proliferation provides a unique tool to evaluate the effect of plasma lipids on atherosclerosis progression and regression.

Coronary angioplasty is now an accepted form of treatment for patients with ischemic heart disease and although the number of PTCA procedures continues to increase (> 250,000 performed in the USA in 1988), restenosis complicates about 30% of cases within 3 to 6 months (18). Excessive intimal proliferation of myocytes has been shown both at necropsy and more recently *in vivo* by means of atherectomy devices. No therapy evaluated to date, including low-dose aspirin and dipyridamole, calcium channel blockers, warfarin, and heparin, has yielded a consistent benefit (18). Lovastatin and simvastatin may decrease the incidence of restenosis by reducing serum-cholesterol levels and by preventing cell proliferation. To address this issue, Gellman et al. (20) has recently shown that lovastatin reduces intimal hyperplasia after balloon angioplasty of the femoral artery in the hypercholesterolemic rabbit. The beneficial effect of lovastatin in this study appears not to be related to its hypocholesterolemic action, in agreement with clinical studies that did not find any correlation between serum cholesterol and restenosis (20). Preliminary results in patients undergoing coronary angioplasty have shown a significant reduction in the incidence of restenosis with lovastatin (21), but validation from a larger study is still required.

Our results indicate that lipophilic statins, but not pravastatin, interfere with arterial myocyte proliferation. This effect could be a component, along with the inhibition of liver and intestine cholesterol biosynthesis, of the effective anti-atherosclerotic action of these drugs.

CONCLUSION

In conclusion, treatment of atherosclerosis may be achieved by directly affecting the processes involved in atheroma formation. This effect may be obtained with compounds already able to modify major risk factors of atherosclerosis such as hypertension and hypercholesterolemia or by new compounds specifically designed as direct anti-atherosclerotic drugs.

REFERENCES

1.   Vanhoutte P.M. and Paoletti R. (1987): *Trends in Pharmacol. Sci,* 8: 4-5.

2.   Nayler W.G. (1988): In: *Calcium Antagonists,* pp. 1-338. Academic Press, New York.

3.   Weinstein D.B. (1988): *J. Cardiovasc. Pharmacol.,* 12(suppl.6): S29-S35.

4.   Bernini F., Catapano A.L., Corsini A., Fumagalli R. and Paoletti R. (1989): *Am. J. Cardiol.,* 64: 129I-134I.

5.   Paoletti R. and Bernini F. (1990): *Am. J. Cardiol.,* 66: 28H-31H.

6.   Brown M.S., Ho Y.K and Goldstein J.L. (1980): *J. Biol. Chem.,* 255(19): 9344-9352.

7.   Brown M.S., Dana S.E. and Goldstein J.L. (1975): *J. Biol. Chem.,* 250(10): 4025-4027.

8.   Stein O. and Stein Y. (1987): *Arteriosclerosis,* 7: 578-584.

9.   Etingin O.R. and Hajjar D.P. (1985): *J. Clin. Invest.,* 75: 1554-1558.

10. Daugherty A., Rateri D.L., Schonfeld G. and Sobel B.E. (1987): *Br. J. Pharmacol.,* 91: 113-118.

11. Bellosta S., Didoni G., Bertulli S.M., Fumagalli R. and Bernini F. (1991): In: *Calcium Antagonists in Cardiovascular Care,* Second International Symposium, Basel, Switzerland, abstract.

12. Ross R. (1986): *New Eng. J. Med.,* 314: 488-500.

13. Goldstein J.L. and Brown M.S. (1990): *Nature,* 343: 425-430.

14. Habenicht A.J., Glomset J.A. and Ross R. (1980): *J. Biol. Chem.,* 255: 5134-5140.

15. Corsini A., Raiteri M., Soma M., Fumagalli R. and Paoletti R.

(1991): *Pharmacol. Res.*, 23: 173-180.

16.   Falke P., Mattiasson L., Stavenow L. and Hood B. (1989): *Pharmacol. Toxicol.*, 64: 173-176.

17.   Booth R.F.G., Martin J.F., Honey A.C., Hassal D.G., Beesley J.E. and Moncada S. (1989): *Atherosclerosis*, 76: 257-268.

18.   Ip J.H., Fuster V., Badimon L., Badimon J., Taubman M.B. and Chesebro J.H. (1990): *J. Am. Coll. Cardiol.*, 15: 1667-1687.

19.   Brown G., Albers J.J., Fisher L.D., Schaefer S.M., Lin J.T., Kaplan C., Zhao X.Q., Bisson B.D., Fitzpatrick V. and Dodge H.T. (1990): *New Engl. J. Med.*, 323: 1289-1298.

20.   Gellman J., Ezekowitz M.D., Sarembock I.J., Azrin M.A., Nochomowitz L.E., Lerner E. and Haudenschild C.C. (1991): *J. Am. Coll. Cardiol.*, 17: 251-259.

21.   Sahni R., Maniet A.R., Voci G. and Banka V.S. (1991): *Am. Heart J.*, 121: 1600-1608.

CALCIUM CHANNEL BLOCKERS AS AGENTS FOR THE TREATMENT OF ATHEROSCLEROSIS

Philip D. Henry and Moises Bucay

Baylor College of Medicine
One Baylor Plaza
Houston, Texas 77030

Therapeutic trials in animals and experiments with cells in culture suggest that calcium blockers (calcium channel blockers, calcium antagonists) may exert anti-atherogenic effects (1-3).

THERAPEUTIC TRIALS IN ANIMALS

Calcium channel blockers have been shown to retard the formation of arteriosclerotic lesions in major experimental models of atherosclerosis (1-3). In fat-fed rabbits (4,5), rats (6), and monkeys (7), dihydropyridine and verapamil derivatives have been repeatedly shown to suppress the formation of foam-cell lesions (fatty streaks). Treatment with calcium blockers retards hypercholesterolemic endothelial dysfunction as reflected by decreased responsiveness to endothelium-dependent vasodilators (8). Atherosclerotic impairment of endothelium-dependent relaxation affects not only large arteries, but also small arteries, vessels not known to develop atheromas (9). Of interest is that nifedipine has been reported to limit, in cholesterol-fed rabbits, accumulation of lipid in microvessels (10). Therapeutic effects of calcium blockers have usually not been associated with reductions in plasma lipids or arterial pressure (1-3).

In another model of arterial disease, lesion formation produced by mechanical arterial trauma, calcium blockers have been shown to suppress proliferative responses. In normocholesterolemic rabbits and rats, treatment with nifedipine or nilvadipine suppresses the formation of proliferative lesions after injury with balloon catheters or periarterial cuffs (11-13).

CELL CULTURE EXPERIMENTS

Experiments with cells in culture have been performed in an attempt to elucidate mechanisms by which calcium blockers might inhibit the formation of foam-cells or the proliferation of vascular smooth muscle in animal models of atherosclerosis.

T. Godfraind et al. (eds.), Calcium Antagonists, 133–138.
© 1993 *Kluwer Academic Publishers and Fondazione Giovanni Lorenzini.*

Etingin and Hajjar (14) have shown that nifedipine in submicromolar concentration promotes efflux of cholesterol from fat-laden rabbit vascular smooth muscle cells in culture. They ascribed the effect to a stimulation of lysosomal cholesteryl ester hydrolase activity. In a subsequent study, these authors collected aortic tissue from patients undergoing open heart operations and demonstrated that cholesteryl ester hydrolase activity in arterial homogenates was several times higher in patients treated preoperatively with calcium blockers compared to untreated control patients (15). Because excess intracellular cholesterol is stored predominantly as an ester and because cholesterol can be discharged from cells mainly as a free sterol, cholesteryl ester hydrolysis represents a pivotal step in reverse cholesterol transport. Schmitz et al., (16) observed that fat-laden mouse macrophages in culture excreted lipid-rich lamellar bodies after addition of nifedipine to the culture medium. The reports by Etingin and Hajjar (14,15) and Schmitz et al. (16) raise the question, whether calcium blockers exert anti-atherosclerotic effects by stimulating reverse cholesterol transport. Effects of calcium blockers on net lipid uptake by nonmuscle cells in culture have been reported also by Stein et al. (17) and Daugherty et al. (18).

In other cell culture experiments, calcium blockers have been shown to suppress the migration (19) and proliferation of vascular smooth muscle (20-22). Nilsson et al. (20) reported that nifedipine inhibited smooth muscle replication stimulated by whole serum or platelet-derived growth factor. Orekhov et al. (21) cultured intimal cells (modified smooth muscle cells) from human aortic lesions and observed that verapamil inhibits thymidine incorporation into DNA. Similarly, Stein et al. (22) showed that verapamil inhibited DNA synthesis by cultured rabbit and bovine aortic smooth muscle cells.

There is increasing evidence that peroxidation of lipoproteins accumulating in arterial walls may play an important role in atherogenesis (23). Agents such as probucol (24) or butylated hydroxytoluene (BHT) (25) may retard atherogenesis by preventing the formation of atherogenic oxidatively modified lipoproteins. Oxidatively modified low-density lipoproteins and lysolipids contained in such lipoproteins exert complex effects including actions on monocyte chemotaxis, macrophage immobilization and lipid uptake by macrophages (23). These changes appear to be associated with altered secretion of cytokines (23) and increased expression of adhesion molecules (athero-ELAM's) (26). It is possible that calcium blockers influence multiple leukocyte functions including those involved in immune responses (27).

Recently, several authors have reported that structurally unrelated calcium blockers (dihydropyridines, verapamil, diltiazem, and others) possess antiperoxidative properties (28-33) (table 1). Therefore, it is

TABLE 1. Antiperoxidative Effects of Calcium Blockers on Membrane Lipids

Author	Oxidation System (Lipid Preparation)	Nifedipine IC_{50} μM[a]	Verapamil IC_{50} μM[a]	Diltiazem IC_{50} 50 μM[a]
Robak et al., 1986 (28) 1988 (29)	Inhibition of NADH[b]-stimulated malonaldehyde formation (liver liposomes)	60	1	21
Janero et al., 1988 (30) 1989 (31)	Xanthine oxidase/hypoxanthine/Fe^{3+}-ADP[c] (cardiac liposomes)	>500 (nisoldipine 80)	>500 1/140 as effective as BHT[d]	510 1/100 as effective as BHT[d]
Ondrias et al., 1989 (32)	Autoxidation in air-exposed electrolyte solution (egg yolk liposomes; lipid/drug molar ration = 100/1)	1/10 as effective as BHT[d] (no IC_{50} value)	1/100 as effective as BHT[d] (no IC_{50} value)	not tested
Mak et al., 1990 (33)	Dihydroxyfumarate/Fe^{3+}-ADP[c] (cardiac liposomes)	38	206	850

[a]IC_{50}=Drug concentration (μM) to produce 50% inhibition of peroxidation; [b]NADH=nicotinamide adenine dinucleotide; [c]ADP=adenosine 5' diphosphate; [d]BHT=butylated hydroxytoluene.

possible that some of the anti-atherosclerotic effects of calcium blockers might be mediated by antiperoxidative effects similar to those postulated for probucol and BHT.

CLINICAL TRIALS

Recently, several controlled angiographic trials on the effects of calcium blockers on the progression of coronary disease in humans have been completed (34-37). Other studies are still in progress (38,39). In the completed trials, dihydropyridine derivatives appeared to retard the formation of new or small coronary lesions developed over a two- to three-year period (34-37). As observed in animal trials, hypolipidemic and arterial hypotensive effects were small or absent. Two of these studies were the largest trials in which coronary lesions were evaluated by computer-assisted quantitative arteriography (34,35). However, in these studies sample sizes ($\approx$400 or less) were too small to provide statistically valid information on important clinical endpoints such as myocardial infarction or total mortality. Therefore, additional large-scale studies are necessary to evaluate potential beneficial effects of calcium blockers for the treatment of atherosclerosis and its complications.

REFERENCES

1. Henry P.D. (1990): *J. Cardiovasc. Pharmacol.*, 16: S12-S15.
2. Henry P.D. (1990): *Arteriosclerosis*, 10: 963-965.
3. Henry P.D. (1990): *Circulation*, 82: 2251-2253.
4. Henry P.D. and Bentley K.I. (1981): *J. Clin. Invest.*, 68: 1366-1369.
5. Rouleau J.L., Parmely W.W., Stevens J., Wilkman-Coffelt J., Sievers R., Mahley R.W. and Havel R.J. (1983): *J. Am. Coll. Cardiol.*, 1:1453-1460.
6. Seuter F. (1989): *Z. Kardiol.*, 78: 117-119.
7. Lichtor T., Davis H.R., Vesselinovitch D., Wissler R.W. and Mullan S. (1989): *Appl. Pathol.*, 7: 8-18.
8. Habib J.B., Bossaller C., Wells S., Williams C., Morrisett J.D. and Henry P.D. (1986): *Circ. Res.*, 58: 305-309.
9. Yamamoto H., Bossaller C., Cartwright J., Jr and Henry P.D. (1988): *J. Clin. Invest.*, 81: 1752-1758.
10. Kirschenbaum M.A., Roh D.D. and Kamanna V.S. (1991): *Atherosclerosis*, (in press).
11. Handley D.A., Van Valen R.G., Melden M.K. and Saunders R.N. (1986): *Am. J. Pathol.*, 124: 88-93.
12. Jackson L.J., Bush R.C. and Bowyer D.E. (1988): *Atherosclerosis*, 69: 115-122.
13. Nomoto A., Hirosumi J., Sekiguchi C., Mutoh S., Yamaguchi I. and Aoki H. (1987): *Atherosclerosis*, 64: 255-261.
14. Etingin O.R. and Hajjar D.P. (1985): *J. Clin. Invest.*, 75: 1554-1558.
15. Etingin O.R. and Hajjar D.P. (1990): *Circ. Res.*, 66: 185-190.
16. Schmitz G., Robenek H. and Beuck M. (1988): *Arteriosclerosis*, 8: 46-56.
17. Stein O., Leitersdorf E. and Stein Y. (1985): *Arteriosclerosis*, 5: 35-44.
18. Daugherty A., Rateri D.L., Schonfeld G. and Sobel B.E. (1987): *Br. J. Pharmacol.*, 91: 113-118.
19. Nomoto A., Mutoh S., Hagihara H. and Yamaguchi I. (1988): *Atherosclerosis*, 7: 213-219.
20. Nilsson J., Sjolund M., Palmberg L. and Von Euler A M. (1985): *Atherosclerosis*, 58: 109-122.
21. Orekhov A.N., Tertov V.V., Khashimov K.A., Kudryashov S.A. and Smirnov V.N. (1986): *J. Hypertens.*, 4: S153-S155.
22. Stein O., Halperin G. and Stein Y. (1987): *Arteriosclerosis*, 7: 585-592.
23. Steinberg D. and Witztum J.L. (1990): *JAMA*, 264: 3047-3052.
24. Carew T.E., Schwenke D.C. and Steinberg D. (1987): *Proc. Natl. Acad. Sci. USA*, 8: 7725-7729.

25. Björkhem I., Henriksson-Freyschuss A.H., Breuer O., Diczfalusy U., Berglund L. and Henriksson P. (1991): *Arteriosclerosis Thromb.*, 11: 15-22.

26. Cybulsky M.I. and Gimbrone M.A. (1991): *Science*, 251: 788-791.

27. Levy R., Dana R., Gold B., Alkan M. and Schlaeffer F. (1991): *Isr. J. Med. Sci.*, 27: 301-306.

28. Robak J. and Duniec Z. (1986): *Pharmacol. Res. Commun.*, 12: 1107-1117.

29. Shridi F. and Robak J. (1988):*Pharmacol. Res. Commun.*, 20: 13-21.

30. Janero D.R., Burghardt B. and Lopez R. (1988): *Biochem. Pharmacol.*, 37: 4197-4203.

31. Janero D.R. and Burghardt B. (1989): *Biochem. Pharmacol.*, 38: 4344-4348.

32. Ondrias, K., Misík V., Gergel D. and Stasko A. (1989): *Biochim. Biophys. Acta*, 1003: 238-245.

33. Mak T. and Weglicki W.B. (1990): *Circ. Res.*, 66: 1449-1452.

34. Lichtlen P.R., Hugenholtz P.G., Rafflenbeul W., Hecker H., Jost S. and Deckers J.W. (1990): *Lancet*, 1: 1109-1113.

35. Waters D., Lespérance J., Francetich M., Causey D., Théroux P., Chiang Y-K., Hudon G., Lemarbre L., Reitman M., Joyal M., Gosselin G., Dyrda I., Macer J. and Havel R.J. (1990): *Circulation*, 82: 1940-1953.

36. Loaldi A., Polese A. and Montorsi P. (1989): *Am. J. Cardiol.*, 64: 433-439.

37. Gottlieb S.O., Brinker J.A. and Mellits D. (1989): *Circulation*, 80: II-228.

38. Kober G., Schneider W. and Kaltenbach M. (1989): *J. Cardiovasc. Pharmacol.*, 13: S2-S6.

39. Furberg C.D., Byington R.P. and Borhani N.A. (1989): *Am. J. Med.*, 86: 37-39.

ANTI-ARTERIOSCLEROTIC EFFECTS OF CALCIUM ANTAGONISTS: DO CALCIUM ANTAGONISTS INHIBIT CHOLESTEROL ACCUMULATION IN CORONARY ARTERIES OF CHOLESTEROL-FED RABBITS?

G. Fleckenstein-Grün, F. Thimm, M. Frey, and A. Fleckenstein

Study Group for Calcium Antagonism
Physiological Institute, University of Freiburg
Hermann-Herder-Str. 7, W-7800 Freiburg i.Br., Germany

INTRODUCTION

Historically, arteriosclerosis was considered an abnormal ossification of heavily calcified vessels (1). But since Anitschkow (2) introduced cholesterol-intoxicated rabbits in 1913, their fatty arterial atheromata were pushed forward as models for human arteriosclerosis and, equally misleading, as targets for the experimental evaluation of anti-arteriosclerotic drug potencies (3,4). However, the development of human arteriosclerosis, notably conventional coronary plaque formation, is characterized by a progressive calcium (Ca) uptake into the arterial walls in contrast to an only modest cholesterol accumulation. Arterial Ca overload could be mimicked in rats under the influence of various risk factors. In these experiments, the anti-arteriosclerotic efficacy of Ca antagonists, first described by Fleckenstein and colleagues for verapamil in 1971 (5), nifedipine in 1980 (6), and diltiazem in 1982 (7), was demonstrated to be based on the specific prevention of progredient mural Ca uptake and its morphological consequences. Contrarily, in cholesterol-fed New Zealand (NZ) rabbits, neither the stenosing coronary cholesterol accumulation nor the generalized lipid storage disease, concerning aortic, myocardial hepatic, and renal tissue as well, could be inhibited by a prophylactic therapy with Ca antagonists (verapamil, nitrendipine, nifedipine). In fact, no significant increase in life expectancy of cholesterol-fed NZ rabbits was observed under Ca-antagonistic treatment.

CONVENTIONAL HUMAN CORONARY PLAQUES ARE CALCIUM DOMINATED

Apart from very rare cases of familiar hypercholesterolemia, the formation of conventional human coronary plaques is governed by a progredient arterial Ca accumulation. Fig. 1 demonstrates our analytical

T. Godfraind et al. (eds.), Calcium Antagonists, 139–148.
© 1993 Kluwer Academic Publishers and Fondazione Giovanni Lorenzini.

data obtained from 43 autopsies aged 41 to 90 years (8). Over the whole lifespan, in healthy coronary arteries (stage 0) the contents of free and total cholesterol predominate over the amount of Ca. This situation changes as soon as arteriosclerotic alterations set in. Accordingly, in "fatty streaks" [type I plaques (WHO classification)] Ca was increased on average 13 times, in type II ("fibrous plaques") 25 times, and in fully developed type III plaques ("complicated lesions") 80 times above normal. Conventional coronary plaques of type III that had produced lethal coronary infarction consisted of almost 50% Ca salts. In contrast, there is no correlation between mural coronary free and total cholesterol content and plaque severity. Accordingly, stenosing coronary type III plaques contained less free and total cholesterol (1.5 and 1.8 times normal) than did fatty streaks. Obviously, these tiny amounts of cholesterol cannot per se be responsible for coronary occlusion. However, products such as oxydized low-density lipoproteins (oLDL) may be involved in the development of arteriosclerotic lesions making use of various messenger and even killer functions of Ca ions (see below).

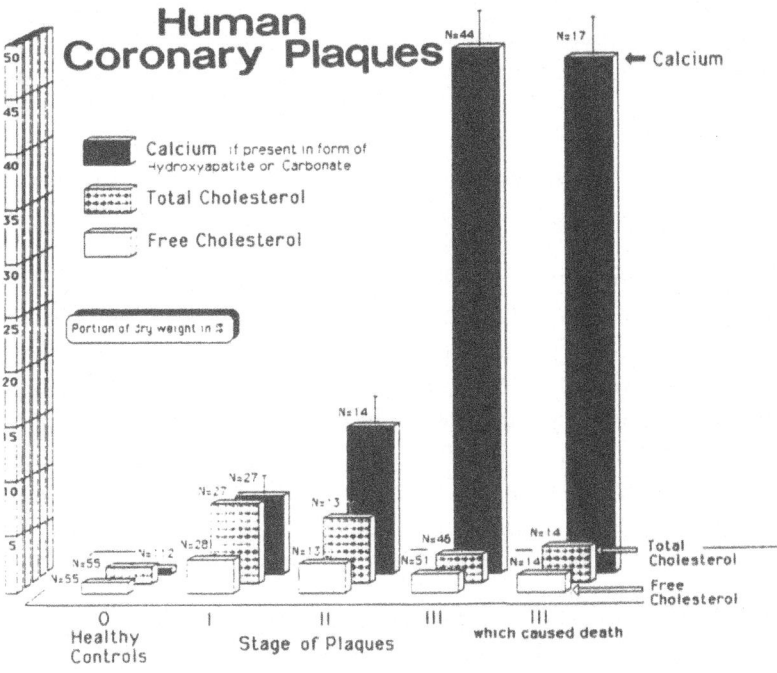

FIG. 1. Chemical composition of conventional human coronary artery plaques. Control values (g/kg dry weight): 2.2±0.3 (Ca); 9.4±1.2 (free cholesterol); and 16.5±1.8 (total cholesterol).

PATHOGENETIC ROLE OF CA IN THE DEVELOPMENT OF ARTERIOSCLEROTIC LESIONS

Ca ions, besides a multitude of vital biological functions, may develop strong pathogenetic potencies, if defectively controlled (9,10). In arteriosclerotic plaque formation, the role of Ca is multifactorial:

1. Ca ions exert various messenger functions particularly in vascular smooth muscle cells (VSMCs), concerning proliferation and migration, secretion of matrix elements, effects of growth factors, endothelin, oLDL, or nicotine to mention just a few. In fact, net Ca uptake into monolayers of cultured aortic media cells (AMCs) of normotensive rats (WKY) was significantly increased by nicotine, endothelin, and oLDL, as well as in AMCs of spontaneously hypertensive rats (SHR) compared to WKY (fig. 2).

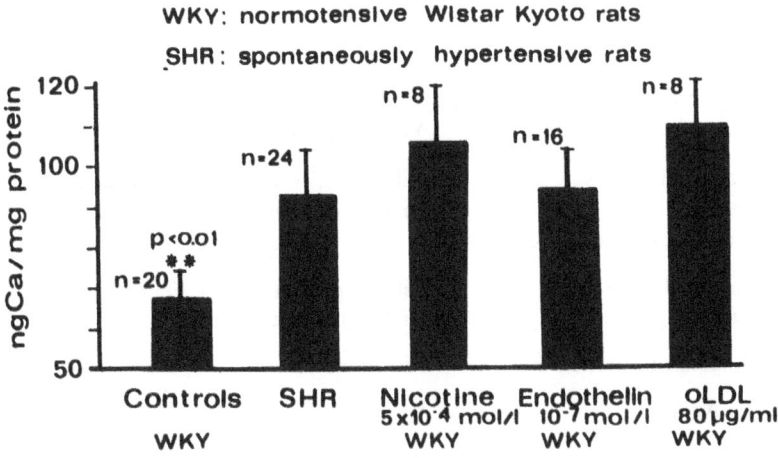

FIG. 2. Monolayers of cultured aortic media cells. Increase of net Ca uptake induced by risk factors (measured as ^{45}Ca incorporation).

2. Ca ions even adopt killer capacities if excessively accumulated in the intracellular space. The deleterious sequelae of Ca overload, first demonstrated for mammalian myocardium by Fleckenstein (9), are excess intracellular Ca signalling and finally cell death and necrosis, brought about by overstimulation of Ca-dependent ATPases, proteases, and lipases, mitochondrial damage, and breakdown of intracellular high energy phosphate. In VSMCs disintegration of mitochondria and contractile elements was demonstrated (10).

3. Ca ions specifically interact with extracellular matrix proteins. Arteriosclerotic calcification of elastic fibers with subsequent decay and stiffness are well-known consequences.

CALCIUM-DOMINATED TYPES OF EXPERIMENTAL ARTERIOSCLEROSIS: ANTI-ARTERIOSCLEROTIC EFFECTS OF CALCIUM ANTAGONISTS

In experiments on rats (table 1), promotion of arteriosclerosis by risk factors (vitamin D_3, hypertension, diabetes, nicotine, advanced age) was found to be correlated with facilitation of mural Ca uptake. Usually, this arterial Ca overload was the most characteristic feature of early lesions, preceded by far the occurrence of manifest arteriosclerosis, and quantitatively corresponded to the severity of lesions. Contrarily, cholesterol accumulation was absent or rather modest (10). However, the decisive experimental milestone that challenged the exclusively cholesterol-oriented hypothesis of atherogenesis was the discovery of Ca antagonists with anti-arteriosclerotic potencies in our laboratory (5-10).

TABLE 1.   Types of Arterial Calcinosis and
                      Arteriosclerosis in Rats

Types	Effective Ca Antagonists Examined
Intoxication with Vitamin D_3	Verapamil, Diltiazem
Hereditary Spontaneous Hypertension (Okamoto Rats; SHRs)	Nifedipine, Nisoldipine, Nitrendipine, Nimodipine, Amlodipine, Verapamil
NaCl-Sensitive Hypertension (Dahl-S Rats)	Nitrendipine, Anipamil, Felodipine, Amlodipine, Verapamil
Renal Goldblatt Hypertension	Nifedipine
Advanced Age	Verapamil
Alloxan Diabetes	Verapamil
Chronic Oral Doses of Nicotine	Nifedipine, Verapamil, Diltiazem
Combined Administration of Vitamin D_3 plus Nicotine	Diltiazem, Verapamil

These drugs counteract the destructive power of excess Ca uptake into arterial walls and thereby protect intracellular structures, notably in VSMCs, as well as elastic fibers from Ca-mediated functional and morphological alterations (11,12; table 1).

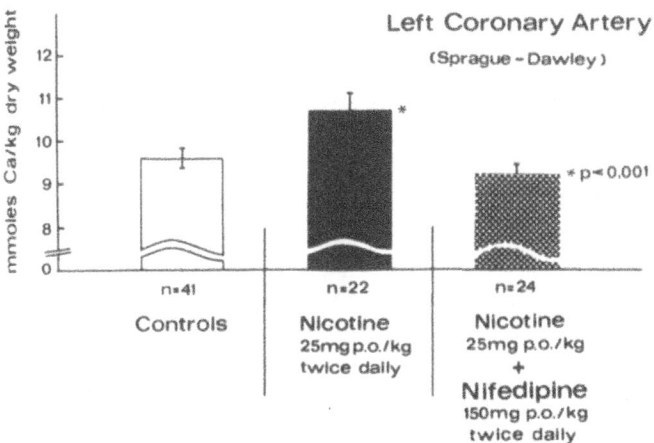

FIG. 3. Restriction of nicotine-induced increase in arterial Ca content of rats by long-term treatment with nifedipine (5 months).

Fig. 3 demonstrates the significant increase in Ca uptake into coronary arteries of rats under the chronic influence of nicotine and its prevention with the Ca antagonist nifedipine. Typically, Ca-dominated arteriosclerotic alterations developed in distal branches of superior mesenteric arteries of 20-21-month-old spontaneously hypertensive rats (SHRs). We classified these lesions according to their severity (fig. 4): type I: meanderous windings and irregular discrete increase in wall diameter; type II: massive wall thickening; type III: excessive wall thickening by a factor of more than 40, with proliferation, endothelial damage, Ca overload, calcification of elastic fibers, lipid accumulation, and at a final state calcified thrombi. In early lesions (type I), net ^{45}Ca incorporation, calculated at % of plasma activity ($=100\%$) was doubled compared to normal; in type II plaques Ca uptake rose by a factor of 4; and in type III by more than 10 times above controls (78.1 ± 1.9). In contrast, net ^{14}C-cholesterol incorporation (in % of plasma activity) into arteriosclerotic walls was not related to plaque severity and only amounted to 1.3 at the most, compared to control values (66.1 ± 1.7). Under the influence of suitable Ca antagonists, for instance nitrendipine, arterial Ca overload could be prevented from the very beginning. Consequently, these drugs completely protected arterial walls of SHRs from any arteriosclerotic alteration and prolonged life expectancy considerably (8).

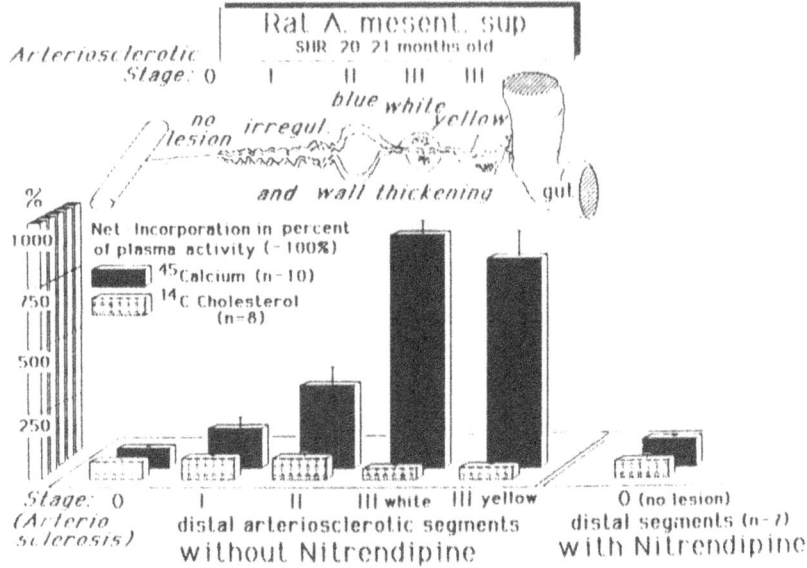

FIG. 4. Net incorporation of ^{45}Ca and ^{14}C-cholesterol in arteriosclerotic plaques of increasing severity [distal branches of A. mesenteric sup. (SHR)].

CALCIUM ANTAGONISTS FAILED TO PREVENT TISSUE CHOLESTEROL ACCUMULATION IN CHOLESTEROL-FED NZ RABBITS

In sharp contrast with these Ca overload-induced forms of experimental arteriosclerosis, feeding of male NZ rabbits with a cholesterol-rich diet (2% cholesterol over 14-17 weeks) produced in our studies a totally different type of arterial alteration. Coronary arteries of these animals showed lethal occlusion by large amounts of accumulated cholesterol in the presence of excess plasma lipid concentrations [(g/dl); n = 15: 2172 ± 171 (total cholesterol); 1151.5 ± 90 very-low-density lipoproteins; 457.4 ± 34.4 low-density lipoproteins; 40.9 ± 3.7 high-density lipoproteins]. These coronary atheromata are characterized by several peculiarities: (1) already early lesions (stage I, fig. 5) were governed by the large deposition of free (3.5- to 7-fold) and total cholesterol (7- to 9-fold) without any concomitant increase in Ca content; (2) coronary occlusion developed by progressive storage of free and total cholesterol up to 20- and 40-fold, respectively, with an only discrete increase in mural Ca [4-fold in fully developed atheromata compared to normal (fig. 6)]; and (3)

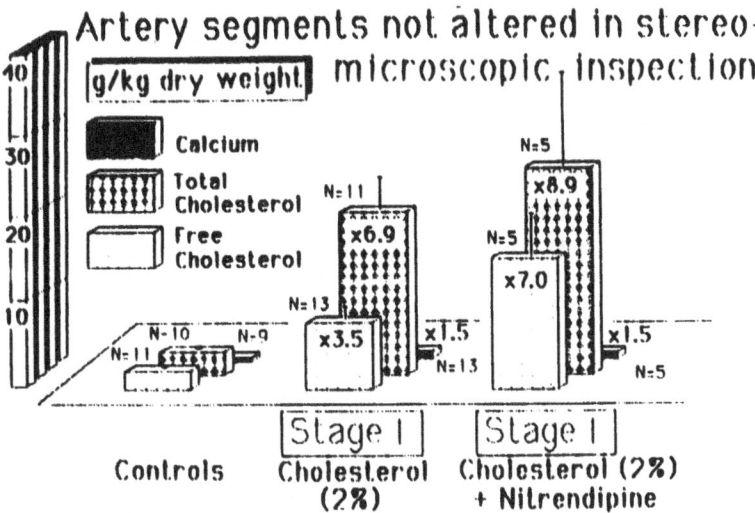

FIG. 5. Chemical composition of early lesions (stage I) in coronary arteries of cholesterol-fed NZ rabbits [control values (g/kg dry weight): 2.2±0.2 (free cholesterol); 2.7±0.4 (total cholesterol); 0.5±0.02 (Ca)]. No inhibition of cholesterol uptake by nitrendipine.

Ca antagonists (nitrendipine and nifedipine: 12-16 mg/kg/die; verapamil: 50 mg/kg/die) applied simultaneously with the cholesterol-rich regimen, did not significantly inhibit this cholesterol deposition in early coronary lesions nor in manifest stenosing plaques. However, they prevented modest Ca accumulation. It has to be emphasized in this context, that NZ rabbits after 14-17 weeks of 2% cholesterol feeding died from coronary occlusion whether treated with Ca antagonists or not. As documented in table 2, free and total cholesterol considerably accumulated in aortas of cholesterol-fed NZ rabbits too, and again a modest increase in Ca content could be demonstrated. Preventative treatment with nitrendipine, in a dosage range sufficient to suppress Ca overload even in this particular experimental model, could not significantly protect aortas from severe cholesterol deposition; nor did nitrendipine prevent the generalized lipid storage disease in myocardial, hepatic, and renal tissue of cholesterol-intoxicated NZ rabbits. Comparable data were obtained with nifedipine and verapamil.

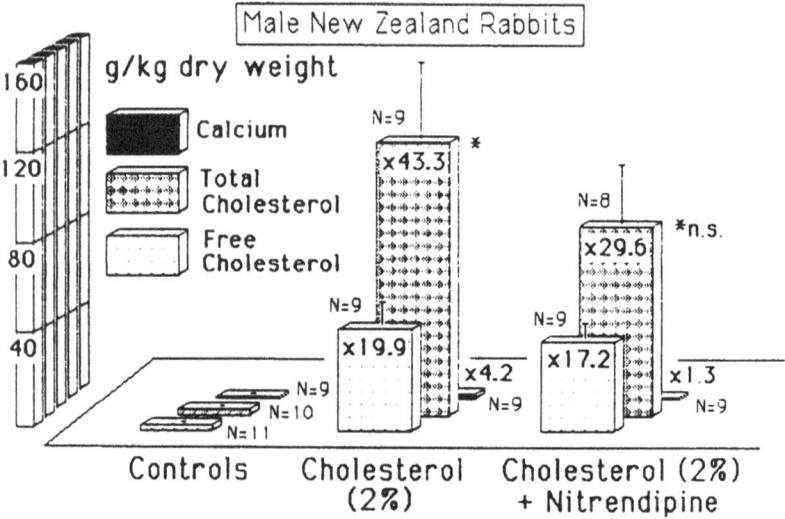

FIG. 6. Chemical composition of stenosing coronary atheromata in cholesterol-fed NZ rabbits. Nitrendipine did not inhibit arterial cholesterol accumulation.

TABLE 2.   Chemical composition of aortic segments
and total aorta of cholesterol-fed NZ rabbits
(Chol=2% cholesterol diet; Nitr = 12-16 mg nitrendipine/kg/die)

		Calcium g/kg d.w.	Free Cholesterolg/k g d.w.	Total Cholesterol g/kg d.w.
Arcus Aortae	Control	0.7± 0.1 n=12	2.0± 0.2 n=12	2.4± 0.2 n=12
	Chol	6.8± 1.6 n=15	43.9± 6.2 n=15	101.9± 8.4 n=15
	Chol+Nitr	1.5± 0.4 n=10	36.6± 5.4 n=10	122.9± 14.6 n=9
Thorac Aortae	Control	0.5± 0.1 n=12	1.7± 0.2 n=12	2.3± 0.3 n=11
	Chol	5.7± 2.1 n=15	24.5± 6.7 n=15	56.5± 13.2 n=15
	Chol + Nitr	0.7± 0.03 n=10	18.3± 4.7 n=10	53.7± 14.1 n=10
Bifurc Aortae	Control	0.5± 0.1 n=6	1.2± 0.2 n=6	4.6± 1.8 n=6
	Chol	1.4± 0.4 n=14	10.0± 2.1 n=14	24.7± 5.3 n=14
	Chol + Nitr	0.7± 0.1 n=12	12.4± 1.8 n=11	42.5± 8.9 n=11
Whole Aortae	Control	0.6± 0.04 n=12	1.7± 0.1 n=12	2.6± 0.3 n=11
	Chol	4.9± 1.3 n=15	27.6± 5.4 n=14	63.7± 7.3 n=15
	Chol + Nitr	10± 0.2 n=10	18.3± 2.8 n=10	65.6± 11.0 n=9

CONCLUSION

The present data indicate the existence of two basically different types of experimental coronary plaques according to their chemical composition, microscopic aspect, and responsiveness to Ca antagonists (fig. 7):

1. The Ca type, developing in vitamin D_3-treated rats (1x300,000 I.U./kg IM). Mesenchymal activation of the arterial wall is caused by an initial Ca-mediated media injury. Progression is characterized by an enormous mural Ca overload without any concomitant cholesterol accumulation. Suitable Ca antagonists completely prevented this type of experimental arteriosclerosis.

2. The cholesterol type, represented by fatty coronary atheromata of cholesterol-fed NZ rabbits. Intimal xanthomatosis, following toxic hypercholesterolemia, reflects the early lesions. Excess cholesterol accumulation, resistant to the prophylactic treatment with Ca antagonists, finally causes coronary occlusion.

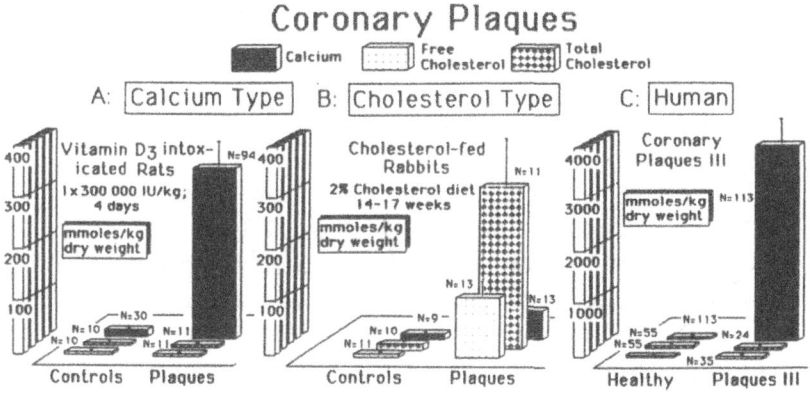

FIG. 7. Chemical composition of different types of experimental coronary artery plaques (A and B) versus conventional human coronary plaques of type III.

Though experimental animal models will never perfectly meet human pathophysiology, both should have at least some characteristic features in common. Coronary atheromata of cholesterol-fed NZ rabbits may be suitable models for coronary heart disease in rare cases of human familiar hypercholesterolemia. The formation of conventional human coronary artery plaques, however, essentially requires a progressive uptake of Ca, thereby representing a Ca-dominated type of arteriosclerosis (fig. 1; C in fig. 7). Ca antagonists specifically inhibited progredient mural Ca uptake in all experimental models of arteriosclerosis tested so far (10). However, neither in atheromatous arteries nor in afflicted organs (myocardium, liver, kidneys) of cholesterol-fed NZ rabbits was any significant prevention of cholesterol accumulation by Ca antagonists found.

REFERENCES

1. Blumenthal H.T. (1967): *Cowdry's Arteriosclerosis: A Survey of the Problem*. Charles C. Thomas, Springfield.
2. Anitschkow N. and Chalatow S. (1913): *Zentralbl. f. Allgemeine Pathol. u. Anat.*, 14: 1-9.
3. Henry D.P. (1990):*J. Cardiovasc. Pharmacol.*, 16(suppl.1): S12-S15.
4. Lichtlen P.R., Hugenholtz P.G., Rafflenbeul W., Hecker H., Jost S. and Deckers J.W. (1990): *Lancet*, 335: 1109-1113.
5. Janke J., Hein B., Pachinger O., Leder O. and Fleckenstein A. (1972): In: *Vascular Smooth Muscle*, edited by E. Betz pp. 71-72. Springer Verlag, Berlin.
6. Fleckenstein A. and Fleckenstein-Grün G. (1980): *Eur. Heart. J.,* 1: 15-21.
7. Fleckenstein A., Frey M. and Leder O. (1983): In: *Drug Development and Evaluation - New Calcium Antagonists - Recent Developments and Prospects* edited by A. Fleckenstein, K. Hashimoto, M. Herrmann, A. Schwartz and L. Seipel pp. 15-31. Gustav Fischer, Stuttgart and New York.
8. Fleckenstein A., Frey M., Thimm F. and Fleckenstein-Grün G. (1990): *Cardiovasc. Drugs and Therapy*, 4: 1005-1014.
9. Fleckenstein A. (1983): *Calcium Antagonism in Heart and Smooth Muscle - Experimental Facts and Therapeutic Prospects,* edited by A. Fleckenstein. John Wiley & Sons, New York.
10. Fleckenstein A., Frey M., Zorn J. and Fleckenstein-Grün G. (1990): In: *Hypertension: Pathophysiology, Diagnosis and Management,* edited by J.H. Laragh and B.M. Brenner pp. 471-509. Raven Press, New York.
11. Fleckenstein-Grün G., Frey M. and Fleckenstein A. (1984): *Trends in Pharmacological Sciences*, 5:283-286.
12. Fleckenstein A., Frey M., Zorn J., and Fleckenstein-Grün G. (1987): *Trends in Pharmacological Sciences,* 8:496-501.

CHOLESTEROL ENRICHMENT DURING DIETARY ATHEROSCLEROSIS ALTERS SMOOTH MUSCLE PLASMA MEMBRANE WIDTH AND STRUCTURE: EVIDENCE FOR REVERSAL BY THE 1,4-DIHYDROPYRIDINE AMLODIPINE

R. Preston Mason, Leo G. Herbette, and Thomas N. Tulenko*

Department of Radiology, Medicine and Biochemistry
Center for Cardiovascular Membrane Research and
the Biomolecular Structure Analysis Center
University of Connecticut Health Center
Farmington, Connecticut 06030

*Department of Physiology and Biochemistry
Medical College of Pennsylvania
3300 Henry Avenue
Philadelphia, Pennsylvania 19129

Free (unesterified) cholesterol is the single most abundant lipid species found in smooth muscle cell (SMC) plasma membrane and appears to have an important role in its membrane structure and physiology (1). In dietary atherosclerosis, the arterial SMC plasma membrane free cholesterol: phospholipid (C:PL) mole ration increased by 83% from 0.38 to 0.78 (1). Cholesterol enrichment in SMCs have been associated with the pathobiology of atherosclerosis, including marked changes in SMC transmembrane movements of cations and SMC proliferation (1-4).

Increased Ca^{2+} transport may be a factor in atherogenesis since Ca^{2+} channel blockers have been suggested to exert anti-atherosclerotic activity (5). For example, cholesterol enrichment as a result of dietary atherosclerosis caused an increase in SMC plasma membrane Ca^{2+} influx which was effectively antagonized by several Ca^{2+} channel blockers (2). Moreover, amlodipine, the 1,4-dihydropyridine (DHP) Ca^{2+} channel blocker, has been shown to inhibit SMC proliferation (6) and migration (7) at concentrations which are lower than required for blocking voltage-sensitive Ca^{2+} channels. However, the physical and chemical basis for amlodipine's potential anti-atherosclerotic activity is not yet understood. In this study, we examined the molecular interactions of amlodipine versus other DHPs with cholesterol enriched model and biological membranes. These data may provide a mechanism for amlodipine's potential anti-atherosclerotic effects.

T. Godfraind et al. (eds.), Calcium Antagonists, 149–155.
© 1993 *Kluwer Academic Publishers and Fondazione Giovanni Lorenzini.*

RESULTS AND DISCUSSION

Cholesterol enrichment alteration of structure and lipid organization of the membrane bilayer

In fig. 1, the electron density profiles of cardiac phospholipid membranes in the absence and presence of a 0.3 C:PL mole ratio are superimposed. The addition of cholesterol resulted in a broad increase in electron density in the membrane hydrocarbon core. This effect of cholesterol is attributed to its electron-rich steroid ring structure lying parallel to the phospholipid acyl chains. In addition, there was a marked increase in the overall membrane bilayer width. This effect of cholesterol on membrane width is due to its planar ring structure decreasing trans-gauche isomerizations in the phospholipid acyl chains (8). Preliminary data using isolated SMC from control and dietary atherosclerotic New Zealand rabbits demonstrate a similar effect of cholesterol-enrichment on plasma membrane structure that is directly proportional to the C:PL mole ratio.

The two peaks of electron density on either side of fig. 1 correspond to the electron-dense phosphate atoms in the membrane headgroups, while the minimum of electron density correlates with the terminal methyl groups in the center of the membrane. The resolution of these profiles is 9 Å. These changes in membrane thickness and lipid organization as a function of dietary atherosclerosis would be expected to alter the tertiary structure of amphipathic membrane proteins which, in turn, could perturb their function.

Fig. 2 illustrates the effect of an increased hydrocarbon core thickness on the conformation and depth of insertion of an amphipathic, α-helical polypeptide. The result of cholesterol enrichment on membrane structure may provide a molecular rationale for altered transmembrane cation transport observed in dietary atherosclerotic SMCs (1-4).

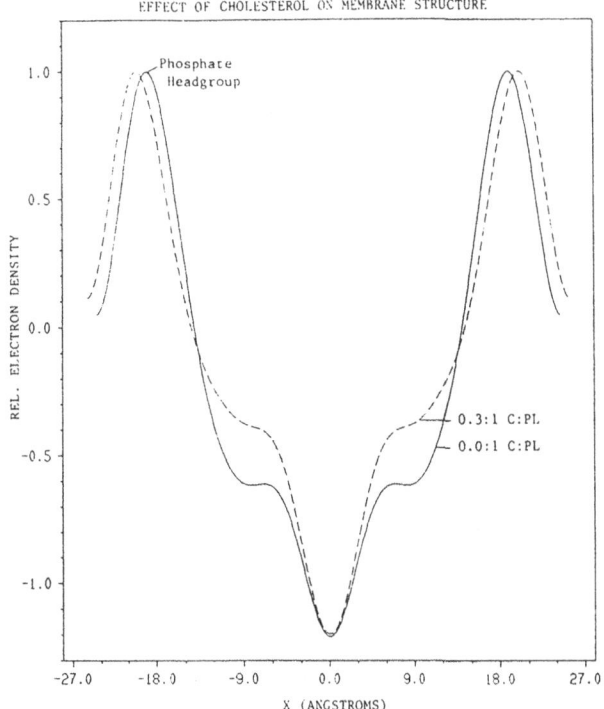

FIG. 1. The effect of cholesterol on membrane structure.

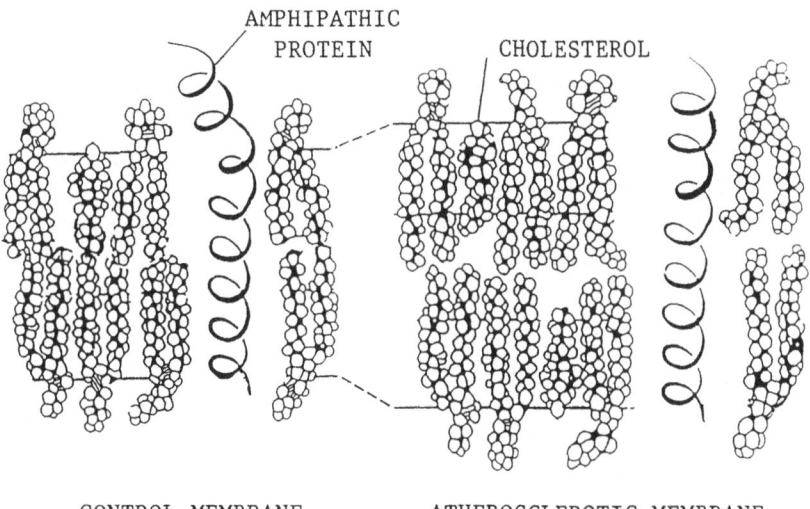

FIG. 2. Comparison of the control membrane with atherosclerotic membrane.

Amlodipine reversal of structural changes in SMC plasma membrane associated with early dietary atherosclerosis

In preliminary studies, we examined the molecular structure of freshly isolated SMC plasma membrane after only 2.5 weeks on a high cholesterol diet (2% by weight). A significant increase in the plasma membrane C:PL mole ration from 0.37 to 0.52 was measured in these membranes along with a 7% increase in the membrane hydrocarbon core width from 43 Å to 46 Å.

Following the addition of amlodipine, the membrane bilayer width appeared to be fully restored to control levels. These data indicate that the lipophilic amlodipine molecule may actually reverse the lipid ordering effect of cholesterol, resulting in a decreased membrane bilayer width. By inserting into the membrane near the hydrocarbon core/water interface, amlodipine may disrupt intermolecular phospholipid packing and charge interactions. This effect of amlodipine, the first ever described for a DHP, may provide a structural rationale for its anti-atherosclerotic activity at the membrane level.

Amlodipine's high affinity for cholesterol-enriched membranes

These potential membrane structure changes mediated by amlodipine may be related to its high affinity for the membrane bilayer. The equilibrium membrane partition coefficient was measured for several DHPs as shown in table 1. Although amlodipine is charged and has a low partition coefficient in octanol/buffer ($K_{p[oct]}=30$), it had a partition coefficient in cholesterol enriched cardiac membranes which was substantially higher than the uncharged isradipine, nitrendipine and nimodipine. These data suggest that amlodipine has distinct chemical and physical properties which affect its interactions with the anisotropic membrane bilayer relative to the other Ca^{2+} channel blockers. These complex interactions clearly could not be predicted based on these drugs' octanol/buffer partition coefficient values alone.

Amlodipine is an amphipathic molecule with a charged, 2-aminoethoxymethyl substituent at the 2-position of the dihydropyridine ring ($pK_a = 9.02$) which, in its crystal structure, extends away from the hydrophobic portion of the molecular (9). Molecular modeling of amlodipine, based on its location in the membrane bilayer and crystal structure, predicts that the charged portion of the drug molecule interacts with the anionic oxygen in the phospholipid headgroup while the hydrophobic dihydropyridine and chlorophenol ring structures penetrate into the hydrocarbon core, as illustrated in fig. 3 which was adapted from Franks (8). The DHP nimodipine, by contrast, has only hydrophobic

TABLE 1.   1,4-dihydropyridine partition coefficients
into control and
cholesterol-enriched phosphatidylcholine
membranes

Drug	Normal Cholesterol Phospholipid Membrane[a]	Cholesterol-enriched Phospholipid Membrane[a]
Amlodipine	19,000	8,000
Nitrendipine	10,000	800
Isradipine	6,000	900
Nimodipine	5,000	500

[a]Drug concentrations were maintained at 5×10^{-10}M. Bovine cardiac phospholipid concentration was 20 $\mu g/ml$ (pH 7.0, 21°C) with a cholesterol:phospholipid ration of 0.3:1 (normal) or 0.6:1 (cholesterol-enriched).

interaction with the membrane bilayer. The combination of ionic and hydrophobic bonding between amlodipine and the phospholipid molecules may be the chemical basis for its high $K_{p[mem]}$ values and slow rate of membrane dissociation or "washout" (9) observed for this drug at normal and elevated cholesterol conditions.   Moreover, amlodipine's high membrane concentration over time may mediate its long duration of activity as a Ca^{2+} channel blocker (9,10,11).

Amlodipine's high affinity for the membrane bilayer may mediate
its potent anti-atherosclerotic effects by reducing calcium influx

The C:PL mole ration in arterial SMC plasma membranes has been shown to be elevated by 83% as a result of dietary atherosclerosis (1). This increase in cholesterol appears to be directly associated with structural and functional changes in the SMC plasma membrane. Structurally, there is a correlation between the C:PL mole ration and the SMC plasma membrane bilayer width. Functionally, the elevated C:PL mole ratio and membrane bilayer width correlates with increased transmembrane Ca^{2+} influx (1). In addition, free cholesterol itself has recently been demonstrated to have potent mitogenic activity in cultured arterial SMC preparations (6).   Increased Ca^{2+} influx and SMC proliferation, in turn, have been considered to play an important role in the pathobiology of atherosclerosis.   Thus, amlodipine's anti-atherosclerotic activity may be related to its ability to inhibit elevated

R. P. MASON ET AL.

transmembrane Ca^{2+} transport, indirectly, by reducing the membrane

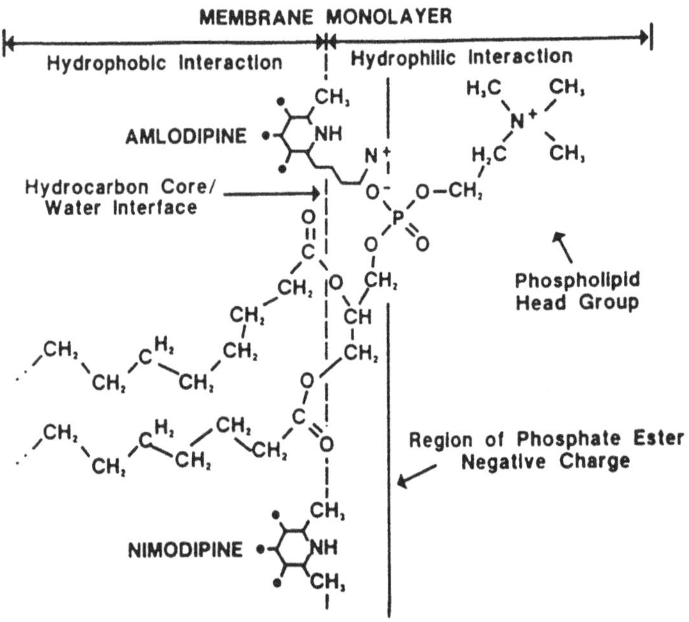

FIG. 3.   Comparison of molecular structures of amlodipine and nimodipine (reproduced by permission of the authors and publisher).

bilayer width; directly, by blocking voltage-sensitive Ca^{2+}; or possibly a combination of both mechanism. It should be emphasized, however, that amlodipine may utilize other cellular mechanisms to produce its potential antiproliferative and antimigrational effects on arterial SMCs.

REFERENCES

1.   Chen M., Mason R.P. and Tulenko T.N. (1991): *FASEB J.,* 5: 531a.
2.   Gleason M.M., Medow M.S. and Tulenko T.N. (1991): *Circ. Res.,* 69: 216-227.
3.   Tulenko T.N., Bialecki R., Gleason M. and D'Angelo G. (1990): In: *Ion Channels, Membrane Lipids and Cholesterol: A Role for Lipid Domains in Arterial Function,* edited by T. Colatsky pp. 187-

203. Alan R. Liss, New York.

4.   Strickberger S.A., Russek L.N. and Phair R.D. (1988): *Circ. Res.*, 62: 75-80

5.   Henry P.D. (1985): *Circulation,* 72: 456-459.

6.   Tulenko T.N. (1991): *J. Amer. Coll. Card.,* 17: 24a.

7.   McMurray H.F. and Chahwala S.B. (1991): *J. Amer. Coll. Card.,* 17: 194A.

8.   Franks N.P. (1976): *J. Mol. Biol.,* 100: 345-358.

9.   Mason R.P., Campbell S., Wang S. and Herbette L.G. (1989): *Mol. Pharm.,* 36: 634-640.

10.   Mason R.P., Rhodes D.G. and Herbette L.G. (1991): *J. Med. Chem.,* 34: 869-877.

11.   Burges R.A., Gardiner D.G., Gwilt M., Higgins A.J., Blackburn K.J., Campbell S.F., Dross P.E. and Stubbs J.K. (1987): *J. Cardiovasc. Pharmacol.,* 9: 110-119.

LACIDIPINE AND ATHEROSCLEROSIS

F. Bernini, A. Corsini, M. Mazzotti, S. Bellosta,
M. Soma, and R. Paoletti

Institute of Pharmacological Sciences, Via Balzaretti 9,
20133 Milan, Italy

Lacidipine is a new calcium antagonist (CA) structurally related to nifedipine. We compared lacidipine with CA prototypes, nifedipine and verapamil, for their effects on two major processes of atherogenesis: cholesteryl ester (CE) metabolism in macrophages and proliferation of myocytes. Lacidipine (5-50 μM) inhibited from 20 to 90% the ability of acetyl low-density lipoprotein (acLDL) to stimulate cholesterol esterification in macrophages. Little or no effect was observed with nifedipine (10-50 μM). Verapamil (50 μM) also prevented (>95%) the capacity of acLDL to stimulate cholesterol esterification. The effect of Ca was also tested on myocytes cultured from rat aorta. Lacidipine decreased myocyte proliferation in a dose-dependent manner with an IC_{50} value of 8 μM.

INTRODUCTION

Major processes involved in the formation of atherosclerotic lesions are deposition of lipids, mainly cholesteryl esters, and arterial smooth muscle cell (myocyte, SMC) proliferation (1). The accumulating cholesterol derives from plasma lipoproteins that contain apo B [LDL, intermediate-density lipoproteins (IDL), and very low-density lipoproteins (VLDL)]. In contrast, apo A_1-containing lipoproteins, high-density lipoproteins (HDL), are believed to be important in the removal of cholesterol from cells within the arterial wall (2). The cholesterol content of the artery, therefore, is the net result of cholesterol influx via LDL, IDL, and VLDL, and cholesterol efflux to HDL. An accumulation of cholesterol in the arterial wall can be expected to occur when there is an imbalance between these two processes, and it may be due to 1) an increase in the number of apo B-containing lipoprotein particles; 2) a modification of the physical-chemical properties of apo B-containing lipoproteins so that intracellular deposition of cholesterol is enhanced; and 3) a decrease of apo A_1-containing lipoprotein particles, with reduced cholesterol efflux from cells.

The atheroma contains two main cell types, smooth muscle cells and

157

T. Godfraind et al. (eds.), Calcium Antagonists, 157–164.

macrophages.   Macrophages derive from circulating monocytes and represent the predominant lipid-loaded cells in the lesions.   The mechanism by which they accumulate lipoprotein cholesterol and develop into foam cells depends mainly upon receptor-mediated processes involving the so-called "scavenger receptor" that recognizes biologically and chemically modified LDL, such as acLDL and malondialdehyde LDL (3). The scavenger receptor does not undergo any regulation, so that the accumulation of cholesterol may proceed to cell death. Recent studies have elucidated the intracellular mechanisms involved in the homeostasis of intracellular cholesterol content. Cholesterol accumulates in macrophages in esterified form which undergoes a continuous cycle of hydrolysis and esterification; these two processes are catalyzed by the enzymes acyl-C_oA-cholesterol acyltransferase (ACAT) and by neutral hydrolase, respectively (4).   Free cholesterol modulates its own synthesis and esterification, as well as LDL receptor expression (5).   Moreover, free cholesterol is excreted (cholesterol efflux) by the cells thus reducing its intracellular content, including the esterified form (3).

Arterial myocytes migrate from the media layer to and proliferate in, the intima layer under the effect of various mitogens; both migration and proliferation are critical events in the development of atheromatous plaques (1).

Calcium antagonists are able to affect the major processes involved in atherogenesis. Verapamil and diltiazem were reported to stimulate the receptor-mediated LDL catabolism (6,7). Interestingly, we observed that only verapamil-like compounds, diltiazem, and amlodipine may stimulate LDL receptor activity.   On the contrary, nifedipine and related compounds (but not amlodipine), and flunarizine are inactive (7,8). Verapamil is also able to reduce cellular cholesterol esterification (9). In addition, CA are able to inhibit both myocyte migration and proliferation in *in vivo*, as well as in *in vitro* experimental models (10-13).

In the present study, we have evaluated in cell culture models, the effect of lacidipine, a new long-lasting calcium antagonist, on processes involved in atherogenesis.

METHODS

Cells

Cultures of smooth muscle cells (SMC) from rat aorta (Sprague Dawley, male) are prepared according to Ross (1). The medium is changed every second day.

Resident peritoneal macrophages are obtained by peritoneal lavage and cultured in Dulbecco's modified Eagle's Medium (DMEM) containing 10% fetal calf serum (FCS) (4).

Cell viability, in the presence or absence of the indicated compounds, is assessed by tripan blue exclusion test.

Preparation of LDL, acLDL, and LPDS

LDL (d 1.019-1.063 g/ml) are isolated by sequential preparative ultracentrifugation from the plasma of clinically healthy normolipidemic volunteers (14), and acetylated by incubation with acetic anhydride (15). Human lipoprotein-deficient serum (LPDS) is prepared by ultracentrifugation of pooled human sera at d = 1.25 g/ml, 40,000 rpm in 50.2 Ti Beckman rotor for 72 hours. LDL undergoes iodination with ^{125}I by the monochloride procedure (16).

Cell proliferation

Vascular myocytes are seeded at a density of $2x10^5$ cells per petri dish (35 mm) and incubated with MEM supplemented with 10% FCS. Eighteen hours later the medium is changed with one containing 0.4% FCS to stop cell growth, and the monolayers incubated for 48 hours. After this time (time 0) the medium is replaced with one containing 10% FCS and known concentrations of lacidipine. At time 0, just before the addition of the drugs to be tested, some petri dishes are used for cell counting. Cell number is determined by Coulter Counter after trypsinization of the monolayers. SMC doubling time is computed according to Elmore and Swift (17). In another set of experiments cell growth is evaluated by nuclear incorporation of [3H] thymidine, incubated with cells (3 μCi/ml medium) for 48 hours. Radioactivity is measured with Lumagel scintillation cocktail.

Cholesterol metabolism in macrophages

Cells are incubated as indicated with acLDL in the presence of the compounds under investigation. The esterification of intracellular cholesterol is measured as the amount of cholesteryl oleate formed in cells incubated for 4 hours with [14C] oleate (4). Each experimental point is evaluated in triplicate. The results are expressed as mean plus or minus standard deviation and analyzed by two-tailed student's t-test. The results are corrected for the protein content of each plate.

RESULTS

Effect on cholesterol esterification

The activity of lacidipine on cholesterol esterification induced by acLDL was tested in mouse peritoneal macrophages incubated with increasing concentration of the drug (1 to 50 μM) for 6 hours. Drug inhibitory effect was observed at concentrations as low as 10 μM (table 1). When the exposition of cells was extended to 24 hours the effect was observed also with lacidipine 3 μM.

The effect of lacidipine on cholesterol esterification in macrophages loaded with acLDL was compared to that of verapamil, nifedipine, and progesterone (table 2). As expected (4,18), both verapamil and progesterone completely inhibited [^{14}C]-oleate incorporation into cholesteryl esters. These effects were similar to those obtained with lacidipine 50 μM. According to previous data of our laboratory, nifedipine did not affect cholesterol esterification (19).

Cellular viability and dish protein content were not significantly affected by lacidipine in our experimental conditions.

TABLE 1.   Effect of lacidipine on cholesterol
            in mouse peritoneal macropages incubated with AcLDL

Drugs (μM)		[^{14}C]-Oleate Incorporation Into Cholesteryl Ester (ng/mg Cell Prot x h)	% Of Control
Control		356 ± 22.03	100 ± 4.4
Lacidipine	1	386 ± 39.9	108.5 ± 11.3
"	5	356 ± 37.8	100 ± 10.7
"	10	281 ± 4.9*	79 ± 4.2*
"	20	94 ± 10.4**	26 ± 2.8**
"	50	10 ± 1.5**	3 ± 0.4**

Cells were preincubated in DMEM containing Fatty Acid Free Albumin 0.1% and drugs for 2 hours. Monolayers underwent a second incubation (4 hours) in the presence of AcLDL (50μg/ml),[^{14}C]-oleate albumin complex, and drugs. Each point is the mean ± SD of triplicate dishes. Student's t-test: * p<0.05; ** p< 0.001.

TABLE 2.   Effects of lacidipine, verapamil, nifedipine
                and progesterone on the cholesterol esterification
        in mouse peritoneal macrophages loaded with AcLDL (50 g/ml)

Drugs (μM)		[^{14}C]-Oleate Incorporation Into Cholesteryl Ester (ng/mg Cell Prot x h)	% Of Control
Control		1555.2 ± 77.4	100.0 ± 4.9
Lacidipine	50	6.6 ± 0.5**	0.4 ± 0.1
Verapamil	1	0.2 ± 4.2**	0.7 ± 0.3
Nifedipine	50	1399.2 ± 50.4*	89.9 ± 3.2
Progesterone	30	32.1 ± 7.9**	2.1 ± 0.5

Experimental conditions as in table 1.

Effect of lacidipine on cellular smooth muscle cell proliferation

The potential antiproliferative action of lacidipine was studied in rat
aortic myocytes (SMC) at drug concentrations ranging between 1 and 20
μM (not significantly affecting cell viability). In a first set of experiments
the effect of the compounds were evaluated by cell counting. Calculation
of cellular doubling time indicate that lacidipine decreased SMC
proliferation and the effect, already detectable at the lowest concentration
tested, became statistically significant at 5 μM. This activity was dose-
dependent with an IC$_{50}$ value of 8 μM; verapamil was less potent in this
respect (table 3).

TABLE 3.   Doubling time for smooth muscle cells cultured
               from rat arota effect of lacidipine and verapamil

Addition μM		Doubling Time (h) x ± SD
None		21.7 ± 0.1
Verapamil	50	31.0 ± 0.6**
Lacidipine	1	23.3 ± 1.1
"	5	23.7 ± 0.6*
"	10	31.5 ± 2.6*
"	20	70.3 ± 11.4*

Doubling time was measured the second day of culture growth; each point
was run in triplicate.   * p< 0.01;   ** p< 0.001

In further experiments, cellular growth was measured as nuclear incorporation of labelled thymidine by SMC; the results paralleled those obtained by cell counting (table 4).

TABLE 4. Effect of lacidipine and verapamil on [³H]-thymidine incorporation in rat aorta myocytes

Addition μM		[³H]-Thymidine Uptake cpm/plate x 10⁻³ x ± SD
None		72.5 ± 7.8
Verapamil	50	57.4 ± 4.9*
Lacidipine	1	72.9 ± 10.3
"	5	60.5 ± 6.8
"	10	56.1 ± 8.0
"	20	51.7 ± 9.3*

[³H]-thymidine was added to the incubation medium along with the reported concentration of drugs. The incorporation was evaluated after 48 hours of incubation: each point represents the mean ± SD of triplicate dishes. * $p < 0.05$

DISCUSSION

The observation that calcium antagonists reduce the severity of experimental atherosclerosis in cholesterol-fed rabbits (20) have raised interest for this class of drugs as possible anti-atherosclerotic agents. A recent report suggests that nifedipine may slow the progression of atherosclerotic plaques in patients affected by coronary heart disease (21).

Among the processes involved in the formation of atherosclerotic lesion a pivotal role is played by the deposition of lipids in the arterial wall, which depends on plasma concentration of circulating lipoproteins, as well as on lipid metabolism of the parietal cells. Several CA are reported to reduce the severity of experimentally induced atherosclerosis in cholesterol-fed animals without affecting plasma lipid levels (22). Studies performed *in vitro* with cell cultures indicate that verapamil-like compounds and diltiazem enhance the LDL receptor-mediated pathway (7). The mechanism involved in this stimulatory activity does not seem to be related to their calcium channel blocking properties. Moreover, selective calcium blockers like nifedipine, although possessing anti-atherosclerotic activity on cholesterol-fed animals, failed to stimulate LDL receptor activity (7). Lacidipine, similarly to nifedipine, did not affect the metabolism of LDL mediated by the specific receptor (data not shown).

CA may also affect intracellular cholesterol metabolism (22). In this study, we tested lacidipine on cholesterol esterification in macrophages. Stimulation of cholesterol esterification by acLDL in these cells involves their receptor-mediated internalization, delivery to and hydrolysis of the cholesteryl esters in lysosomes, and release of free cholesterol formed to the esterification site of ACAT present on the endoplasmic reticulum. In the cytoplasm, cholesterol undergoes a continuous cycle of esterification and de-esterification which implies a constant activation of ACAT (4). In the absence of a cholesterol acceptor in the extracellular space this cycle remains activated even after removal of acLDL (4).

The results show that lacidipine affects intracellular cholesterol homeostasis by fully inhibiting cholesterol esterification induced by acLDL. According to previous data obtained in this laboratory, verapamil but no nifedipine was effective on cholesterol esterification (19).

Among the processes involved in the formation of atherosclerotic lesion a relevant role is played by initiation of smooth muscle cells proliferation in the medial layer, their migration and further proliferation into the intima of the arterial wall (23). It has been reported that CA inhibit the proliferation of smooth muscle cells both *in vivo* and in cultured cells (10-13).

We assessed the effect of the calcium antagonist lacidipine on proliferation. The present results indicate that lacidipine is effective in reducing proliferation of arterial myocytes. The antiproliferative effect of the drug was dose-dependent, with an IC_{50} value of 8 μM. At this regard, lacidipine was able to inhibit neointimal proliferation induced *in vivo* in hypercholesterolemic rabbits (data not shown). Neointimal proliferation was obtained by perivascular positioning of a hollow silastic collar as described by Booth et al. (24).

The mechanism by which CA affects SMC growth is unknown. Cytosolic free calcium is involved in cell division cycle (25) and CA could effect cell proliferation by modulating extracellular calcium influx. This interpretation, however, is not supported by the high concentrations of CA required to display the antiproliferative effect, which largely exceed calcium channel blockade. Whatever the mechanism, CA seem to possess an apparent selectivity for arterial myocyte growth based on the observation that in balloon-catheterized rats and rabbits, these drugs markedly reduced the incorporation of thymidine into aortic DNA without affecting DNA synthesis in other proliferating tissues (13).

In conclusion, these *in vitro* studies support the potential anti-atherosclerotic properties of lacidipine. The results reported herein provide experimental support to a possible antiatherosclerotic action of lacidipine through effects on mechanisms involved in atherogenesis.

REFERENCES

1. Ross R. (1986): *N. Engl. J. Med.*, 314: 488-500.
2. Picardo M., Massey J.B., Kuhn D.E., Gotto A.M., Gianturco S.H. and Pownall H.J. (1986): *Arteriosclerosis*, 6: 434.
3. Brown M.S. and Goldstein J.L. (1983): *Ann. Review. Biochem.*, 52: 223-261.
4. Brown M.S., Ho Y.K. and Goldstein J.L. (1980): *J. Biol. Chem.*, 255: 9344-9352.
5. Brown M.S. and Goldstein J.L. (1986): *Science*, 232: 34-47.
6. Stein O., Leiterzsdorf E., Stein Y. (1985): *Arteriosclerosis*, 5: 35.
7. Paoletti R., Bernini F., Fumagalli R., Allorio M. and Corsini A. (1988): *Ann. N.Y. Acad. Sci.*, 522: 390-398.
8. Paoletti R. and Bernini F. (1990): *Am. J. Cardiol.*, 66: 28H-31H.
9. Daugherty A., Rateri D.L., Schonfeld G., and Sobel B.E. (1987): *Br. J. Pharmacol.*, 91: 113-118.
10. Betz E., Hammerle H. and Viele D. (1984): *Inter. Angio.*, 3: 33-42.
11. Heinle H. and Reich A. (1985): *Arzneim. Forsch. Drug. Res.*, 35: 1811-1812.
12. Orekhov A.N., Tertov V.V., Khashimov K.A., Kudryashov S.A. and Smirnov V.N. (1986): *J. Hyperten.*, 4(suppl.6): S153-S155.
14. Jackson C.L., Bush R.C. and Bowyer D.E. (1988): *Atherosclerosis*, 69: 115-122.
15. Basu S.K., Goldstein J.L., Anderson R.G.W., and Brown M.S. (1976): *Proc. Natl. Acad. Sci. USA.*, 79: 1712-1716.
16. McFarlane, A.S. (1958): *Nature*, 182: 53.
17. Elmore E. and Swift M. (1976): *J. Cell. Physiol.*, 187: 229-34.
18. Stein O., Halperin G. and Stein Y. (1986): *Arteriosclerosis*, 6: 70-78.
19. Bernini F., Bellosta S., Didoni G., and Fumagalli R. (1991): *J. Cardiovasc. Pharmacol.* (in press).
20. Henry P.D. and Bently K.I. (1981): *J. Clin. Invest.*, 68: 1366.
21. Lichtlen P., Hugenholtz P., Rafflenbeul W., Hecker H., Jost S., and Deckers J.W. (1990): *Lancet*, 335: 1109-1113.
22. Bernini F., Catapano A.L., Corsini A., Fumagalli R., and Paoletti R. (1989): *Am. J. Cardiol.*, 64: 129I-134I.
23. Clowes A.W., Clowes M.M., Fingerle J., and Reidi M.A. (1989): *J. Cardiovas. Pharmacol.*, 14: S12-S15.
24. Booth R.F.G., Martin J.F., Honey A.C., Hassal D.G., Beesley J.E. and Moncada S. (1989): *Atherosclerosis*, 76: 257-268.
25. Harris P.J. (1981): In: *Mitotis/Cytokinesis, Cell Biology*, edited by A.M. Zimmerman and A. Forer pp. 29-57. Academic Press, New York.

POTENTIATION OF THE ANTIPLATELET EFFECT OF ASPIRIN BY THE CALCIUM CHANNEL ANTAGONIST AMLODIPINE

John D. Folts, Jonathan W. Hamilton
and Charles K. Stone

Department of Medicine, Cardiology Section
University of Wisconsin
Medical School
Madison, Wisconsin 53792

The primary role of platelets in the precipitation of unstable angina, myocardial infarction, and sudden death is generally accepted (1). Platelet thrombi are also thought to be significant in transient ischemic attacks and some strokes (2). To protect patients from these thrombotic events, aspirin (3-5) has been primarily utilized in clinical trials. The use of aspirin has been the only successful agent and has been recommended for patients with unstable angina and myocardial infarction (1).

However, protection with aspirin has not been as complete as desired. One possible problem with aspirin as a platelet inhibitor or antithrombotic has been suggested. Human platelets studied *ex vivo* and inhibited with aspirin, can be reactivated by incubating them with physiologic concentrations of epinephrine (6). In addition, epinephrine reverses the inhibitory effect of aspirin on human platelet-vessel wall interactions (7). It has been suggested that one of the effects of epinephrine on platelets is to enhance calcium uptake by the platelet (8,9,10). This is thought to be due in part to alpha 2 adrenergic receptor stimulation (10).

We have developed a model of stenosed canine coronary arteries with intimal damage in which periodic acute platelet thrombus formation occurs and produces cyclical reductions in coronary blood flow (CFRs) and myocardial ischemia (11). This has been called the "Folts" model and is considered to be a good "on line" bioassay of *in vivo* platelet activity (12). This acute periodic thrombus formation and the resulting CFRs can be abolished by aspirin but are renewed by physiologic infusions of epinephrine (13). This study was undertaken to determine if the combination of aspirin and the calcium channel antagonist amlodipine would be more efficacious than aspirin alone in preventing epinephrine-mediated renewal of acute platelet thrombus formation and CFRs in our canine model.

T. Godfraind et al. (eds.), Calcium Antagonists, 165–171.

METHODS

Ten dogs were premedicated with morphine sulfate 3 mg/kg followed 1/2 hour later by sodium pentobarbital anesthesia (20 mg/kg) with supplemental anesthesia as needed. An IV line was placed in the cephalic vein and also in the femoral vein. A Milar pressure transducer was passed retrograde through the femoral artery and localized just superior to the aortic valve for measuring arterial blood pressure. A left thoracotomy was performed and the heart was suspended in a pericardial cradle. A 2 cm segment of the left circumflex coronary artery was dissected out, tying small side branches where necessary. An appropriately sized Statham-Gould electromagnetic flow probe was placed on the proximal portion of the artery. The vessel was clamped lightly with a vascular clamp to induce moderate intimal damage and expose subintimal structures such as collagen, known to stimulate platelet adhesion and aggregation (11). Then, an external constricting cylinder was placed around the outside of the damaged segment of the coronary artery encircling and constricting it as previously described (11-13). The size of the constricting cylinder was chosen so that a "critical" 70% coronary artery stenosis was produced (11). A pair of 20 mHz ultrasonic crystals was placed in the endocardial layer of the left anterior descending coronary artery bed to measure changes in myocardial segment length (MSL) for a measure of myocardial function as previously described (14). Following the surgical preparation, the dogs were observed for 30 minutes. When CFRs had been observed for at least 30 minutes, each animal was given 5 mg/kg aspirin slowly IV. Fifteen minutes after the aspirin was administered, with confirmation that aspirin had indeed abolished CFRs, an IV epinephrine infusion, 0.2 g/kg/min was administered for a 20-minute period.   Fifteen minutes after completion of the epinephrine infusion, 0.4 mg/kg of amlodipine was administered over a 10-minute period. Coronary flow, arterial pressure, heart rate, and MSL were continuously monitored. Sixty minutes following completion of the drug infusion, a second epinephrine infusion of 0.2 g/kg/min was performed for 30 minutes while all physiologic parameters were monitored.

RESULTS

CFRs due to periodic acute platelet thrombus formation followed by embolization occurred in all 10 dogs (fig. 1). Coronary blood flow averaged 47 ± 10 ml/min and there was a mean frequency of cyclical flow reductions of 7 ± 2 CFR/30 min.

The administration of 5 mg/kg of aspirin abolished the CFRs in all dogs (fig. 1). With infusion of 0.2 g/kg/min epinephrine after aspirin, the

CFRs were transiently renewed at a frequency of 9 ± 3 CFR/min (fig. 1). Furthermore, administration of epinephrine 0.2 g/kg/min resulted in a mean arterial pressure of 14 ± 7 mm Hg ($p < 0.05$). When the epinephrine effect had worn off and blood pressure had returned to control levels, the amlodipine was given, and there was a decrease in arterial blood pressure of 10 ± 5 mm Hg ($p < 0.05$) and an increase in heart rate of 18 ± 6 beats/min ($p < 0.05$).

There was no significant change in (MLS) due to the amlodipine in any dog (fig. 2). When the epinephrine infusion was repeated 60 minutes after amlodipine there was no renewal of CFRs and the arterial blood pressure did not rise a significant amount (fig. 3). If the epinephrine infusion was repeated 30 minutes after the amlodipine was given small CFRs returned in 5 out of the 10 dogs. This suggests that 30-60 minutes are required for the amlodipine to achieve maximal effect.

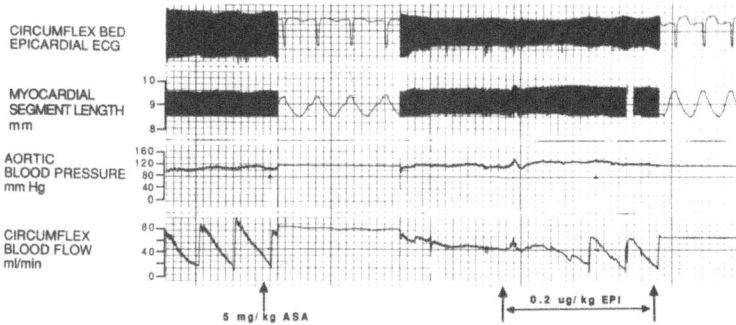

FIG. 1. Cyclic flow reductions (CFRs) are shown on the lower left and they are abolished by 5 mg/kg of aspirin (ASA). On the lower right the CFRs are renewed by infusing epinephrine (EPI).

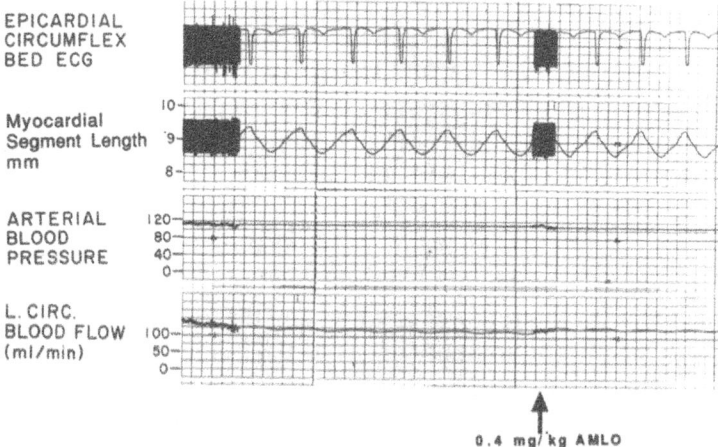

FIG. 2. Arterial blood pressure falls ≈ 19 mm Hg but there is no change in the myocardial segment length or contractility due to amlodipine.

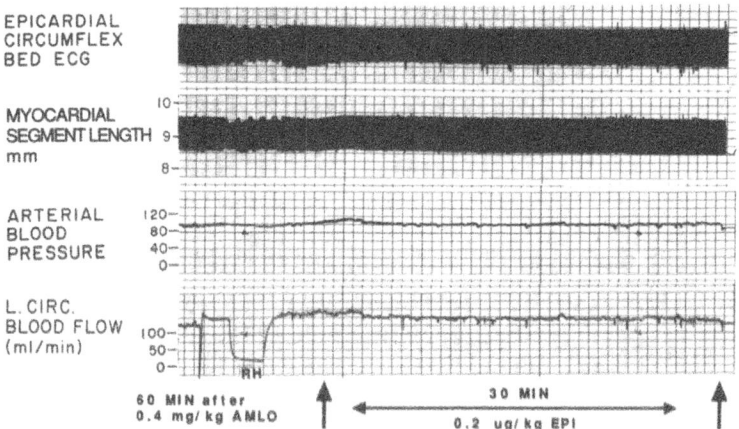

FIG. 3. After amlodipine (AMLO) treatment an infusion of epinephrine (EPI) fails to renew the CFRs.

DISCUSSION

We have shown that CFRs in this model could be abolished with aspirin 5 mg/kg (13). However, that study also demonstrated that acute platelet thrombus formation and CFRs could be renewed with IV epinephrine infusions. Rao et al. have shown that human platelets inhibited with aspirin can be reactivated with epinephrine (6,7) and human platelets are more sensitive to epinephrine than dog platelets (15). This ability of catecholamines to act synergistically with other common platelet stimuli occurs at catecholamine levels normally achieved in the circulation. Thus for elevated catecholamines the effect will be even greater (16). This concern over the "epinephrine challenge" problem has recently been raised with regard to the possible clinical relevance for patients taking aspirin (17).

The mechanism of the epinephrine effect on aspirin-treated platelets is not clear. It is believed to be due in part to Ca^{2+} mobilization from both intra- and extracellular sources. Owen et al. have demonstrated that human platelets exposed to epinephrine are induced to take Ca^{2+} (8). Rao et al. have also shown that epinephrine-induced calcium uptake in human platelets can be blocked by both α antagonists and calcium channel blockers (10). Thus, despite the role of prostaglandins in platelet hemostasis, there might exist a salvage pathway independent of the prostaglandin pathway which not only allows platelets to function normally, but also relies heavily on cytoplasmic calcium concentrations.

Our hypothesis for this study was based on the supposition that the epinephrine effect on aspirin-treated platelets could conceivably be blocked by using calcium channel blockers. This assumes that the effect of epinephrine on platelets is due in part to internal and external calcium mobilization which has been shown to be inhibited by the calcium channel blockers (18).

The calcium channel blockers, nifedipine, verapamil, and diltiazem, inhibit *ex vivo* platelet aggregation in usual therapeutic doses (18). Platelet aggregability increased in man during exercise, but was decreased by premedication with 10 mg of nifedipine orally (18). In our studies, amlodipine alone (0.4-0.8 mg/kg) decreased the size and frequency of CFRs but did not abolish them (unpublished observations). In our model the exacerbation of CFRs by epinephrine infusion after aspirin therapy was abolished by amlodipine. The amlodipine appears to have a long onset of action, as complete protection against the epinephrine challenge does not occur until 45 to 60 minutes after administering the drug. These results suggest that the induction of CRFs by epinephrine in aspirin-treated platelets *in vivo* are mediated by changes in platelet intracellular Ca^{2+}.

The clinical implications of these results are twofold: calcium channel blockers may be a useful antiplatelet adjuvant in patients with persistent unstable angina in spite of aspirin therapy. It is possible that some of the efficacy of sublingual nifedipine in unstable angina may be due to its effect on calcium flux in platelets. Secondly, calcium channel antagonists may be useful in preventing arterial thrombosis in these patients with elevated Ca^{2+} levels (19). It has been shown that patients with arterial thrombosis have elevated resting platelet cytoplasmic Ca^{2+} levels and their platelets are hypersensitive to aggregating agents (19). In addition, it has been demonstrated that elevated Ca^{2+} levels in patients return to normal with divided doses of nifedipine from 40 to 80 mg/day for 10 days (20). Amlodipine combined with aspirin may well provide better protection against platelet-mediated thrombosis in patients with elevated catecholamine and increased intraplatelet Ca^{2+} levels than aspirin alone.

REFERENCES

1. Hjemdahl-Monsen C.E., Lewis D., Cairns J., Chesebro J.H. and Fuster V. (1986): *J. Am. Coll. Cardiol.*, 8(suppl.6): 67B-75B.
2. Kerson L.A. and Olmos-Lau N. (1984): *Cardiovasc. Rev. Rep.*, 5: 227-239.
3. Cairns J.A., Gent M. and Singer J. (1985): *N. Engl. J. Med.*, 313: 1369-1375.
4. Lewis H.D., David J.W. and Archibald D.G. (1983): *N. Engl. J. Med.*, 309: 396-403.
5. Preliminary Report (1988): *N. Engl. J. Med.*, 318(suppl.4) 262-264.
6. Rao G.H.R., Johnson G.J. and White J.G. (1980): *Prost. Med.*, 5: 45-58.
7. Rao G.H.R., Escolar G. and White J.G. (1986): *Thrombo. Res.*, 44: 65-74.
8. Owen N.E., Fanberg H. and Lebreton G.C. (1980): *Am. J. Physiol.*, 239: H483-H438.
9. Rao G.H.R., Reddy K.R. and White J.G. (1981): *Prostagland. Med.*, 6: 75-90.
10. Rao G.H.R. (1988): *Thromb. Res.*, 50: 510-516.
11. Folts J.D. (1991): *Circulation,* 83(suppl.4): 3-14.
12. Bush L.R. and Shebuski R.J. (1990): *FASEB J.*, 4: 3087-3098.
13. Folts J.D. and Rowe G.G. (1988): *Thromb. Res.*, 50: 5107-516.
14. Bertha B.G. and Folts J.D. (1985): *Cardiovasc. Res.*, 8: 495-506.
15. Mason R.G. and Read M.S. (1967): *Exp. Mol. Pathol.*, 6: 370-375.
16. Ardlie N.G., McGuiness J.A. and Garrett J.J. (1985): *Atherosclerosis,* 58: 251-259.
17. Folts J.D., Rowe G.G. and Rao G.H.R. (1988): *Lancet,* 1: 937-938.
18. Ahn Y.S. and Harrington W.J. (1987): *Adv. Intern. Med.*, 32: 137-154.

19.   Shanbaky N.M., Ahn Y., Jy W., Harrington W., Fernandez L. and Haynes D.H. (1987): *Thrombosis and Hemostasis,* 57: 1-10.
20.   Ahn Y.S., Jy W., Harrington W.J., Shanbaky N., Fernandez L.F. and Haynes D.H. (1987): *Thrombo. Res.,* 45: 135-143.

THE EFFECTS OF LACIDIPINE ON INSULIN SENSITIVITY

R. Donnelly, A.D. Morris, J.M.C. Connell*, and J.L. Reid

Department of Medicine and Therapeutics
and the *MRC Blood Pressure Unit
Gardiner Institute, Western Infirmary
Glasgow, G11 6NT Scotland

Clinical and pathophysiological associations between obesity, noninsulin-dependent diabetes mellitus (NIDDM), and essential hypertension have raised important questions about the role of insulin resistance in both the etiology and clinical course of hypertension. Loss of tissue sensitivity to the actions of insulin, i.e., insulin resistance, leads to a rise in circulating insulin levels and, when this compensatory hyperinsulinemic response is insufficient, to impaired glucose tolerance or type II diabetes. Insulin resistance is a well-recognized feature of obesity and NIDDM, but more recently evidence has shown that lean nondiabetic essential hypertensives are also insulin resistant and hyperinsulinemic compared with well-matched normotensive controls (1). Moreover, treated hypertensives may remain insulin resistant which may contribute to the apparent deficiencies of antihypertensive treatment in reducing coronary heart disease mortality. Dyslipidemia is an accompanying metabolic feature of insulin resistance and generally there is a positive correlation between plasma insulin and triglyceride concentrations and an inverse relationship between insulin and high-density lipoprotein cholesterol (2). In addition to these adverse lipid effects, hyperinsulinemia itself may be an independent risk factor for coronary heart disease (3) and this epidemiological evidence is supported by numerous experimental studies showing that insulin accelerates atherosclerosis (4). Interestingly, insulin resistance and hyperinsulinemia also occur in classic rodent models of hypertension, for example, the spontaneously hypertensive rat (SHR) and the Dahl salt-sensitive animal (5). In addition, Reaven showed some time ago that replacing the usual carbohydrate in rat chow with fructose led to a state of insulin resistance, hyperinsulinemia and hypertension within 21 days (6). Moreover, exercise training and somatostatin administration (interventions which enhance insulin sensitivity and lower plasma insulin concentrations) attenuate the rise in blood pressure associated with this dietary intervention (7).

The mechanism of the insulin resistance associated with hypertension remains unclear. The resistance seems to be located mainly in skeletal

T. Godfraind et al. (eds.), Calcium Antagonists, 173–177.

muscle and to be limited to the nonoxidative (i.e., glycogen synthetic) pathways of intracellular glucose utilization (8). In contrast, obesity and NIDDM are associated with a more widespread defect in glucose transport.

The precise role of insulin resistance in the initiation and maintenance of high blood pressure remains unclear. For example, there is debate about whether abnormalities in glucose transport are the cause or consequence of hypertension. However, there is evidence that insulin resistance and hyperinsulinemia occur in patients with primary essential hypertension but not in those with secondary forms of hypertension (9). In addition, a recent epidemiological study showed that men who later became hypertensive had impaired glucose tolerance up to 18 years before the onset of raised blood pressure (10). Similarly, animal studies have shown that insulin resistance is a feature of primary genetic models of hypertension (e.g., the SHR) whereas secondary hypertension in the Goldblatt animal is not associated with insulin resistance.

Whether insulin resistance is involved in the initiation of high blood pressure is debatable, but this triad of metabolic abnormalities (namely insulin resistance, hyperinsulinemia and dyslipidemia) undoubtedly affects the long-term clinical course of hypertension and coronary heart disease. Accordingly, the impact of long-term antihypertensive treatment on these metabolic parameters may be of clinical importance.

CALCIUM ANTAGONISTS AND INSULIN SENSITIVITY

Antihypertensive drugs have differential effects on insulin sensitivity and treated hypertensives may remain insulin resistant. Thiazide diuretics and beta blockers often decrease insulin sensitivity and glycemic control but there has been conflicting evidence about the effects of calcium antagonists on glucose tolerance, insulin secretion, and peripheral insulin action. For example, previous studies have reported adverse, neutral, and even beneficial effects on metabolic parameters. Many of these studies, however, were uncontrolled and used surrogate measurements of insulin sensitivity, e.g., glucose tolerance tests and fasting insulin levels, which have contributed to the conflicting evidence. In addition, since calcium antagonists are associated acutely with reflex increases in plasma catecholamines and renin activity, there may be differences between single dose and chronic administration.

The purpose of this study was to evaluate the acute and chronic effects of lacidipine on whole-body-insulin sensitivity in normotensive male volunteers using the euglycemic hyperinsulinemic clamp (11).

METHODS

Twelve healthy male volunteers, age range 18-29 years, completed a double-blind, placebo-controlled, crossover study to evaluate the effects of single and multiple doses (14 days) of lacidipine 4 mg on whole-body-insulin sensitivity. Each subject attended four 4-hour study mornings in the Clinical Pharmacology Research Unit to evaluate acute and chronic lacidipine and the corresponding placebo administrations. On each occasion, following an overnight fast, lacidipine or placebo was administered orally with 100 mls of water and then whole-body-insulin sensitivity was measured using the euglycemic hyperinsulinemic clamp (11). In brief, this method consists of a primed constant rate infusion of Actrapid insulin (1.5 mU/kg/min) for 3 hours and a variable rate infusion of 20% dextrose to maintain euglycemia (5.2 mmol/l). Venous cannulae were inserted into an antecubital vein for administration of insulin and dextrose, and retrogradely into a dorsal hand vein for removal of blood samples. The dorsal hand vein was surrounded by a heated box (55 °C) in order to "arterialize" the venous blood. At a steady state, the rate of infusion of dextrose can be related to whole-body uptake of glucose from the circulation. Thus, following a standard "space correction," the dextrose infusion rate is equated with M which has units of milligrams per kilogram body weight per minute (11). Measurements are expressed as mean ± SD and M values were compared by repeated measures analysis of variance.

RESULTS

Lacidipine was generally well tolerated and there were no significant adverse events or withdrawals from the study. Measurements of insulin-stimulated glucose uptake were 8.9±2.0 and 9.6±2.1 mg/kg/min following acute and chronic lacidipine, respectively, compared with 9.1±1.7 and 9.7±1.5 mg/kg/min after placebo (table 1). The intrasubject coefficient of variation in insulin sensitivity during the placebo phase was 9%. The profiles of serum insulin on each of the 4 study days were not significantly different.

TABLE 1. <u>Insulin Sensitivity Values (± SD) in mg/kg/min</u>

	Lacidipine	Placebo
Acute	8.9±2.0	9.1±1.7
Chronic	9.6±2.1	9.7±1.5

DISCUSSION

This controlled study has shown that lacidipine has no significant effect on insulin sensitivity in healthy male volunteers. Previous uncontrolled studies have reported adverse, neutral, and beneficial effects of dihydropyridine compounds on glycemic control and peripheral insulin sensitivity. However, the interpretation of these studies has been hampered by poor methodology: for example, many studies have used parallel group comparisons without adequate placebo data and attempted to infer differences in insulin sensitivity from oral glucose tolerance tests or fasting plasma insulin levels. The euglycemic hyperinsulinemic clamp is well recognized as the "gold standard" for measuring insulin-stimulated metabolic responses and clearly a crossover design is far more appropriate for examining drug effects on insulin sensitivity.

Dihydropyridine calcium antagonists are associated acutely with reflex sympathetic activation which tends to diminish with chronic dosing. Thus, transient increases in plasma catecholamine levels and plasma renin activity would be expected to have an insulin antagonist effect. In this study, measurements of insulin sensitivity were only marginally lower after acute, compared with, chronic lacidipine, which is probably indirect evidence for minimal reflex stimulation on first dose administration.

CONCLUSION

In conclusion, this controlled study has shown that lacidipine has no adverse effects on insulin-stimulated glucose uptake, which is entirely consistent with most previous studies which have shown that dihydropyridine drugs have a neutral metabolic profile. Further work is in progress to confirm these effects in a more insulin-resistant group of hypertensive patients.

REFERENCES

1.   Ferranini, E., Buzzigoli G. and Bonadona R. (1987): *N. Engl. J.*

Med., 317: 350-357.
2. Garg A., Helderman J.H., Koffler M., Ayuso R., Rosenstock J. and Raskin P. (1988): *Metabolism,* 37: 982-987.
3. Pyorala K. (1979): *Diabetes Care,* 2: 131-141.
4. Stout R.W. (1990): *Diabetes Care,* 13: 631-654.
5. Mondon C.E. and Reaven G.M. (1989): *Am. J. Physiol.,* 257: E4 91-E498.
6. Hwang I.S., Ho H., Hoffman B.B. and Reaven G.M. (1987): *Hypertension,* 10: 512-516.
7. Reaven G.M., Ho H. and Hoffman B.B. (1988): *Hypertension,* 12: 129-132.
8. Natali A., Santoro D., Palombo C., Cerri M., Ghione S. and Ferrannini E. (1991): *Hypertension,* 17: 170-178.
9. Marigliano A., Tedde R., Sechi L.A., Pala A., Pisanu G. and Pacifico A. (1990): *Am. J. Hyperten.,* 3: 521-526.
10. Salomaa V.V., Strandberg T.E., Vanhanen H., Naukkarinen V., Sarna S. and Miettinen T.A. (1991): *Br. Med. J.,* 302: 493-496.
11. DeFronzo R., Tobin J.D. and Andres R. (1979): *Am. J. Physiol.,* 237: E214-E233.

CALCIUM ANTAGONISTS, HYPERTENSION, AND ATHEROSCLEROSIS

C. Carpi

Glaxo Research Laboratories
Via A. Fleming 4
37100 Verona, Italy

Calcium antagonists are now a well-established class of drugs in widespread clinical use. They comprise a large variety of chemical structures in which several subclasses have been identified: the dihydropiridines, with nifedipine as the classical reference compound and lacidipine as a drug representing the latest generation of calcium antagonists; the phenylalkylamines, represented by verapamil; the benzothiazepines, with diltiazem, and a small group represented by compounds like bepridil and flunarizine.

Although they possess very different chemical structures, all share the same mechanism of action, namely the inhibition of calcium influx through the voltage-dependent calcium channels (1). By means of this mechanism they have proved to be particularly effective in several cardiovascular diseases including angina and hypertension. Indeed, calcium antagonists are now widely accepted as first line antihypertensive agents together with beta-blockers, diuretics, and ACE-inhibitors (2).

However one of the most intriguing findings which have emerged from large-scale clinical trials of antihypertensive drugs, is that treatment decreased the incidence of stroke by about 40% but had insignificant effects on myocardial infarction (3). The most likely explanation is that they are very different diseases: stroke includes both thrombotic and hemorrhagic events, the later being more directly related to the level of blood pressure, whereas myocardial infarction is purely thrombotic (4). Thus, for stroke, blood pressure is the most important risk factor, while for myocardial infarction atherosclerosis assumes a major role (5).

Atherosclerosis is a slowly progressive disease of arterial walls that begins at a young age but becomes manifest only in the middle age. It is a multifaceted metabolic disorder, impairing circulation by reducing the elasticity of the arterial wall and occluding the lumen with the classic atheromatous lesions. The lesion protrudes into the lumen, often with an associated thrombotic cap, and reduces the effective cross-sectional area of the blood vessel available for the circulation.

The pathogenesis of atherosclerosis is the result of a complex network

T. Godfraind et al. (eds.), Calcium Antagonists, 179–181.
© 1993 *Kluwer Academic Publishers and Fondazione Giovanni Lorenzini.*

of different metabolic and cellular events whose complete description is still unclear (6,7). Nevertheless, many of the cellular steps that contribute to the formation of atheroma are calcium dependent and, therefore, potentially sensitive to calcium entry blockers (8).

Recently, knowledge has become available that opens up a much wider field for new therapeutic applications for specific calcium antagonists. One of the most interesting is the antiatherosclerotic activity of these drugs. An increasing body of experimental evidence suggests that calcium antagonists possess considerable antiatherosclerotic potential (9,10). The protection that calcium antagonists exert in the initial stages of developing atheroma has been extensively documented in different *in vivo* and *in vitro* models.

So far a number of calcium antagonists of different classes has been tested in the cholesterol-fed rabbit model. These compounds, without affecting lipid metabolism, decrease the area of the lesion to varying extents (11,12,13). The mechanism of this action is undoubtedly complex. However, many *in vitro* experimental models suggest a significant activity on cellular migration and proliferation and on matrix secretion from smooth muscle cells (14,15,16,17).

For all these reasons, a drug that simultaneously normalizes blood pressure and inhibits atherogenesis would be of great benfit for the treatment of cardiovascular diseases. The main objective of this Symposium is to present lacidipine as a new long-lasting calcium entry blocker which effectively controls hypertension and shows promising antiatherogenic properties (18).

REFERENCES

1.   Fleckenstein A. (1983): *Calcium Antagonism in Heart and Smooth Muscle.* John Wiley & Sons, New York.
2.   Zanchetti A. (1987): *Am. J. Card.*, 59: 130B-136B.
3.   MacMahon S.W., Cutler J.A., Furberg C.D. and Payne G.H. (1986): *Prog. Cardiovasc. Dis.*, 29(suppl.1): 99-118.
4.   Strandgaard S. and Haunso S. (1987): *Lancet*, 2: 658-660.
5.   Welin L., Svardsudd K., Wilhemsen L., Larsson B. and Tibblin G. (1987): *New Engl. J. Med.*, 317: 521-526.
6.   Ross R.R. (1986): *New Engl. J. Med.*, 314: 488-500.
7.   Murno J.M. and Cotran R.S. (1988): *Laboratory Invest.*, 58: 249-261.
8.   Kiowski W., Eme P. and Bühler F.R. (1989): *Clin. Exp. Hyper. Theory and Practice*, A11(5-6): 1085-1096.
9.   Parmley W.W., Blumlein S. and Sievers R. (1985): *Am. J. Cardiol.*, 55: 165B-171B.
10.   Weinstein D.B. (1988): *J. Cardiovasc. Pharmacol.*, 12(suppl.6): S29-S35.
11.   Kramsc D.M., Aspen A.J. and Rotzler L.J. (1981): *Science*, 213:

1511-1512.

12. Mutoh S., Nomoto A., Sekiguchi C. and Yamaguchi I. (1988): *Atherosclerosis*, 73: 181-189.

13. Kolbuchi Y., Sakai S., Miura S., Ono T., Shibayama F. and Ohtsuka M. (1989): *Atherosclerosis*, 79: 147-155.

14. Kjieldsen K. and Stender S. (1989): *Proc. Soc. Exp. Biol. Med.*, 190: 219-228.

15. Nomoto A., Mutoh S., Hagihara H. and Yamaguchi I. (1988): *Atherosclerosis*, 72: 213-219.

16. Jackson C.L., Raymond C.B. and Bowyer D. (1989): *Atherosclerosis*, 80: 17-26.

17. Betz E. (1988): *Ann. N.Y. Acad. Sci.*, 522: 399-410.

18. Micheli D., Ratti E., Toson G. and Gaviraghi G. (1991): *J. Cardiovasc. Pharmacol.*, 17(suppl.4): S1-S8.

EXPRESSION, LOCALIZATION, AND FUNCTION OF CALCIUM CHANNELS IN NGF-TREATED PC12 CELLS

H. Reuter, B.F.X. Reber, M.M. Usowicz, and H. Porzig

Department of Pharmacology, University of Bern
Friedbühlstrasse 49, 3010 Bern, Switzerland

The rat pheochromocytoma (PC12) cell line is a widely used model system for nerve growth factor (NGF)-induced neuronal differentiation (1). Morphological differentiation is characterized by enlargement of cell bodies and extension of neurites. The mechanism of action of NGF is incompletely understood (2). However, in addition to morphological differentiation, several properties of PC12 cells change during exposure to NGF. Among others, densities of Ca^{2+} channels in the surface membrane are increased (3-6). This may have important functional consequences, for example for depolarization-evoked catecholamine release from the cells. It has been shown by several groups (7-9) that, in the course of differentiation, there is a change in sensitivity of catecholamine release to pharmacological interventions. The purpose of this paper is to summarize the evidence for this NGF-induced change.

EXPRESSION OF L- AND N-TYPE CALCIUM CHANNELS DURING NERVE GROWTH FACTOR EXPOSURE

In our PC12 cell clone there are at least three pharmacologically distinct types of high-threshold Ca^{2+} channels. The L-type channels are characterized by their sensitivity to 1,4-dihydropyridines (DHPs) and the N-type channels are blocked by ω-conotoxin (10-12). A third component of high-threshold Ca^{2+} current remains when saturating concentrations of DHP and ω-conotoxin are applied together (12,13). This DHP- and ω-conotoxin-resistant current component has not yet been characterized. Low-threshold (T-type) Ca^{2+} channels (4) were absent in our cells.

We have studied in detail the expression of L- and N-type Ca^{2+} channels in PC12 cells after 5-6 days exposure of the cells to NGF (6). We have found that Ba^{2+} current through L-type Ca^{2+} channels increases roughly in proportion to the increase in surface area. Membrane surface area was measured as membrane capacitance (pF). The L-type current component was measured from a holding potential of -30 mV, where most of the N-type current was inactivated (fig. 1). The density of this current component increased only slightly from an average of 3.8 pA/pF

T. Godfraind et al. (eds.), Calcium Antagonists, 183–188.
© 1993 Kluwer Academic Publishers and Fondazione Giovanni Lorenzini.

in 55 native cells to 4.4 pA/pF in 47 NGF-exposed cells (6). Most of this current could be blocked by the DHP isradipine. By contrast, the Ba^{2+} current density measured from a holding potential of -90 mV increased by a larger amount after NGF exposure (fig. 1), from an average of 22 pA/pF (n = 40) in native cells to 32 pA/pF (n = 39) in NGF-treated cells (6).

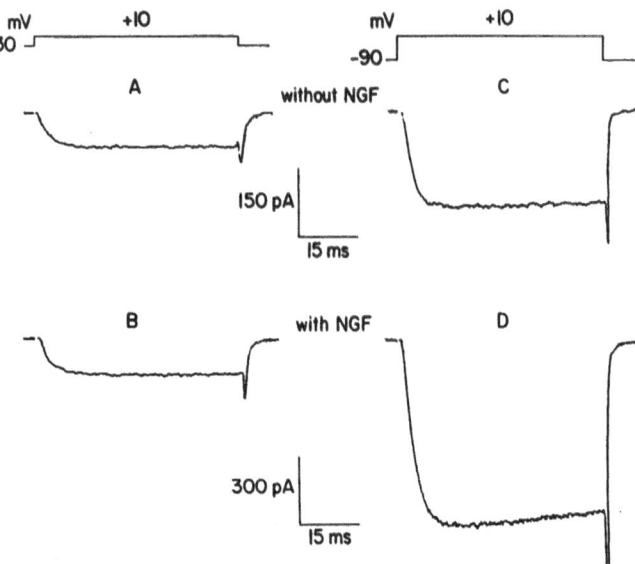

FIG. 1. Comparison of Ca^{2+} channel currents in two PC12 cells cultured for 5-6 days in the absence (A,C) or presence (B,D) of NGF. Whole-cell Ba^{2+} currents evoked by depolarizing voltage steps from -30 mV (A,B) or -90 mV (C,D) to + 10 mV are shown. Currents in the NGF-treated cell (B,D) are much larger than in the untreated cell (A,C), particularly that evoked from a holding potential of -90 mV. Note difference in current scale. For details of the methods see (6).

Fig. 2 illustrates that the current measured from a holding potential of -90 mV was greatly inhibited by ω-conotoxin, while that recorded from -30 mV was not. In agreement with the results obtained by Plummer et al. (5) we conclude that there is a preferential expression of N-type over L-type Ca^{2+} channels in PC12 cells after exposure to NGF. This conclusion is supported by our radioligand binding studies which also showed a larger increase in [125I]-ω-conotoxin binding than in [3H]-isradipine binding in NGF-treated PC12 cells (6).

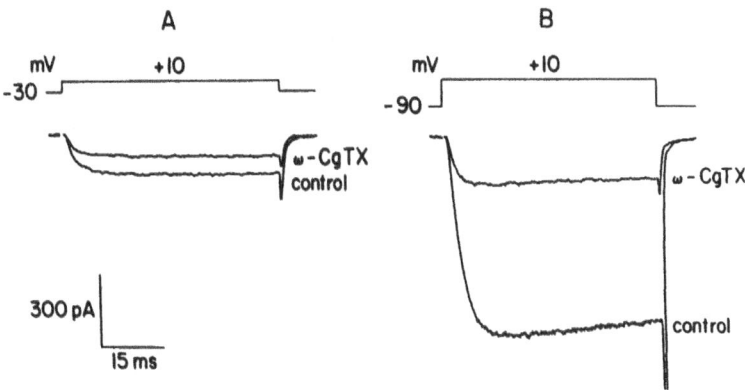

FIG. 2. The whole-cell Ba^{2+} current component that can be blocked by ω-conotoxin (500 nM) in a NGF-treated cell (same cell as in fig. 1) is larger when the holding potential is set at -90 mV than at -30 mV. At -90 mV most N-type channels are available to open during the depolarizing voltage step to +10 mV, while at -30 mV most of these channels are inactivated.

LOCALIZATION OF L- AND N-TYPE CALCIUM CHANNELS

We used fura-2 microfluorimetry in single cells (14), in order to assess the location of L- and N-type Ca^{2+} channels in PC12 cells before and after NGF treatment (13). Fig. 3 shows the results. The cells were depolarized by elevating the external K^+-concentration ($[K^+]_o$) from 5 mM to 70 mM. Depolarization opens the voltage-dependent Ca^{2+} channels which causes a rise in the intracellular Ca^{2+}-concentration ($[Ca^{2+}]_i$) that could be measured with fura-2. In bodies of native cells, not treated with NGF (fig. 3a), a large fraction of the K^+-depolarization-induced rise in $[Ca^{2+}]_i$ could be blocked by the DHP nifedipine and a smaller, additional fraction by ω-conotoxin. Even after application of saturating concentrations of both drugs together, elevation of $[Ca^{2+}]_i$ was not blocked completely, arguing for a third type of Ca^{2+} channels.

The effects of nifedipine and ω-conotoxin were not much different in cell bodies of differentiated PC12 cells after treatment with NGF (fig. 3b). However, in growth cones of differentiating cells, there was only a small inhibition by nifedipine and a much larger one by ω-conotoxin (fig. 3c). A rather large component of the rise in $[Ca^{2+}]_i$ could not be blocked at all by these drugs. From these results we conclude that the ration of distribution of L- and N-type Ca^{2+} channels in cell bodies is not much affected by NGF-induced differentiation. By contrast, in growth cones a much larger fraction of N-type channels seems to be expressed. Thus, growth cones seem to be the structures in differentiated PC12 cells where

most of the additional N-type Ca^{2+} channels, described in the previous
section, are localized.

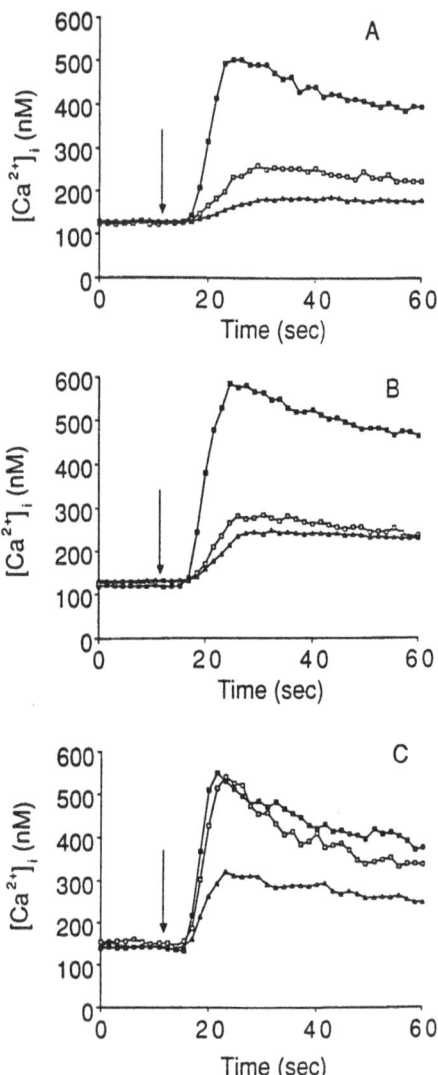

FIG. 3. Effects of nifedipine ($10\mu M$) and of ω-conotoxin (500 nM) on
changes in $[Ca^{2+}]_i$ evoked by K^+-depolarization. Cells were depolarized
by elevating $[K^+]_o$ to 70 mM (arrows), $[Ca^{2+}]_i$ was measured with fura-2;
controls (■), $10\mu M$ nifedipine (□), 10 μM nifedipine plus 500 nM ω-
conotoxin (▲). Panel A: combined results of 6 undifferentiated PC12 cell
bodies. Panel B: combined results of 14 cell bodies from NGF-treated (6
days) cells. Panel C: combined results of 7 growth cones from NGF-
treated (6-7 days) cells. For details of the method, see (13).

[³H]-DOPAMINE RELEASE

Catecholamine release from undifferentiated PC12 cells is very sensitive to blockade of L-type Ca^{2+} channels by DHP. This sensitivity decreases after NGF-induced differentiation of the cells (7,8). We have measured [³H]-dopamine release from PC12 cells before and after differentiation by NGF, and its inhibition by isradipine and ω-conotoxin (9). [³H]-dopamine release was evoked by K^+-depolarization of the cells. The inhibition of the evoked release by the drugs is shown in fig. 4. Isradipine inhibited [³H]-dopamine release much more strongly in non-differentiated (-NGF) than in differentiated (+NGF) PC12 cells. The opposite was true for ω-conotoxin which had a higher efficacy in blocking the release from differentiated cells. Since ω-conotoxin sensitive N-type Ca^{2+} channels are preferentially expressed in these cells after NGF exposure, the results in fig. 4 suggest that these channels become predominant in the K^+-depolarization evoked [³H]-dopamine release. By contrast, in agreement with earlier studies (7,8), L-type Ca^{2+} channels become less important for transmitter release during differentiation of PC12 cells. Our results also argue for a switch of the site of [³H]-dopamine release from cell bodies to growth cones in the course of differentiation. L-type channels are only sparsely expressed in growth cones while N-type channels are dominant (fig. 3c). It is of interest to note that K^+-evoked [³H]-dopamine release is not completely inhibited by saturating concentrations of both drugs. Therefore, the unidentified type of Ca^{2+} channel also seems to play a role in transmitter release.

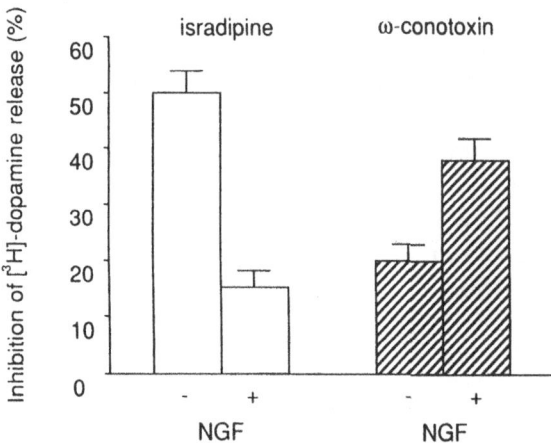

FIG. 4. Inhibition by isradipine (10 μM) and ω-conotoxin (200 nM) of [³H]-dopamine release from PC12 cells cultured in the absence (-NGF) or presence (+NGF) of NGF. Means ± SEM of 3-4 experiments. For details of the method, see (9).

ACKNOWLEDGEMENT

We wish to thank Ms. C. Becker for her unfailing help with the cell cultures and the transmitter release experiments. Financial support by the Swiss National Science Foundation (Grant No. 3.042-0.87) is gratefully acknowledged.

REFERENCES

1.   Greene L.A. and Tischler A.S. (1982): *Adv. Cell Neurobiol.*, 3: 373-414.

2.   Levi A. and Alemà S. (1991): *Annu. Rev. Pharmacol. Toxicol.*, 31: 205-228.

3.   Streit J. and Lux H.D. (1987): *Pflügers Arch.*, 408: 634-641.

4.   Garber S.S., Hoshi T. and Aldrich R.W. (1989): *J. Neurosci.*, 9: 3976-3987.

5.   Plummer M.R., Logothetis D.E. and Hess P. (1989): *Neuron.*, 2: 1453-1463.

6.   Usowicz M.M., Porzig H., Becker C. and Reuter H. (1990): *J. Physiol.*, 426: 95-116.

7.   Takahashi M., Tsukui H. and Hatanaka H. (1985): *Brain Res.*, 341: 381-384.

8.   Kongsamut S. and Miller R.J. (1986): *Proc. Natl. Acad. Sci. USA*, 83: 2243-2247.

9.   Reber B.F.X., Porzig H., Becker C. and Reuter H. (1990): *Neurochem. Int.*, 17: 197-203.

10. Tsien R.W., Lipscombe D., Madison D.V., Bley K.R., and Fox A.P. (1988): *Trends Neurosci.*, 11: 431-438.

11. Carbone E., Sher E. and Clementi F. (1990): *Pflügers Arch.*, 416: 170-179.

12. Regan L.J., Sah D.W.Y. and Bean B.P. (1991): *Neuron.*, 6: 269-280.

13. Reber B.F.X. and Reuter H. (1991): *J. Physiol.*, 435: 145-162.

14. Grynkiewicz G., Poenie M. and Tsien R.Y. (1985): *J. Bio. Chem.*, 260: 3440-3450.

RECEPTOR-MEDIATED INHIBITION OF CALCIUM CURRENTS AND SYNAPTIC TRANSMISSION IN CULTURED HIPPOCAMPAL PYRAMIDAL NEURONS

Kenneth P. Scholz and Richard J. Miller

Department of Pharmacological and Physiological Sciences
University of Chicago
947 E. 58th St.
Chicago, Illinois 60637

In peripheral neurons, there are many examples of neurotransmitters and exogenous agents that inhibit calcium channels or currents (1). Clearly, these examples have been crucial in advancing our understanding of the role of calcium channels in neuronal activity and synaptic transmission. In contrast, there is a relative scarcity of information about neurotransmitters that inhibit calcium channels in the central nervous system. Furthermore, while the action of calcium channel antagonists on transmitter release have been studied biochemically (2), there are only a handful of studies that have utilized single-cell electrophysiological approaches to study blockers of calcium channels in central neurons.

We have begun a series of investigations into the modulation of calcium channels in central neurons and have utilized cultured hippocampal pyramidal cells as a model. While these cells have the problem that they are sometime difficult to voltage-clamp, they provide the benefit that they form synapses in culture. In this condition, the synaptic regions are accessible for local application of pharmacological agents. Thus, we have set out to test the inhibition of calcium currents and channels in these cells by neurotransmitters and pharmacological agents. In addition, parallel studies have been conducted to test the effects of these same compounds on synaptic transmission using several experimental approaches.

ADENOSINE

The endogenous compound adenosine is a neuromodulator with some unique properties. While there is no evidence that adenosine acts like a classical neurotransmitter (3), it has been shown that endogenous levels of adenosine are elevated in the hippocampus as a result of neuronal activity (4). Furthermore, adenosine acts at presynaptic terminals of hippocampal pyramidal neurons to inhibit release of glutamate (5). In

T. Godfraind et al. (eds.), Calcium Antagonists, 189–194.

this way, adenosine serves in a negative feedback capacity to limit excitatory transmitter release. The mechanism by which adenosine exerts this action is not well understood. Some reports have concluded that adenosine does not reduce calcium currents in hippocampal neurons (6,7), and have thus concluded that presynaptic inhibition is mediated by activation a K^+ channel present in the presynaptic terminal. However, there are many problems with studying the action of neurotransmitters on calcium currents in the brain slice preparation. For this reason, we have re-examined this issue in cultured pyramidal neurons.

Adenosine A1 receptor agonists were found to inhibit high-threshold Ca currents in the soma of cultured hippocampal pyramidal neurons (8). This action was associated with a decrease in the rate of activation of high-threshold current. Inhibition was observed in the presence of 2 mM Ba^{2+} as the charge carrier indicating that the decrease in current was not merely a result of activation of K^+ currents in peripheral regions of the cell. These results support the previous findings of Proctor and Dunwiddie (9) who reported that adenosine reduced Ca-dependent action potentials in pyramidal neurons in the slice.

The role of guanine-nucleotide binding proteins (G-proteins) in the receptor-mediated inhibition of Ca currents by adenosine was studied by two approaches. First, the non-hydrolyzable GTP analog GTP-γ-S was perfused into individual cells via the patch pipette. GTP-γ-S induced an irreversible inhibition of I_{Ca} upon exposure of the cells to the A1-selective adenosine receptor agonist cyclopentyladenosine (CPA). This result implicates G-proteins in the mechanism of transduction from the adenosine receptor to the calcium channel. In addition, pre-incubation of the cells with pertussis toxin, which induces the ADP-ribosylation and inhibition of Gi and Go subclasses of G-proteins (10), abolished the actions of CPA.

Since we had characterized the adenosine receptor that inhibited I_{Ca} in these cells, we wished to determine if the same or a different receptor was involved in presynaptic inhibition. It was found that presynaptic inhibition followed the exact same profile of agonists as did inhibition of I_{Ca}. In addition, the A1 specific antagonist, cycoopentyl theophylline (CPT), inhibited the actions of adenosine agonists with an apparent affinity that was essentially identical to that found for inhibition of I_{Ca}. Thus, based on currently available ligands, the receptors linked to inhibition of I_{Ca} and synaptic transmission are indistinguishable (8).

GABA

The neurotransmitter substance GABA has been shown to act on two very different types of receptor (11). The $GABA_A$ type directly gates a Cl- permeable ion channel. This leads to fast acting inhibition in most cases. This type of action has also been linked to presynaptic inhibition at some synapses, for example at the crayfish neuromuscular junction (12) and at some spinal cord synapses (13). It is generally thought that this form of presynaptic inhibition requires axo-axonic synapses.

A second type of GABA receptor has been labeled the $GABA_B$ receptor (11). In all cases that have been found to date, this receptor is not directly associated with an ion channel. In contrast, the $GABA_B$ receptor is a G-protein binding molecule and appears to transmit its signal through the activation of a G-protein. In the hippocampus, $GABA_B$ receptors have been shown to activate a K^+ current (14). In addition, $GABA_B$ receptors have been shown to cause presynaptic inhibition (15). However, this form of presynaptic inhibition may not require axo-axonic synapses. Indeed, there is convincing evidence that presynaptic terminals that release GABA also express $GABA_B$ receptors on their surface membrane. In this way, GABA release may inhibit further release from the same terminal (16).

As with adenosine, the presynaptic inhibitory actions of baclofen have been linked to activation of K^+ channels (14). However, the mechanism of action is far from being understood. We have shown recently that baclofen can inhibit I_{Ca} in the soma of pyramidal neurons and that this action involves activation of a $GABA_B$ receptor (17). Thus, inhibition of Ca currents becomes another putative mechanism by which baclofen can inhibit transmitter release.

Intracellular perfusion of GTP-γ-S was found to potentiate the actions of baclofen and to make the response irreversible, as was found with adenosine. Of interest was the finding that saturation of the response to baclofen in the presence of intracellular GTP-γ-S also led to the occlusion of the response to adenosine. This indicates that the transduction signals for the two receptor-effector systems converge (17). It is unclear whether this happens at the level of the G-protein or the Ca channels.

The presynaptic inhibitory actions of baclofen have been reported to be insensitive to block by pertussis toxin (PTX) in the hippocampal slice preparation (18,19). However, there is evidence that the action of toxin, or its dispersion, may not be particularly efficient under these conditions (20). When we re-examined this issue in cultured pyramidal cells, we found that preincubation with pertussis toxin abolished the actions of baclofen on transmitter release and on calcium currents. Similar results were found for the actions of pertussis toxin on presynaptic inhibition mediated by adenosine receptors. Recent studies of Van de Ploeg et al. (20) have indicated that the discrepancy between our results and those published previously may arise from the incomplete dispersion of PTX in the slice preparation when the toxin is injected into the third ventricle of the rat brain.

In neurons, ω-CgTX sensitive (N-type) Ca channels have received a great deal of attention, probably resulting from the fact that many examples of Ca-channel modulation in neurons to date involve this type of channel (1,21). We have examined the subtypes of Ca channels inhibited by baclofen by using ω-CgTX fraction GVIA and dihydropyridine Ca channel antagonists, compounds that appear to have selectivities for different classes of Ca channels (22). Both ω-CgTX and nimodipine partially occluded the actions of baclofen, although the properties of block by the two antagonists were very different (17). This suggested that baclofen may inhibit both ω-CgTX-sensitive channels and dihydropyridine-sensitive (L-type) channels. This conclusion was supported by the finding that the combination of ω-CgTX and nimodipine were unable to block all high-threshold currents in these cells, as has been observed by Regan et al. (23) and by Mogul and Fox (24). While it seems that this antagonist-insensitive current is also insensitive to inhibition by baclofen, the properties of this high-threshold current remain to be examined.

It has been shown that the best available technique to assess the role of L-type channels during whole-cell recording is to use a dihydropyridine Ca channel agonist to prolong the Ca current tails (22). In this way, a component of the tail current can be attributed to L-type channels without ambiguity. However, this type of experiment is very difficult to perform in our cells due to space-clamp considerations. Therefore, we turned to cell-attached recordings of single L-type Ca channels to investigate the effects of baclofen on the channels in more detail. When L-type channels are active during voltage-clamp pulses from -40 to +10 mV, their mean open time is greatly enhanced by dihydropyridine Ca channel agonists. In this way, L-type channels can be identified. We have found that addition of baclofen to the bathing solution outside of the patch pipette leads to the gradual inhibition of L-type channel activity in the patch. This indicates that baclofen may inhibit L-type channels by acting through a second messenger. Interestingly, ω-CgTX-sensitive channels were not modulated in the patch during application of baclofen outside of the patch. These results suggest that baclofen may inhibit N-type channels by a membrane-delimited action that does not require a soluble second messenger, but may inhibit L-type channels through a change in the concentration of a cytoplasmic second messenger.

SUMMARY

Now that the inhibition of Ca currents by adenosine and baclofen has been established in pyramidal cells, it will be important to determine the role that this mechanism plays in presynaptic inhibition. We have begun experiments looking at the spontaneous release of transmitter quanta from these cells and are beginning to assess the sensitivity of presynaptic Ca channels to these transmitter substances. In addition, we have begun to use new Ca channel toxins from spiders and cone snails to compare the pharmacological profile of somatic Ca channels and presynaptic Ca channels.

In summary, the inhibition of Ca channels by two neurotransmitter pathways that act in the hippocampus to inhibit transmitter release from pyramidal neurons has been established. The receptors involved, the types of Ca channels inhibited, and the role of G-proteins in this process have been investigated. Further, experiments will be aimed at determining the role of the Ca channels in the inhibition of transmitter release.

REFERENCES

1.   Tsien R.W., Lipscombe D., Madison D.V., Bley K.R.and Fox A.P. (1988): *Trends in Neurosci.*, 11: 431-438.
2.   Huston E., Scott R.H. and Dolphin A.C. (1991): *Neurosci.* (in press).
3.   Snyder S.H. (1985): *Ann. Rev. Neurosci.*, 8: 103-124.
4.   Haas H.L. and Greene R.W. (1988): *Naunyn-Schmiedeberg's Arch. Pharm.*, 337: 561-565.
5.   Dunwiddie T.V. and Haas H.L. (1985): *J. Physiol.*, 369: 365-377.
6.   Halliwell J.V. and Scholfield C.N. (1984): *Neurosci. Lett.*, 50: 13-18.
7.   Gerber U., Greene R.W., Haas H.L. and Stevens D.R. (1989): *J. Physiol.*, 417: 567-578.
8.   Scholz K.P. and Miller R.J. (1991): *J. Physiol.*, 435: 373-393.
9.   Proctor W.R. and Dunwiddie T.V. (1983): *Neurosci. Lett.*, 35: 197-201.
10.   Ui M., Katada T., Murayama T., Kurose T., Yajima M., Tamura M., Nakamura T. and Nogimori K. (1984): *Adv. Cyclic. Nucl. Res.*, 17: 145-151.
11.   Bowery N. (1989): *Trends Pharm. Sci.*, 10: 401-407.
12.   Takeuchi A. and Takeuchi N. (1966): *J. Physiol.*, 183: 418-432.
13.   Eccles J.C., Kostyuk P.G. and Kuffler S.W. (1962): *J. Physiol. (Lond)*, 161: 237-257.
14.   Gahwiler G.H. and Brown D.A. (1985): *Proc. Natl. Acad. Sci. USA*, 82: 1558-1562.
15.   Lanthorn T.H. and Cotman C.W. (1981): *Brain Res.*, 225: 171-178.
16.   Davies C.H., Davies S.N. and Collingridge G.L. (1990): *J. Physiol.*, 424: 513-531.
17.   Scholz K.P. and Miller R.J. (1991): *J. Physiol.* (in press).
18.   Colmers W.F. and Pittman Q.J. (1989): *Brain Res.*, 498: 99-105.

19.   Dutar P. and Nicoll R.A. (1988): *Neuron.*, 1: 585-591.
20.   Van der Ploeg I., Cintra A., Altiok N., Askelof P., Fuxe K. and Fredholm B.B. (1991): *Neurosci.*, 44: 205-214.
21.   Miller R.J. (1990): *FASEB J.*, 4: 3291-3299.
22.   Plummer M.R., Logothetis D.E. and Hess P. (1989): *Neuron.*, 2: 1453-1463.
23.   Regan L.J., Sah D.W.Y. and Bean B.P. (1991): *Neuron.*, 6: 269-280.
24.   Mogul D.J. and Fox A.P. (1991): *J. Physiol.*, 433: 259-281.

CALCIUM STORES AND CALCIUM CHANNELS
OF CEREBELLUM PURKINJE CELLS

Alessandra Nori, Adelina Martini, and Pompeo Volpe

Centro di Studio per la Biologia e Fisiopatologia
Muscolare del CNR, Istituto di Patologia Generale
Università di Padova, via Trieste 75
Padova Italy

Calcium ions are involved in both specialized cellular functions, such as contraction, excitability, signal transduction, and secretion, and basic metabolism, e.g., cell division, protein synthesis, and regulation of gene expression. The key role of Ca^{2+} in such a variety of physiological processes implies that the cytosolic concentration of free Ca^{2+} ($[Ca^{2+}]_i$) is strictly controlled, and that specific mechanisms exist to gauge the concentration changes in the time scale required for each function (1,2). Many cellular activities are stimulated by transient changes of $[Ca^{2+}]_i$ in the range 0.1-1 μM. Calcium sources that can be used for eliciting these responses include import of Ca^{2+} from the extracellular space and mobilization from intracellular Ca^{2+} stores (1-3). Intracellular Ca^{2+} stores, for example, release Ca^{2+} and initiate contraction in skeletal muscle fibers: Ca^{2+} release from the sarcoplasmic reticulum (SR) is mediated by a membrane-bound channel, also referred to as ryanodine receptor (Rya-R), because of its sensitivity to the plant alkaloid ryanodine; Ca^{2+} is transported via a Ca^{2+}-pump from the cytosol back into the SR lumen, where it is largely stored in bound form to calsequestrin (CS), an intraluminal high-capacity, low-affinity Ca^{2+} binding protein (4). The architecture and overall development of the SR are ideally suited to control transient $[Ca^{2+}]_i$ changes during contraction relaxation cycles (2,4), i.e., the SR of striated muscle fibers is the prototype of a rapidly-exchanging Ca^{2+} store.

In nonmuscle cells, the endoplasmic reticulum (ER), in particular subcompartments of the smooth-surfaced ER, and membrane compartments other than ER might have properties, functions, and structural features analogous, at least in part, to those of the SR of striated muscles (1,3).

INTRACELLULAR CALCIUM STORES

Some of the proteins functionally important in Ca^{2+} storage organelles

T. Godfraind et al. (eds.), Calcium Antagonists, 195–203.

are Ca^{2+}-pump(s), Ca^{2+} release channels, and intraluminal, high-capacity, low-affinity Ca^{2+} binding proteins (1,3).

Three types of SR/ER Ca^{2+}-pumps have been described (5): SERCA1, in fast-twitch skeletal muscle; SERCA2a, the prevalent form in cardiac and slow-twitch skeletal muscles; and SERCA2b, the predominant form in cells other than striated muscles. *In situ* hybridization studies have confirmed that in cerebellar Purkinje cells, the expression of the Ca^{2+}-pump isoforms is quite different: SERCA2a is expressed at very low levels, whereas SERCA2b is expressed at high levels (6).

The second component of rapidly-exchanging Ca^{2+} stores is represented by high-capacity, low-affinity Ca^{2+} binding proteins which should serve as a Ca^{2+} buffer inside the store (1,3). In mammalian cells, at least four intraluminal proteins with these properties have been described: cardiac type and fast-twitch skeletal muscle type CS (7,8), calreticulin (9) and endoplasmin [also referred to as grp94/ERp99 (9)]. In mammals, CS is restricted to skeletal and cardiac muscles (3,7,8), whereas in chickens it has also been identified in cerebellum, as a putative component of the inositol 1,4,5-trisphophate (IP_3)-sensitive Ca^{2+} store (10).

Calcium-release channels constitute the third class of key proteins needed to define an intracellular Ca^{2+} store. Thus far, two types of channels have been isolated and characterized: at least two isoforms of ryanodine-sensitive Ca^{2+} channels in heart (11), skeletal muscle (12), and brain (11, 13), and IP_3-sensitive Ca^{2+} channels in cerebellum (14,15) and smooth muscle (16).

Several hormones, neurotransmitters, and growth factors interact with plasma membrane receptors of target cells, and activate a signal transduction mechanism which appears to be almost ubiquitous. Transduction of many external signals is, in fact, mediated by breakdown of phosphatydil inositol 4,5-bisphosphate into diacylglycerol and IP_3 (17). IP_3 acts as a soluble cytosolic messenger and triggers Ca^{2+} release from intracellular stores upon interaction with specific IP_3 receptors [IP_3-R (17,18)]. IP_3-sensitive Ca^{2+} stores are not the only source of rapidly-exchanging Ca^{2+}, because Ca^{2+}-, caffeine-, and ryanodine-sensitive Ca^{2+} stores have been described in neuronal adrenal cromaffin, pancreatic acinar, and smooth muscle cells (19). Intracellular messengers responsible for Ca^{2+} release from caffeine-ryanodine-sensitive Ca^{2+} stores have not been identified yet. The physiological role of ryanodine-, caffeine-sensitive Ca^{2+} stores is not yet clear either: these stores are probably involved in Ca^{2+}-induced Ca^{2+} release (CICR) (20) and might participate in the generation of Ca^{2+} oscillations (21).

In nonmuscle cells, many organelles accumulate Ca^{2+}; some of them (lysosomes, mitochondria, secretory granules) are slow exchangers and cannot support transient changes of $[Ca^{2+}]_i$ (1-3). The ER has been proposed as the rapidly-exchanging Ca^{2+} store (18). However, subcellular fractionation of various tissues and cell types, and studies of distribution of biochemical and immunological markers have revealed a marked heterogeneity of the microsomal fractions composed mainly, but not exclusively, of ER membrane fragments (3). These experiments do not rule out a role of ER in Ca^{2+} storage and Ca^{2+} mobilization, but are not consistent with the hypothesis that the whole ER is a Ca^{2+} store. Moreover, studies on ER-poor cell lines, such as HL-60, have led to the suggestion that rapidly-exchanging Ca^{2+} stores other than the ER exist (22). In HL-60 cells and hepatocytes, smooth-surfaced vacuoles and small tubules (collectively named "calciosomes") have been identified, by immunogold labeling, as those subcellular organelles expressing low-affinity, high-capacity Ca^{2+} binding proteins (3,22,23). Our present working hypothesis is that ER subcompartments and/or separate organelles exist, and that their molecular, structural, and functional heterogeneity is the basis for compartmentalization and regulation of distinct rapidly-exchanging Ca^{2+} stores.

Neuronal cells and, in particular, cerebellum Purkinje cells (PCs) express high levels of both IP_3-sensitive Ca^{2+} channels and ryanodine-sensitive Ca^{2+} channels (24,25), Ca^{2+}-pump(s) (6,26) and, in chicken at least, the high-capacity, low-affinity Ca^{2+} binding protein CS (10,27,28). PCs are a valuable model to study the subcellular localization of Ca^{2+} stores, as well as the physiological properties and interrelations between caffeine-, ryanodine-sensitive Ca^{2+} stores, on one hand, and the IP_3-sensitive Ca^{2+} stores, on the other.

The availability of specific antibodies for the putative components of Ca^{2+} stores and the implementation of cryosection immunogold labeling, have allowed an impressive step forward toward the identification of relevant subcellular structures. In chicken PCs the Ca^{2+} pump has a widespread, although not homogeneous, distribution throughout the smooth ER (27,28).

CS, which by immunoflourescence has been observed in axons, perikarya, and dendrites (fig.1), is localized throughout the smooth ER and heavily concentrated in pleiomorphic vacuoles with a moderately dense core (calciosomes) (27,29). The IP_3-R is concentrated in peculiar stacks of ER cisternae, and present at lower levels in individual smooth ER cisternae (27,28,29), and in some of the CS-rich compartments (28,29); IP_3-R labeling is present in perikarya, dendrites up to dendritic spines and axons down to the synaptic terminals. As to the subcellular distribution of Rya-R, preliminary observations indicate a low-density, homogeneous labeling throughout the smooth ER and lack of labeling of dendritic spines (25,30). Immunogold labeling of membrane subfractions obtained by sucrose gradient centrifugation of chicken cerebellum microsomes, has further shown that only some of the CR-rich vesicles express the IP_3-R (29). Interestingly, heavy subfractions, enriched in [^{3}H]

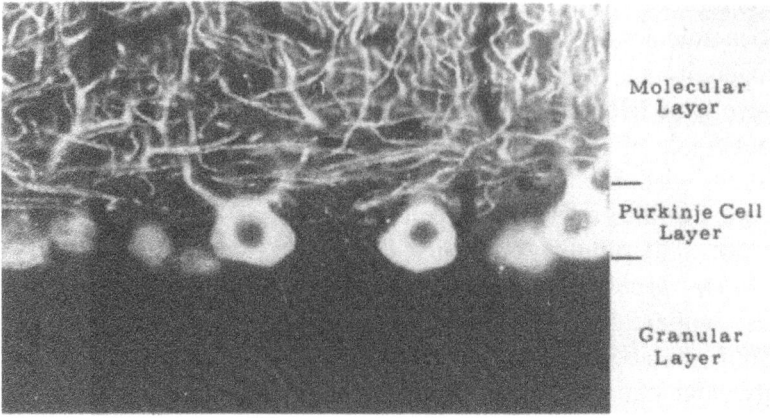

FIG. 1. Immunofluorescence localization of CS in frozen sections of chicken cerebellum cortex. Courtesy of K. Takei and P. De Camilli.

ryanodine binding, contain calciosomes which are not labeled with anti-IP$_3$-R antibodies (29). These observations, taken together, clearly indicate that the smooth-surfaced ER, in spite of its overall uniformity in conventional thin-section electron microscopy, is instead structurally and molecularly heterogeneous. It remains to be ascertained whether IP$_3$-R-rich ER stacks, calciosomes with and without IP$_3$-R, are ER subcompartments or, at least in part, distinct cytological entities. Differences in membrane composition and shape, along with immunoprecipitation techniques, might favor the isolation of purified organelles with specific and diversified functions, a task we are actively pursuing at the present time.

INTRACELLULAR CALCIUM CHANNELS

The IP$_3$-R has been isolated (14,31) and cloned (32,32). The IP$_3$-R is a homotetramer composed of 300 KDa subunits that contain, at the COOH-terminus, multiple transmembrane regions showing fragmentary homology with the skeletal muscle Rya-R (32,33). Scatchard analysis of IP$_3$ binding to the purified receptor yields a K_d of 83 nM and a B_{max} of 2.1 pmol/ug of protein, i.e., one binding site per receptor subunit (31). Expression of mutated IP$_3$-R clones in COS cells has suggested that transmembrane regions of the IP$_3$-R play a role in subunit association but not in ligand binding (34). The IP$_3$ binding site appears to be located within sequences of the large hydrophilic NH$_2$-terminal region protruding into the cytoplasm. However, a monoclonal antibody, reacting with the last 12 amino acid residues of the COOH-terminus, inhibits IP$_3$-induced Ca^{2+} release from cerebellum microsomal fractions and increases the affinity for IP$_3$ (35). This suggests that also the hydrophilic tail of the

COOH-terminus, protruding into the cytoplasm, may be involved in ligand binding and/or regulation of channel opening.

The IP_3-R gene is expressed in a tissue-specific manner. Distinct neuronal and nonneuronal forms differing in phosphorylation and length, and deriving from alternative splicing, have been identified in the rat (36); the long form with a 120 bp insert appears to be exclusively neuronal.

The link between IP_{34} binding and Ca^{2+} release has been clearly established by transfection and stable expression of the cerebellar IP_3-R in a fibroblast L cell line (37). Electrophysiological properties of the IP_3-sensitive channel have been investigated upon reconstitution of the purified cerebellar IP_3-R in lipid bilayers (15). In biionic recording solutions, the most frequent amplitude reveals a slope conductance of 26 pS, whereas in asymmetric NaCl solution, the slope conductance is 21 pS. Thus, IP_3-R behaves, in lipid bilayer at least, as a Na^+ and/or Ca^{2+} permeable channel. The purified channel is activated by micromolar ATP, and is modified to attain a larger conductance state. The ATP binding site has been postulated, by consensus sequence analysis for atypical nucleotide binding site, within the NH_2-terminal cytoplasmic domain (32), and has been revealed by photoaffinity labeling of cerebellar microsomes (15). Scatchard analysis of ATP binding to the purified IP_3-R yields B_{max} and K_d values of 2.3 pmol/ug and 17 μM, respectively. Since B_{max} for ATP and IP_3 are very similar, it seems that there is one ATP binding site per receptor subunit. The physiological role of ATP binding to the IP_3-R has not yet been clarified.

Ca^{2+} itself affects IP_3-induced Ca^{2+} release and IP_3 binding. At first, it appeared that Ca^{2+} was a straight inhibitor of both IP_3-induced Ca^{2+} (38) release and IP_3 binding to the IP_3-R (39). More recently, the Ca^{2+} influence on the IP_3-R has been reinvestigated upon incorporation of microsomal vesicles in planar lipid bilayers (40). The open probability of the channel increases when the free Ca^{2+} varies between 10 nM and 250 nM, i.e., within physiological concentration ranges, but decreases at concentrations above 250 nM. The biphasic Ca^{2+}-dependence of IP_3-gated channel may be biologically relevant, and constitute an important mechanism in Ca^{2+} homeostasis.

Another Ca^{2+} release channel has been identified in neuronal cells of mammalian and avian brain (13,25,29,41): the Ca^{2+}-, caffeine- and ryanodine-sensitive receptors (Rya-R). Striated muscle Rya-R has been isolated, cloned, and functionally expressed (11,12,42). A high M_r protein, binding with high-affinity [^{3}H]ryanodine, has been solubilized from rabbit brain membranes and biochemically characterized (43): the purified protein sediments through sucrose gradients as a large tetrameric complex (about 30 S), like the skeletal and cardiac muscle counterparts, and is immunoprecipitated by polyclonal antibodies specific for the skeletal muscle Rya-R (43). Electrophysiological properties of the brain Rya-R are very similar to those of the muscle isoforms. The brain Rya-R, when reconstituted in lipid bilayers, forms a caffeine- and ryanodine-sensitive Ca^{2+} channel with a 107 pS slope conductance (13). The structural, electrophysiological, and pharmacological similarities between the brain Rya-R and the striated muscle isoforms, suggest a similar role of the channel in Ca^{2+} release from intracellular stores. Thus, the Rya-R acts as a second Ca^{2+} release channel in PCs (13).

Although IP_3 appears to have a major role in the redistribution of Ca^{2+} and regulation of $[Ca^{2+}]_i$ within PCs, physiological and pharmacological evidences indicate the presence of a caffeine-sensitive Ca^{2+} pool that is distinct from the IP_3-sensitive pool (44). The role of both IP_3-sensitive Ca^{2+} stores and caffeine- and ryanodine-sensitive Ca^{2+} stores relies upon the kinetic competence to sustain Ca^{2+} transients *in situ*. Thus, an important experimental question is how fast is Ca^{2+} release mediated by either type of channel, regardless of the possible and mutual interactions which occur inside the cell. With the aid of a stopped-flow apparatus, this issue has been addressed by determining the kinetics of Ca^{2+} release from canine cerebellum microsomes, which are sensitive to both IP_3 and ryanodine. Fig. 2A shows the effect of saturating concentration of IP_3, whereas fig. 2B shows the effect of caffeine (trace a) and of Ca^{2+} jump (trace b). In either case, rat constants of about $100\ s^{-1}$ have been measured, and half-maximal releases have been obtained within 200 msec. These properties indicate that redistribution of Ca^{2+} from intracellular Ca^{2+} stores is not a kinetic constraint to the generation of Ca^{2+} transients *in situ*.

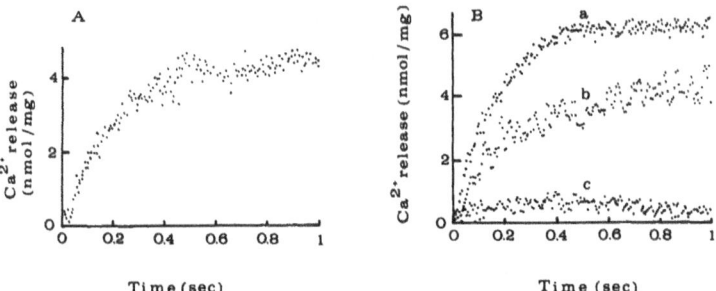

FIG 2. Stopped-flow measurements of calcium release from canine cerebellum microsomes. Panel A: Release evoked by 5 μM IP_3. Panel B: Release evoked by 6 mM caffeine (trace a), 20 μM Ca^{2+} (trace b), and 6 mM caffeine and 4 μM La^{3+} (trace c). For experimental details see reference 41.

REFERENCES

1. Pietrobon D., Di Virgilio F. and Pozzan T. (1990): *Eur. J. Biochem.*, 193: 599-622.
2. Tsien R.W. and Tsien R.J. (1990): *Annu. Rev. Cell Biol.*, 6: 715-760.
3. Volpe P., Pozzan T. and Meldolesi J. (1990): *Sem. Cell Biol.*, 1: 297-304.
4. Fleischer S. and Inui M. (1989): *Annu. Rev. Biophys. Bipohys. Chem.*, 18: 333-364.
5. Lytton J., Zarain-Herzberg A. and MacLennan D.H. (1989): *J. Biol. Chem.*, 264: 7059-7065.
6. Plessers L., Eggermont J.A., Wuitack F. and Casteels R. (1991): *J. Neurosci.*, 11: 650-656.
7. Scott B.T., Simmerman H.K.B., Collins J.H., Nadal-Ginard B. and Jones L.R. (1988): *J. Biol. Chem.*, 263: 8858-8864.
8. Fliegel L., Ohnishi M., Carpenter M.R., Khanna V.H., Reinhart A.F., Reithmeier R.A.F. and MacLennan D.H. (1987): *Proc. Natl. Acad. Sci.*, 84: 1167-1171.
9. Koch G.L.E. (1987): *J. Cell Sci.*, 87: 491-492.
10. Volpe P., Alderson-Lang B.H., Madeddu L., Damiani E., Collins J.H. and Margreth A. (1990): *Neuron,* 5: 713-721.
11. Otsu K., Willard H.F., Khanna V.K., Zorzato F., Green N.M. and MacLennan D.H. (1990): *J. Biol. Chem.*, 265: 13472-13483.
12. Takeshima H., Nishimura S., Matsumoto T., Ishida H., Kangawa K., Minamino N., Matsuo H., Ueda M., Hanaoka M., Hirose T. and Numa S. (1989): *Nature,* 339: 439-445.
13. McPherson P.S., Kim Y.-K., Valdivia H., Knudson C.M., Takekura H., Franzini-Armstrong C., Coronado R. and Campbell K.P. (1991): *Neuron,* 7: 17-25.
14. Supattapone S., Worley P.F., Baraban J.M. and Snyder S.H. (1988): *J. Bio. Chem.*, 263: 1530-2534.
15. Maeda N., Kawasaki T., Nakade S., Yokota N., Taguchi T., Kasai M. and Mikoshiba K. (1991): *J. Biol. Chem.*, 266: 1109-1116.
16. Mayrleitner M., Chadwick C.C., Timerman A.P., Fleischer S. and Schindler H. (1991): *Cell Calcium,* 12: 505-514.
17. Berridge M.J. and Irvine R.F. (1989): *Nature,* 341: 197-205.
18. Streb H., Irvine R.F., Berridge M.J. and Schulz I. (1983): *Nature,* 306: 67-69.
19. Fasolato C., Zottini M., Clementi E., Zacchetti D., Meldolesi J. and Pozzan T. (1991): *J. Biol. Chem.,* (in press).
20. Lipscombe D., Madison D.V., Poenie M., Reuter H., Tsien R.W. and Tsien R.J. (1988): *Neuron,* 1: 355-365.
21. Berridge M.J. (1990): *J. Biol. Chem.*, 265: 9583-9586.

22. Volpe P., Krause K.-H., Hashimoto S., Zorzato F., Pozzan T., Meldolesi J. and Lew D.P. (1988): *Proc. Natl. Acad. Sci.*, 85: 1091-1095.

23. Hashimoto S., Bruno B., Lew D.P., Pozzan T., Volpe P. and Meldolesi J. (1988): *J. Cell Biol.*, 107: 2524-2531.

24. Satoh T., Ross C.A., Villa A., Supattapone S., Pozzan T., Snyder S.H. and Meldolesi J. (1990): *J. Cell Biol.*, 111: 615-624.

25. Ellisman M.H., Deerinck T.J., Ouyang Y., Beck C.F., Tanksley S.J., Walton P.D., Airey J.A. and Sutko J.L. (1990): *Neuron*, 5: 135-146.

26. Kaprielian Z., Campbell A.M. and Fambrough D.M. (1989): *Mol. Brain Res.*, 6: 55-60.

27. Villa A., Podini P, Clegg D.O., Pozzan T. and Meldolesi J (1991): *J. Cell Biol.*, 113: 779-791.

28. Takei K., Stukenbrook H., Metcalf A., Mignery G.A., Sudhof T.C., Volpe P. and De Camilli P. (1992): *J. Neurosci.*, 12: 488-505.

29. Volpe P., Villa A., Damiani E., Sharp A.H., Podini P., Snyder S.H. and Meldolesi J. (1991): *EMBO J.*, 10: 3183-3185.

30. Walton P.D., Airey J.A., Sutko J.L., Beck C.F., Mignery G.A., Sudhof T.C., Deerinck T.J. and Ellisman M.H. (1991): *J. Cell Biol.*, 112: 1145-1157.

31. Maeda N., Niinobe M. and Mikoshiba K. (1990): *EMBO J.*, 9: 61-67.

32. Furuichi T., Yoshikawa S., Miyakawa A., Wada K., Maeda N. and Mikoshiba K. (1989): *Nature*, 342: 32-38.

33. Mignery G.A., Newton C.L., Archer B.T. and Sudhof T.C. (1990): *J. Biol. Chem.*, 265: 12679-12685.

34. Mignery G.A. and Sudhof T.C. (1990): *EMBO J.*, 9: 3883-3898.

35. Nakade S., Maeda N. and Mikoshiba K. (1991): *Biochem. J.*, 277: 125-131.

36. Danhoff S.K., Ferris C.A., Donath C., Fischer G., Munemitsu S., Ullrich L.A., Snyder S.H. and Ross C.A. (1991): *Proc. Natl. Acad. Sci.*, 88: 2951-2955.

37. Miyakawi A., Furuichi T., Maeda N. and Mikoshiba K. (1990): *Neuron*, 5: 11-18.

38. Joseph S.K. and Williamson J.R. (1989): *Arch. Biochem. Biophys.*, 273: 1-15.

39. Worley P.F., Baraban J.M., Supattapone S., Wilson V.F. and Snyder S.H. (1987): *J. Biol. Chem.*, 262: 12132-12136.

40. Bezprozvanny I., Watras J. and Ehrlich B.E. (1991): *Nature*, 351: 751-754.

41. Meszaros L.G. and Volpe P. (1991): *Am. J. Physiol.*, 261: C1048-C1054.

42. Penner R., Neher E., Takeshima H., Nishimura S. and Numa S (1989): *FEBS Lett.*, 259: 217-225.

43.   McPherson P.S. and Campbell K.P. (1990): *J. Biol. Chem.*, 265-18454-18460.
44.   Brorson J.R., Bleakman D., Gibbons S.J. and Miller R.J. (1991): *J. Neurosci.*, (in press).

CALCIUM CHANNEL SIGNALING IN ASTROCYTES

Leif Hertz and William E. Code

Departments of Pharmacology and Anaesthesia
University of Saskatchewan
Saskatoon, Saskatchewan
Canada S7N 0W0

It has been known for more than a century that the brain contains not only neurons but also non-neuronal cells. Most of the non-neuronal cells belong to the two glial cell types, oligodendrocytes and astrocytes. In many species, including man, they vastly outnumber neurons (1). At the beginning of this century several attempts were made to study the role(s) of glial cells. Many theories were suggested but little solid evidence was obtained, leading Ramon and Cajal (2) to the visionary conclusion that the functions of glial cells were unknown and would remain so for a long time to come, because no methodologies were available to study these functions. With the remarkable development in electrophysiological techniques during the following decades, neuronal characteristics were elucidated in ever greater detail, whereas the possibility of a major role of glial cells in brain function was more or less disregarded. This situation began to change in the 1950s. With the aid of histological techniques it was shown that oligodendrocytes synthesize myelin and that radial glial cells (immature astrocytes) serve a guiding role during neuronal migration. Purified samples of either neurons or glial cells were first obtained by Hyden (3), using microdissection; subsequently, neurons and different types of glial cells were prepared by centrifugation or culturing techniques. Another development of decisive importance for the interest in glial cell studies was the development of microelectrodes which made it possible to demonstrate huge alterations in extracellular concentrations of cations (primarily potassium and calcium) during both normal function of the CNS and pathological states, e.g., seizures, hypoglycemia, and ischemia. This new knowledge paved the way for the concept that neurons and astrocytes share a common extracellular space, sheltered behind the bloodbrain barrier and, in conjunction, regulate the local concentration of neuroactive compounds like potassium ions and glutamate. Finally, the development of patch-clamp techniques to study ion channels and of imaging techniques to determine intracellular concentrations (activities) of calcium (4) and, more recently, of sodium and potassium in visually selected individual cells and/or microareas of cells, are as well suited to

T. Godfraind et al. (eds.), Calcium Antagonists, 205–213.
© 1993 *Kluwer Academic Publishers and Fondazione Giovanni Lorenzini.*

measure characteristics of glial cells as of neurons.

Most information pertains to astrocytes; the astrocyte is the predominant glial cell in grey matter, although it is also found in white matter. Recently, a distinction has been made between "type-1" and "type-2" astrocytes, which in many respects behave differently. The astrocytes in the brain cortex are "type-1" astrocytes, which here will be referred to simply as "astrocytes."

VOLTAGE-REGULATED CALCIUM CHANNELS

Astrocytes are exquisitely well suited to react to voltage-affecting signals due to their high potassium conductance (5) which allows them to function as perfect potassium electrodes. They possess a multitude of voltage-dependent channels, including calcium channels (6). The voltage-sensitive L-channel, which is selectively inhibited by low concentrations of dihydropyridines (7), is present in primary cultures of astrocytes, but only if the cells have been treated with dibutyryl cyclic AMP (dBcAMP) or exposed to similar differentiating procedures. This difference between dBcAMP-treated and nontreated cells is also obvious from binding characteristics for the dihydropyridine nitrendipine which in both treated and untreated cells binds to a low-affinity site but only in the treated cells to an additional high-affinity binding site (fig. 1). In the following, only dBcAMP-treated astrocytes will be discussed unless otherwise indicated. In such cultures, uptake studies, using ^{45}Ca, have confirmed a voltage-dependent calcium uptake which is inhibited by nimodipine with an IC_{50} of 2.4 nM (8), i.e., a concentration quite similar to the K_D value for the high affinity binding of nitrendipine.

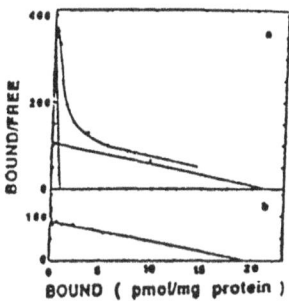

FIG. 1. Skatchard plot of nitrendipine binding to dBcAMP-treated (a) and nontreated (b) cultured astrocytes. Note the two components [high-affinity uptake (B_{max} 0.9 pmol/mg; K_D 2.5 nM) and low-affinity uptake (B_{max} 22.4 pmol/mg; K_D 209 nM)] in (a) versus only one component (low-affinity uptake) in (b).

FREE INTRACELLULAR CALCIUM IN ASTROCYTES

The correlation between extracellular potassium concentration and free intracellular calcium concentrations in primary cultures of astrocytes, measured by the aid of the fluorescent drug Indo-1, is shown together with the almost complete inhibition of the voltage-induced increase in free intracellular calcium concentration by a submicromolar concentration of nifedipine in fig. 2. These observations suggest that the major source for the increased intracellular calcium concentration is calcium accumulation via an L-channel.

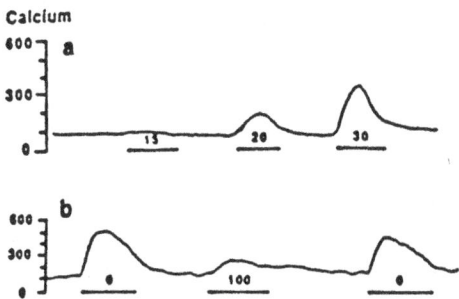

FIG. 2. Free intracellular calcium concentration (nM) in cultured astrocytes exposed to control medium (5 mM potassium) and to 15, 20, or 30 mM of potassium during the intervals indicated (a) or to 20 mM potassium in the absence (0) and presence of 100 nM nifedipine.

An interaction has been suspected between the L-channel and the "peripheral-type" (9), including the astrocytic (10), benzodiazepine receptor. Benzodiazepine concentrations in the micromolar range may inhibit uptake of ^{45}Ca into primary cultures of astrocytes (11). In contrast, as can be seen from fig. 3, nanomolar levels of the benzodiazepine midazolam [which has approximately similar affinity for neuronal and astrocytic benzodiazepine receptors (12)], cause an increase of the effect of 20 mM potassium, i.e., a submaximum potassium concentration (fig. 2). They have, however, no effect at 50 mM potassium which, by itself, causes a maximum increase in free intracellular calcium concentration. This effect, which occurs at a pharmacologically relevant drug level, is abolished by the "peripheral-type" benzodiazepine antagonist PK-11195 (13) which, even at the high concentration of 1 μM, selectively inhibited the enhancement of the potassium-induced increase in free intracellular calcium concentration without touching the effect of the elevated potassium concentration. This mode of action is different from that of dihydropyridines which abolish both the response to the depolarization as such and its enhancement by the benzodiazepine (13).

The response to benzodiazepines described above is not limited to midazolam but probably universal for benzodiazepines showing some affinity for the "peripheral-type" benzodiazepine receptor site, i.e., virtually all clinically used benzodiazepines. This includes the "prototype" benzodiazepine diazepam (Valium), although peak effect is seen at a somewhat higher concentration (around 0.5 μM). However, since the benzodiazepine effect can be regarded as a modulation of L-channel activity, it must be critically dependent upon the expression of L-channels in the astrocytes. It is in agreement with this conclusion that H.S. White (personal communication), using astrocytes which had not been pretreated with dBcAMP did not find any potentiation of potassium effect on free intracellular calcium concentration when the cultures were exposed to diazepam. They did observe a response to potassium in these cells, but this increase in free intracellular calcium concentration was abolished by omega-conotoxin, suggesting the involvement of an N-channel (14). Since these cultures in all other respects were similar to those used in the present work and since the potassium-induced increase in free intracellular calcium concentration in treated cells is abolished by nifedipine, it is possible that the astrocytic N-channel disappears simultaneously with the appearance of L-channel activity.

TRANSMITTER-OPERATED CALCIUM CHANNELS IN ASTROCYTES

Astrocytes are endowed with receptors for a multitude of transmitters (including noradrenaline, serotonin, glutamate, adenosine, and ATP), many of which affect the free intracellular calcium concentration (15-17). They vary between individual astrocytes and probably also on the same cell at different times (K.D. McCarthy, personal communication). Many transmitters also lead to an increase in phosphoinositide turnover, indicating an increased formation of inositol triphosphate (IP_3) and diacylglycerol (DAG), which in turn leads to stimulation of protein kinase C (PKC) and release of intracellularly bound calcium. Although this aspect of calcium signaling also appears to play a major role in astrocytic functions, it will not be dealt with here. However, administration of ATP leads not only to an increase in free intracellular calcium concentration (16) but also to increased uptake of ^{45}Ca into primary cultures of astrocytes (18), suggesting an effect on transmitter-operated calcium channels. Also, the putative neurotransmitter histamine increases free intracellular calcium concentration in astrocytes both *via* a transmitter-operated channel and *via* the phosphoinoside second messenger system (19). In contrast, glutamate which potently increases free intracellular calcium concentration in astrocytes, does not do so in the absence of extracellular calcium (Z. Zhao, L. Peng, and L. Hertz, unpublished experiments).

CALCIUM CHANNEL SIGNALING AS A REGULATOR OF ENERGY METABOLISM

Calcium signaling is ubiquitous, involved in cell development and in normal cell function as well as in many aspects of abnormal function. These include at least some aspects of metabolic regulation. Thus, glycogenolysis in brain slices is increased in the presence of elevated potassium concentrations in a calcium dependent manner (20). Since both glycogen and the glycogenolytic enzyme phosphorylase normally are located in astrocytes but not in neurons (21,22), the glycogenolytic effect must ultimately be exerted on astrocytes although it cannot *a priori* be excluded that the elevated potassium concentration might cause neuronal release of a compound (e.g., a transmitter) which, in turn stimulates glycogenolysis in adjacent astrocytes. In order to distinguish between a direct potassium effect on astrocytes versus a neuronally mediated effect, we have studied potassium effects on primary cultures of astrocytes (containing no neurons). From fig. 4 it can be seen that elevated potassium concentrations do evoke glycogenolysis in homogenous astrocytic cultures (23), but only provided the cultures have been treated with dBcAMP, i.e., express L-channels. Moreover, the effect is inhibited by low concentrations of dihydropyridine L-channel blockers (K.V. Subbarao and L. Hertz, unpublished experiments). This finding strongly suggests that L-channel mediated calcium signaling has a regulatory effect on astrocytic metabolism. This should not be taken as an indication that no other factors regulate glycogenolysis in astrocytes since transmitters such as noradrenaline (fig. 4) or vasoactive intestinal peptide also enhance glycogenolysis (24). There is no additive effect of maximally effective concentrations of noradrenaline and potassium, and adrenergic antagonists inhibit the noradrenergic but not the potassium-induced stimulation of glycogenolysis in primary cultures of astrocytes (23). Other aspects of glucose metabolism in astrocytes, but not in neurons, are also enhanced by elevated concentrations of extracellular potassium, e.g., pyruvate carboxylation (25), but it has not been investigated whether voltage-dependent calcium channels are involved in these responses.

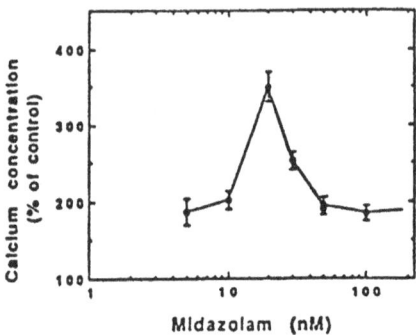

FIG. 3. Dose-response curve for midazolam effect on potassium-induced increase in free intracellular calcium evoked by 20 mM potassium in cultured astrocytes.

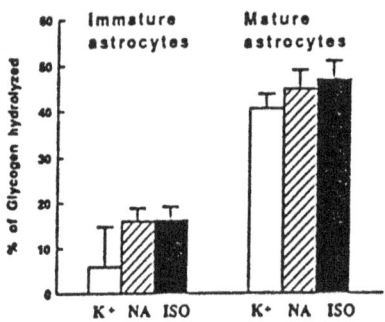

FIG. 4. Effect of elevation of extracellular potassium concentration to 50 mM (K^+), noradrenaline 25 μM (NA) and isoproterenol 25 μM (ISO) on glycogenolysis in immature (non dBcAMP-treated) or mature (dBcAMP-treated) astrocytes.

The important difference between dBcAMP-treated and nontreated cultures with respect to metabolic regulation raises the question whether dBcAMP-treated astrocyte cultures or nontreated cultures more closely resemble their *in vivo* counterparts. The finding that elevated potassium concentrations enhance glycogenolysis (an astrocytic phenomenon) in brain slices from adult rats (20) and mice, but not in slices from new-born mice (K.V. Subbarao and L. Hertz, unpublished experiments) strongly suggests that the treated cultures mimic their differentiated *in vivo* counterparts whereas untreated cultures display characteristics of immature animals. This is in keeping with the concept that dBcAMP treatment of cultures, which routinely are prepared from new-born mouse or rat pups, substitutes for a neuronal signal (probably from noradrenergic locus coeruleus fibers) reaching the cerebral hemispheres postnatally during normal *in vivo* development, but obviously not in homogenous cultures of astrocytes (26,27).

OTHER REGULATORY EFFECTS OF CALCIUM SIGNALING

It is highly unlikely that signaling via calcium channels should not effect other parameters than energy metabolism. Quandt and MacVicar (28) have reported that potassium channel activity in primary cultures of astrocytes is enhanced as a result of L-channel activation. It should, however, be noted that the potassium channel conductance is extremely low in their preparation, i.e., only one tenth of the potassium channel activity observed in mouse astrocytes by Berwald-Netter and coworkers (29). Since a high potassium conductance is a key characteristic for astrocytes (5) the correlation between L-channel activity and potassium channel activity should be reinvestigated before any firm conclusions can be made.

CONCLUDING REMARKS

It is rapidly becoming obvious that channel-operated calcium signaling is of functional importance in astrocytes. This may especially be the case for voltage-regulated L-channels which have been found to be of remarkably low activity (or even absent) in several neuronal preparations, although L-channel activity is quite pronounced in whole brain preparations (3). Thus, effects of drugs interacting with channel-mediated signaling in brain may to a large extent be primarily exerted on astrocytes and only secondarily affect neuronal output as a result of neuronal astrocytic-interactions (e.g., 31,32) in brain metabolism and function.

ACKNOWLEDGEMENTS

Mrs. E. Habbick is thanked for typing this manuscript and Mr. Z. Zhong for help in preparing figures. Financial support by Saskatchewan Health Research Board to W.E.C. and the Medical Research Council of Canada to L.H. is gratefully acknowledged.

REFERENCES

1.   Pope A. (1978): In: *Dynamic Properties of Glia Cells,* edited by E. Schoffeniels, G. Franck, L. Hertz and D.B. Tower pp. 13-20. Pergamon, Oxford.
2.   Ramon Y. and Cajal S. (1909): *Histologie du Systeme Nerveux de l'Homme et des Vertebres* vol. 1, p. 246. Paris.

3. Hyden H. (1959): *Nature,* 184: 433-435.
4. Tsien R.W. (1988): *TINS,* 11: 419-424.
5. Walz W. (1989): *Prog. Neurobiol.,* 33: 309-333.
6. Barres B.A., Chun L.L.Y. and Corey D.P. (1990): *Annu. Rev. Neurosci.,* 13: 441-474.
7. Janis R.A., Silver P.J. and Triggle D.J. (1987): *Adv. Drug Res.,* 16: 309-591.
8. Hertz L., Bender A.S., Woodbury D. and White S.A. (1989): *J. Neurosci. Res.,* 22: 209-215.
9. Cantor E.H., Kennessey A., Semenuk G. and Spector S. (1984): *Proc. Natl. Acad. Sci. USA,* 81: 1549-1552.
10. Bender A. and Hertz L. (1985): *Eur. J. Pharmacol.,* 110: 287-288.
11. Code W.E. and Hertz L. (1989): *Can. J. Anaesth.,* 36: S151.
12. Bender A.S. and Hertz L. (1987): *J. Neurosci. Res.,* 18: 366-372.
13. Code W.E., White H.S. and Hertz L. (1991): *N.Y. Acad. Sci.,* 625: 430-432.
14. White H.S., Skeen G., Howell M.C. and Litzinger M.A. (1991): *Trans. Am. Soc. Neurochem.,* 22(1): 101.
15. Enkvist M.O., Holopainen I. and Akerman K.E. (1989): *Glia,* 2: 397-402.
16. Salm A.K., Lerea L., Castros H. and McCarthy K.D. (1990): In: *Differentiation and Functions of Glial Cells,* edited by G. Levi pp. 275-288. Alan R. Liss, New York.
17. Cornell-Bell A.H., Finkbeiner S.M., Cooper M.S. and Smith S.J. (1990): *Science,* 247: 470-473.
18. Neary J.T., Van Breemen C., Laskey R., Blicharska, Norenberg L.-O.B. and Norenberg M.D. (1991): *Ann. N.Y. Acad. Sci., USA,* 603: 473-475.
19. Arbones L., Picatoste F. and Garcia (1990): *Mol. Pharmacol.,* 37: 921-927.
20. Hof P.R., Pascale E. and Magistretti P.J. (1988): *J. Neurosci.,* 8: 1922-1928.
21. Ibrahim M.Z.M. (1975): *Adv. Anat. Cell Biol.,* 52: 5-89.
22. Pfeiffer B., Elmer K., Roggendorf W., Reinhart P.H. and Hamprecht B. (1990): *Histochemistry,* 94: 73-80.
23. Subbarao K. and Hertz L. (1990): *Brain Res.,* 536: 220-226.
24. Magistretti P.J. (1988): *Diabete & Metabolisme,* 14: 237-246.
25. Kaufman E.E. and Driscoll (1990): *Trans. Amer. Soc. Neurochem.,* 21: 289.
26. Hertz L. (1990): In: *Molecular Aspects of Development and Aging in the Nervous System,* edited by A. Privat, E. Giacobini, P. Timiras and A. Vernadakis pp. 227-243, Plenum, New York.
27. Meier E., Hertz L. and Schousboe A. (1990): *Neurochem. Int.,* 19: 1-15.

28.   Quandt, F.N. and MacVicar B.A. (1986): *Neuroscience,* 19: 29-41.
29.   Nowak L., Ascher P. and Berwald-Netter Y. (1987): *J. Neurosci,* 7: 101-109.
30.   Miller R.J. (1987): *Science,* 235: 46-52.
31.   Hertz L. (1989); In: *Regulatory Mechanisms of Neurons to Vessel Communication in Brain,* edited by S. Govoni, F. Battaini and M.S. Mangoni pp. 271-305. Springer, Heidelberg.
32.   Hertz L., Code W.E., Shokeir O., Shargool M., Woodbury D.M. and White M.S.: In: *Neuroglia,* edited by A. Roitbak (in press). Tbilisi, Georgia.

STUDIES ON SYNAPTOSOMES:
NEW INSIGHTS ON CALCIUM ANTAGONISTS

Stanley L. Cohan, Rena Getz, and Mei Chen

Georgetown University School of Medicine
Washington, D.C., 20007

The use of cerebral synaptosomes permits investigation of certain aspects and consequences of calcium metabolism by the brain not readily accomplished in other biological samples. Since alterations in brain calcium homeostasis appear to play a critical role in cell damage following anoxic/ischemic insult (1,2) use of synaptosomes may provide an important investigative tool for the *in vitro* study of this aspect of brain anoxia and ischemia. Neuronal cell culture permits study of some mechanisms of calcium toxicity as well as the neurotoxic potential of excitatory amino acids (EAA) in ischemia, but does not readily permit study of antecedent presynaptic events. Microdialysis techniques are able to monitor EAA release following ischemia (3), but the poor time resolution of the measurements precludes analysis of the time course or kinetics of EAA release, nor can the cell of origin of the compounds be readily discerned.

SYNAPTOSOMES AS *IN VITRO* MODELS

Ultrastructural studies demonstrate that important changes in calcium metabolism appear to take place in the presynaptic compartment, particularly in the first few hours following ischemia (4) and, although able to demonstrate the time course of both presynaptic and postsynaptic calcium accumulation following brain ischemia, these methods do not readily permit quantitative measurement. Use of cerebral synaptosomes permits investigation of mechanisms regulating early postischemic presynaptic calcium accumulation relatively free of the contaminating influences of the postsynaptic and glial intracellular milieu found in other preparations. In as much as presynaptic calcium uptake following ischemia may in large measure be responsible for the excessive release of EAA neurotransmitters from the presynaptic compartment (5,6), use of synaptosomes also permits the quantitative and kinetic study of the interrelationship between calcium uptake and EAA release in the same preparation. The study of mechanisms regulating EAA release is of great interest since these neurotransmitters appear to mediate postsynaptic

215

T. Godfraind et al. (eds.), Calcium Antagonists, 215–220.
© 1993 *Kluwer Academic Publishers and Fondazione Giovanni Lorenzini.*

calcium-induced cell damage (7,8). Additionally, the presynaptic compartment may be an important therapeutic target and provide an opportunity for pharmacologic agents to block the toxic effects of calcium in the presynaptic compartment by inhibiting anoxia/ischemia-induced, voltage-regulated excessive presynaptic calcium uptake and thereby also block excessive calcium uptake-dependent EAA release into the synaptic cleft. Thus, synaptosomes may be used to study the relation of the synaptic EAA release by voltage-dependent presynaptic calcium uptake as well as a test system to assess the potential efficacy of therapeutic agents in blocking both presynaptic calcium uptake and EAA release in *in vitro* models of brain anoxia/ischemia.

Postischemic Presynaptic Calcium Accumulation

Synaptosomes provide an *in vitro* model which is suitable for addressing important questions about calcium metabolism during ischemia. Depolarization follows cerebral anoxia or ischemia and is primarily the result of increasing extracellular potassium (9,10). Depolarization of synaptosomes with increased concentration of potassium mimics this event and permits study of presynaptic voltage-regulated calcium channels and the effect of pharmacologic agents on ischemia-induced presynaptic calcium accumulation. We have measured cerebral cortical synaptosomal calcium accumulation in the gerbil, using prolonged depolarization (5-60 mm K^+), using the fluorescent calcium indicator FURA 2 (11) to quantify synaptosomal calcium concentration, $[Ca]_{is}$. There is a continuous and progressive increase in $[Ca]_{is}$ that reaches approximately 1200 nM 45 minutes after depolarization with 60 mM potassium. This calcium accumulation is dependent upon influx from the extracellular space and not from mobilization of intracellular stores since depolarization-induced increase in $[Ca]_{is}$ is completely blocked by use of calcium-free buffers (12). Calcium does not enter via sodium channels, since use of the sodium-channel inhibitor tetrodotoxin, even in sodium-free buffer, has no effect on depolarization-induced $[Ca]_{is}$ (12).

The Effect of Calcium Antagonists

The L-type channel blockers, nimodipine, nifedipine, and verapamil, have no effect on $[Ca]_{is}$, nor did the T-channel blocker, nickel (12). Flunarizine, a type IV calcium antagonist, was a potent blocker of depolarization-induced presynaptic calcium influx (table 1). with an Ed_{50} of $10^{-8}M$ (12). The N-channel blocker cadmium, also blocked depolarization-induced calcium accumulation, as did the N-channel blocker omega-conotoxin (13), but to a far lesser extent than either flunarizine or cadmium (table 1). These results suggest that presynaptic voltage-regulated calcium uptake occurs at least in part through N-type calcium channels, although it is also evident that voltage-dependent presynaptic calcium uptake also occurs through channels other than the L, N, and T types described in studies utilizing single channel recordings (14,15).

TABLE 1.   Effect of flunarizine, cadmium,
        omega conotoxin and nimodipine on depolarization
        induced (K+) synaptosomal calcium ([Ca]$_{is}$).

Duration of Incubation	1 min.		45 min.	
[K+]	5 mM	60 mM	5 mM	60 mM
Control	214 ± 37	852 ± 64	350 ± 68	1246 ± 212
Flunarizine 50 nM	218 ± 48	*375 ± 62	306 ± 40	#350 ± 76
Cadmium 50 uM	168 ± 37	*340 ± 68	318 ± 70	#532 ± 72
ω-conotoxin 1 uM	237 ± 38	ᶦ498 ± 77	344 ± 52	*870 ± 112
Nimodipine 1 uM	218 ± 38	*546 ± 68	621 ± 104	1190 ± 124

[Ca]$_{is}$ in nM ± SD. Each value is mean of 12 experiments *P < .01, ᶦP < .005, and #P < .001 when compared to control for significance.

Postischemic Synaptic EAA Release

Cerebral synaptosomes may be used to measure calcium-dependent EAA release as a means of studying postischemic synaptic release of EAA, particularly glutamate. The addition of glutamic acid dehydro-genase and NADP to the extrasynaptosomal medium permits continuous monitoring of glutamate concentrations in the extrasynaptosomal space by fluorospectrophotometric measurement of NADPH production by the enzyme (16,17). Because glutamate concentration can be obtained every 2 seconds, the time course of glutamate release following sustained strong depolarization, such as that which occurs during anoxic/ischemic insult (9,10), can be studied in ways not possible by microdialysis sampling of whole brain following *in vivo* ischemia (5). Not only may the kinetic properties of EAA release be studied, but the effect of calcium antagonists on these properties of EAA release can be studied as well. Our results demonstrate that flunarizine, cadmium, and omega-conotoxin all inhibit calcium-dependent glutamate release following depolarization but to a varying degree (13). Fig. 1 demonstrates glutamate release occurring at two rates, the more rapid studied, but the effect of calcium antagonists on these properties of EAA release can be studied as well. Our results demonstrate that flunarizine, cadmium, and omega-conotoxin all inhibit calcium-dependent glutamate release following depolarization but to a varying degree (13). Fig. 1 demonstrates glutamate release occurring at two rates, the more rapid being calcium uptake-dependent which cannot be demonstrated when using calcium-free buffers and similar to that previously described (18).

The calcium-dependent rate of glutamate release is impaired by flunarizine, omega-conotoxin and cadmium but note should be made of the differences in the effective concentration of these compounds. Nimodipine, nifedipine, verapamil and nickel have no effect on calcium-dependent glutamate release (data not shown). The slow rate is a calcium-independent glutamate leak, which has been previously described (17,19) and is unaffected by these calcium antagonists.

The Effect of Flunarizine, Cadmium and Omega Conotoxin
on Ca^{2+} - Dependant Glutamate Release from Synaptosomes

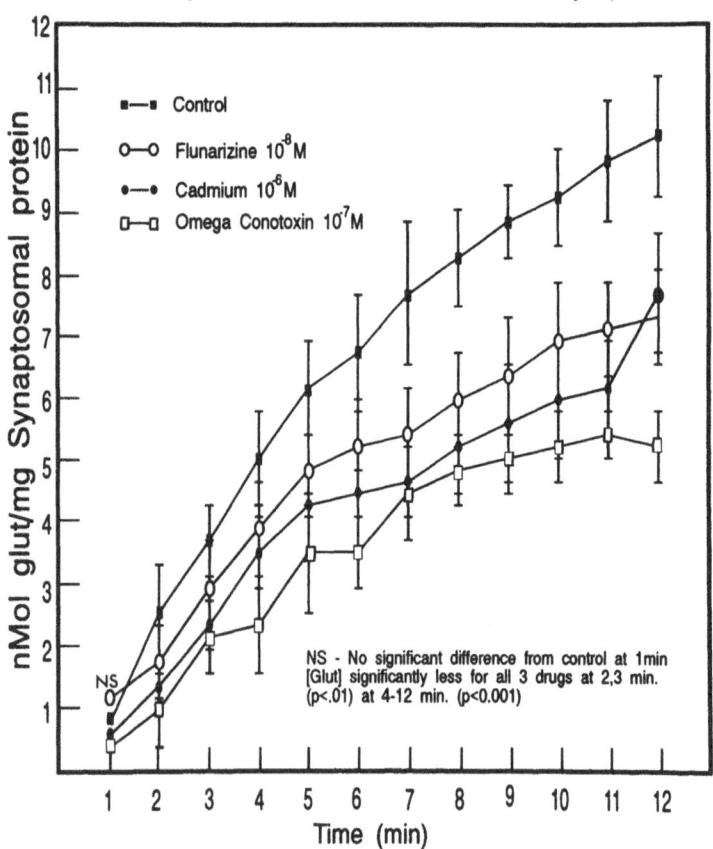

FIG. 1

SUMMARY

Our results suggest that N-type calcium channels are an important, but not sole, venue of voltage-regulated presynaptic calcium uptake following prolonged strong depolarization such as that accompanying ischemia. This is demonstrated in part by the difference between omega-conotoxin and flunarizine inhibition of $[Ca]_{is}$ increases (table 1). Although it appears that presynaptic calcium entry via N-type channels is important in the regulation of neurotransmitter release, a significant degree of voltage-dependent presynaptic calcium accumulation occurs via voltage sensitive channels other than L-type, omega conotoxinsensitive N-type or T-type calcium channels, which are probably not coupled to neurotransmitter release. Flunarizine has demonstrated cytoprotective effects in studies of cerebral ischemia (20-22) and it has been assumed that these protective effects are the direct result of its action as a calcium antagonist. Although this is probably true to a great extent, the cytoxic effects of EAA release may also be diminished by flunarizine and thus provide a second cytoprotective mechanism. The ability of a pharmacologic agent such as flunarizine to block the synaptic release of glutamate and probably other EAAs, provides a potential alternative to pharmacologic agents which block calcium influx without affecting EAA release or which block postsynaptic ion channel-regulating toxic side effects *in vivo* that may ultimately preclude their use as therapeutic agents, particularly in stroke. Thus, synaptosomes may be used to study factors regulating the homeostasis of calcium and EAAs, each of which may play critical roles in the evolution of cell death following anoxia and ischemia, and may be used to assess the affect pharmacologic agents have on calcium and EAA metabolism, and their potential use as therapeutic agents in stroke.

REFERENCES

1. Farber J.L. (1981): *Life Sci.*, 29: 1289-1295.
2. Siesjo B.K. (1981): *J. Cereb. Blood Flow Metab.*, 1: 155-185.
3. Benveniste H., Drejer J., Schousboe A. and Diemer N.H. (1984): *J. Neurosci.*, 43: 1369-1374.
4. Van Reempts J. and Borgers M. (1985): *Ann. Emerg. Med.*, 14: 736-742.
5. Nicholls D.G. (1989): *J. Neurochem.*, 52: 331-341.
6. McMahon H.T. and Nicholls G. (1991): *J. Neurochem*, 56: 86-94.
7. Rothman S. (1984): *J. Neurosci.*, 4: 1885-1891.
8. Choi D.W. (1987): *J. Neurosci.*, 7: 369-379.
9. Harris R.J., Symon L., Branston N.M. and Bayhan M. (1981): *J. Cereb. Blood Flow Metab.*, 1: 203-209.
10. Holler M., Dierking H., Dengler K., Tegtmeyer F. and Peters T. (1986): In: *Acute Brain Ischemia.*, edited by N. Battistine, R. Coubier, P. Fiorani, F. Plum and C. Fieschi pp. 229-233. Raven Press, New York.
11. Verhage M., Besselsen E., Lopes da Silva F.H. and Ghijsen W.E.J.M. (1988): *J. Neurochem.*, 51: 1667-1674.
12. Stanley E.F., Nowycky M.C. and Triggle D.J. (1991): *Ann. N.Y. Acad. Sci.*, 635: 397-399
13. Cohan S.L., Getz R. and Chen M. (1991): *Ann. Neurol.*, 30: 235.
14. Fox A.P., Nowycky M.C. and Tsien R.W. (1987): *J. Physiol.*, 394: 149-172.
15. Fox A.P., Nowycky M.C. and Tsien R.W. (1987): *J. Physiol.*, 394: 173-200.
16. Nicholls D.G. and Sihra T.S. (1986): *Nature*, 321: 772-773.
17. Nicholls D.G., Sihra T.S. and Sanchez-Prieto J. (1987): *J. Neurochem.*, 49: 50-57.
18. McMahon H.T. and Nicholls D.G. (1990): *J. Neurochem.*, 54: 373-380.
19. Pockock J.M., Murphie H.M. and Nicholls D.G. (1988): *J. Neurochem.*, 50: 745-751.
20. Van Reempts J., Borgers M., Van Dael L., Van Eyndhoven L. and Van de Ven M. (1983): *Arch. Int. Pharmacodyn. Ther.*, 262: 76-88.
21. Deshpande J.K. and Wieloch T. (1986): *Anesthesiol.*, 64: 215-224.
22. Alps B.J., Calder C., Hass W.K. and Wilson A.D. (1988): *Br. J. Pharmacol.*, 93: 877-883.

THE REGULATION OF NEURONAL CALCIUM CHANNELS

R. Bangalore, J. Ferrante*, M. Hawthorn, W. Zheng,
A. Rutledge, M. Gopalakrishnan**, and D.J. Triggle

School of Pharmacy, State University of New York
at Buffalo, Buffalo, New York 14260

*Laboratory of Biochemical Genetics,
National Institutes of Health, Bethesda, Maryland 20892

**Department of Physiology, Baylor College
of Medicine, Houston, Texas 77030

Voltage-gated Ca^{2+} channels may be regarded as pharmacologic receptors. As such, they possess the following general properties:

1. Drug binding sites for activators and antagonists with defined structure-activity relationships including stereoselectivity.
2. Drug binding sites linked to channel permeation and gating machinery.
3. Association with G proteins.
4. Regulated by homologous and heterologous factors.
5. Expression and function altered during experimental and clinical disease states.

Voltage-gated Ca^{2+} channels constitute a major family with at least four classes, L,T,N, and P (1) and there likely exists within each class, several subclasses derived from the products of separate genes and through alternative splicing. Each of the channel classes has its own distinct pharmacology, but it remains to be determined how closely channel classifications derived from pharmacological and molecular biologic classifications will agree. In any event, it is the L class of channel that has been the best characterized by pharmacological criteria since it carries binding and functional sites for the first generation Ca^{2+} channel antagonists of the 1,4-dihydropyridine (nifedipine and Bay K 8644), phenylalkylamine (verapamil), and benzothiazepine (diltiazem) classes. These drugs have achieved major prominence both as therapeutic agents and as molecular probes for the channel (2). Regulatory studies of Ca^{2+} channels have thus far principally focused on the L class of channel (reviewed in 3).

T. Godfraind et al. (eds.), Calcium Antagonists, 221–229.
© 1993 Kluwer Academic Publishers and Fondazione Giovanni Lorenzini.

CALCIUM CHANNEL REGULATION

Calcium channels are regulated species. They are regulated by chronic drug administration, by chronic activation or antagonism, by hormone influence, by neurotoxins including lead and ethanol, by pathology including ischemia, cardiomyopathy, autoimmune disorders, and hypertension, and they are altered in number and function during aging. These studies have been reviewed recently by Ferrante and Triggle (3) and provide clear evidence that at least the L class of voltage-gated Ca^{2+} channel behaves as a pharmacologic receptor.

Regulation of Neuronal Calcium Channels

Our interest in the regulation of Ca^{2+} channels has focused on the effects of chronic activation and the antagonism, on the effects of neurotoxins and on the effects of aging. We have found conditions where significant changes in channel number and function occur and which suggest that these may be of physiological and pathological significance.

Chronic activation and antagonism.

The work of DeLorme and McGee (4) and DeLorme et al. (5) showed that chronic depolarization of PC12 cells down-regulated the 1,4-dihydropyridine-sensitive Ca^{2+} channels. Depolarization is both a physiologic and pathologic signal for Ca^{2+} channels and it is of interest that in the same cell line chronic channel occupancy by nifedipine or Bay K 8644 produces a corresponding up- and down-regulation of channel numbers and function (6).

Extension of these studies to chick retinal neurons reveals that the 1,4-dihydropyridine-sensitive Ca^{2+} channels are similarly down-regulated by chronic depolarization in dose-dependent fashion (1), that this is a Ca^{2+}-dependent process since it is mimicked by the ionophore A23187, and that Ca^{2+} entry through the L class of channels is involved since the down-regulation is protected by co-treatment with D600 (7). The specific mechanisms of these depolarization-induced events are not established but activation of immediate early genes is important to the ability of cells, particularly neurons, to respond to signals by long-term modulation. The proto-oncogene c-fos is activated by membrane depolarization and may serve to initiate gene expression programs (8). It is of interest, however, that chronic depolarization of cardiac myocytes fails to regulate the channels although the pharmacologic behavior of the channels in both preparations is very similar (7). Thus, significant differences apparently exist in channel regulation phenomena.

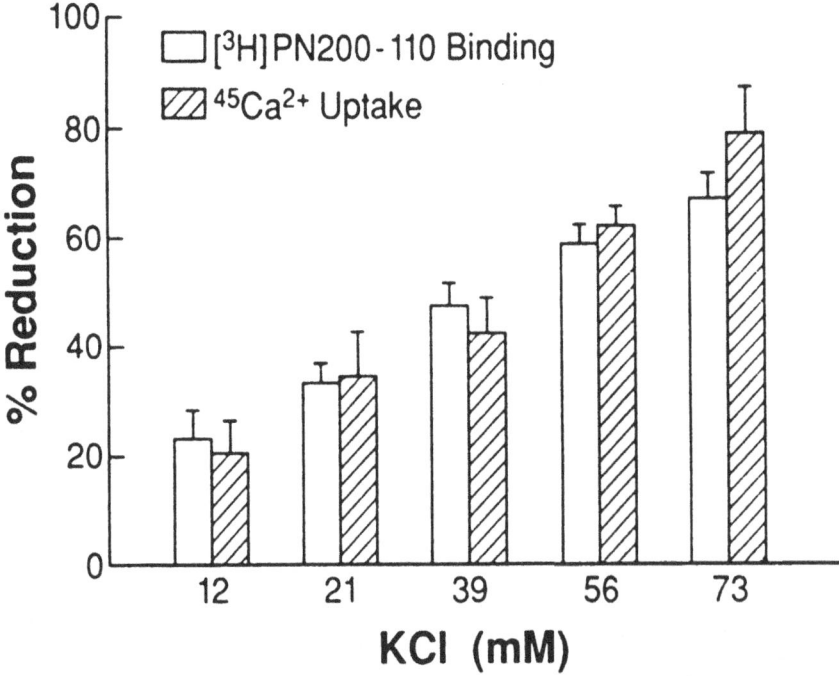

FIG. 1. Comparison between reduction of [3H]isradipine binding and $^{45}Ca^{2+}$ uptake in cultured chick neural retina cells treated with the indicated concentrations of K^+ for 4 days [data from Ferrante and Triggle (7)].

Voltage-dependent binding of 1,4-dihydropyridines.

It is well established that the interactions of 1,4-dihydropyridine antagonists at L channels is voltage-dependent, ligand affinity increasing with increasing depolarization (9,10). These voltage-dependent interactions have been studied principally in the cardiovascular system. We have measured the voltage-dependent binding of 1,4-dihydropyridine antagonist and activators to cardiac myocytes. Antagonist binding is voltage-dependent with affinity increasing with decreasing membrane potential, whereas activator binding is essentially independent of membrane potential (11). This can be seen also in chick retinal neurons (fig. 2) where there exists, for a limited series of 1,4-dihydropyridines, a 1:1 relationship for activator binding and pharmacologic activity, whereas a significant difference exists between these two measures of affinity for the antagonist ligands(12).

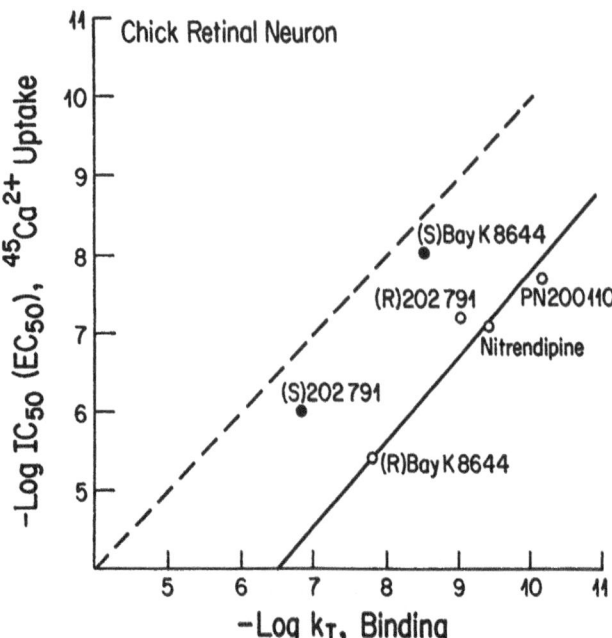

FIG. 2. Correlation between radioligand binding and pharmacologic activities ($^{45}Ca^{2+}$ uptake) in chick retinal neurons. (S)202 791 and (S)Bay K 8644 are activators and (R)Bay K 8644, (R)202 791, nitrendipine, and PN 200110 are antagonists. Data from Wei et al. (12).

In neonatal rat cerebellar granule cells in primary culture high affinity 1,4-dihydropyridine binding sites can be quantitated in membrane preparations with [^{3}H]isradipine and by competition with both antagonist and activator 1,4-dihydropyridine ligands (table 1). Radioligand binding in intact cells at 5 and 50 mM K^+, representing polarized and depolarized states, respectively, reveals that, as anticipated, binding under depolarized conditions exhibits a similar affinity to that observed in membranes and that the binding affinity in polarized cells is lower than that in depolarized cells by approximately 15-fold (table 2). These observations parallel those made in other systems particularly cardiac and vascular smooth muscle (9-11). Of particular interest, however, is that this voltage-

TABLE 1. 1,4-Dihydropyridine binding in rat cerebellar
granule cell membranes

Ligand	K_D, x 10^{-10}M	B_{max} fmoles/mg	n_H
[³H]Isradipine	5.57 ± 0.43	101.9	1.04
R (+) Nimodipine	6.00 ± 0.55	-	0.92
S (−) Nimodipine	1.30 ± 0.03	-	1.02
R (+) Bay K 8644	56.1 ± 0.55	-	0.96
S (−) Bay K 8644	36.1 ± 0.54	-	1.01

TABLE 2. 1,4-Dihydropyridine binding in rat cerebellar
granule cells

Ligand	K^+, mM	K_D, x 10^{-10}M	B_{max} fmoles/mg	n_H
[³H]Isradipine	5.8	3.75 ± 0.62	364.2 ± 95.5	0.95
[³H]Isradipine	50	0.25 ± 0.05	70.6 ± 14.4	0.97
R(+) Nimodipine	5.8	9.35 ± 7.22	-	0.64
R(+) Nimodipine	50	1.33 ± 0.49	-	0.78
S(−) Nimodipine	5.8	7.63 ± 4.02	-	0.69
S(−) Nimodipine	50	0.94 ± 0.42	-	0.72
R(+) Bay K 8644	5.8	8.10 ± 1.04	-	0.63
R(+) Bay K 8644	50	9.59 ± 5.83	-	1.05
S(−) Bay K 8644	5.8	11.3 ± 4.70	-	0.97
S(−) Bay K 8644	50	2.68 ± 2.43	-	1.12

dependent behavior includes a significant down-regulation under the conditions, 60-120-minute depolarization with elevated K^+, employed in these experiments. This suggests that the L-type channels in these neuronal cells are rapidly regulated and that this may contribute to neuronal growth and development and also to changes under pathological conditions.

The neurotoxin iminodipropionitrile.
3,3'-Iminodipropionitrile (IDPN) is a neurotoxin producing a

characteristic syndrome of excitation with choreiform and circling movements (13). The symptoms resemble those of certain human movement disorders including Huntington's disease and Tourette's syndrome. The biochemical basis of the defect is not established, but virtually all of the transmitter systems appear to be involved including the striatal dopaminergic system (13,14). The symptoms of IDPN neurotoxicity are relieved by nifedipine (15) indicating that changes in L-type Ca^{2+} channels may be involved. This is supported by observations that in IDPN-treated mice, there is a significant increase in striatal density of $[^3H]$1,4-dihydropyridine binding of approximately 50%, with no changes in other brain areas or in ω-conotoxin binding (16; table 3). These observations suggest that the movement disorders produced by IDPN neurotoxicity are associated with an overexpression of L-type channels in the striatum. It remains, however, to be determined whether this change is a primary defect.

TABLE 3.   Effect of IDPN on $[^3H]$isradipine binding in mouse brain areas

Group	Cortex		Striatum	
	K_D, pM	B_{max}	K_D, pM	B_{max}
Saline	68 ± 10	563 ± 51	38 ± 5	157 ± 6
IDPN*	45 ± 5	485 ± 41	33 ± 2	237 ± 31**

* 1g/kg ip, 3 days.   ** p < 0.05

Aging and calcium channel function.

The effects of age on a variety of neurotransmitter receptors in rat brain have been relatively well studied. A majority of reports indicate that there is an age-dependent decrease in density of a number of other receptors including dopamine, muscarinic, opiate, N-methyl-D-aspartate (NMDA), and serotonin (inter alia: 17,18).

In marked contrast to the study of receptors, little information is available concerning the regulation with age of Ca^{2+} channels in the brain although there are well-expressed hypotheses indicating that changes in central-nervous-system Ca^{2+} homeostasis are associated with pathological features of the aging process (19). Thus, Leslie et al. (20) reported that synaptosomal Ca^{2+} uptake decreased with aging and, consistent with this, some reports indicate a reduced density of 1,4-dihydropyridine binding sites with age (21,22). However, other reports indicate no changes in these channels or even increases (23,24). Clearly, clarification through a systematic analysis is needed to resolve this issue. In Fisher 344 rats, a systematic regional survey of Ca^{2+} and other ion channel distribution and function reveals large and heterogenous changes with aging (fig. 3). ω-Conotoxin binding is reduced at 30 months by almost 75% in the striatum and 1,4-dihydropyridine binding is reduced by approximately 50%. In contrast, there is no change in 1,4-dihydropyridine binding site density in the cerebellum and the heart reveals no changes in the density of L-type channels or ATP-dependent K^+ channels from six to thirty months.

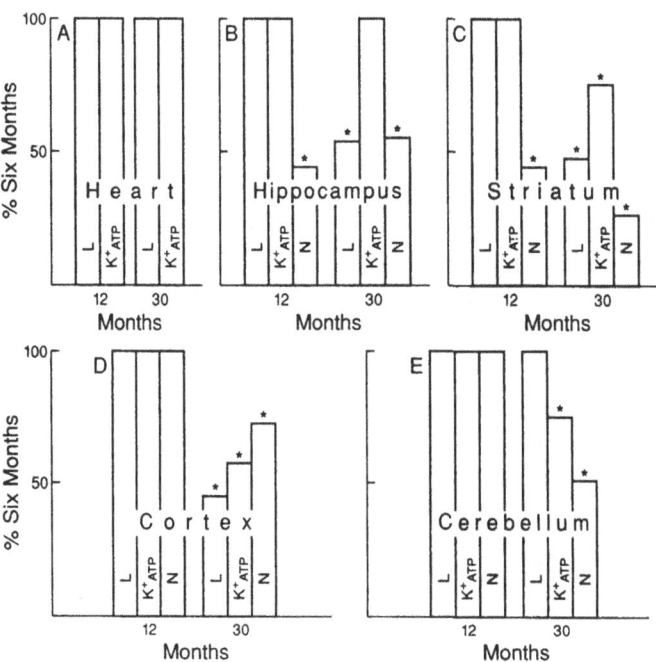

FIG. 3.   The densities of binding sites for L channels (1,4-dihydropyridine), N channels (ω-conotoxin), and sulfonylurea (ATP-dependent K^+ channels) in the heart and brain regions of Fisher 344 rats at 12 and 30 months of age expressed as a percentage of values at six months.   Values are the means of six individual experiments and * indicates significance at the 0.05 level.

SUMMARY

The voltage-gated neuronal Ca^{2+} channel is a regulated species. This regulation likely has important physiological, pharmacological, and pathological consequences. These include the regulation of ligand affinity through membrane potential, the down-regulation of channel number by chronic activation or occupancy, the alteration in receptor number by a neurotoxin, and the regional changes in channel number upon aging. Much work remains to be done to determine the consequences of channel regulation and the extent to which differential regulation of the several channel classes and subclasses occurs.

ACKNOWLEDGMENTS

This work was supported by a grant from the National Institutes of Health. Additional support from Miles Inc. and Pfizer Inc. is gratefully acknowledged.

REFERENCES

1.   Tsien R.W. and Tsien R.Y. (1990): *Ann. Rev. Cell. Biol.*, 6: 715-760.
2.   Janis R.A., Silver P. and Triggle D.J. (1987): *Adv. Drug Res.*, 16: 307-591.
3.   Ferrante J. and Triggle D.J. (1990): *Pharmacol. Rev.*, 42: 29-44.
4.   DeLorme E.M. and McGee R. (1986): *Brain Res.*, 397: 189-192.
5.   DeLorme E.M., Rabe C.S. and McGee R. (1988): *J. Pharmacol. Exp. Therap.*, 244: 838-843.
6.   Skattebol A., Brown A.M. and Triggle D.J. (1989): *Biochem. Biophys. Res. Comm.*, 160: 929-936.
7.   Ferrante J., Triggle D.J. and Rutledge A. (1991): *Can. J. Physiol. Pharmacol.*, 69: 914-920.
8.   Morgan J.I. and Curran T. (1986): *Nature*, 322: 552-555.
9.   Sanguinetti M.C. and Kass R.S. (1984): *Circ. Res.*, 55: 336-348.
10. Hondeghem L.M. and Katzung B.G. (1984): *Ann. Rev. Pharmacol. Toxicol.*, 24: 387-423.
11. Wei X.Y., Rutledge A. and Triggle J.D. (1989): *Mol. Pharmacol.*, 35: 541-552.
12. Wei X.Y., Rutledge A., Zhong Q., Ferrante J. and Triggle D.J. (1989): *Can. J. Physiol. Pharmacol.*, 76: 506-514.
13. Selye H. (1957): *Rev. Can. Biol.*, 16: 1-73.
14. Cadet J.L., Braun T. and Freed W.J. (1987): *Exp. Neurol.*, 96: 594-600.
15. Cadet J.L., Taylor E. and Freed W.J. (1988): *Pharmacol. Biochem. Behav.*, 29: 381-385.
16. Bangalore R., Hawthorn M. and Triggle D.J. (1991): *J. Neurochem.*, 57: 550-555.
17. Maggi A., Schmidt M.J., Ghetti B. and Enna S.J. (1979): *Life Sci.*, 24: 367-374.
18. Govoni S., Memo M., Saiani L., Spano P.F. and Trabucchi M.

(1980): *Mech. Aging Dev.*, 12: 39-46.

19.   Gibson G.E. and Peterson C. (1987): *Neurobiol. of Aging*, 8: 329-343.

20.   Leslie S.W., Chandler L.J., Barr E.M. and Farrar R.P. (1985): *Brain Res.*, 329: 177-183.

21.   Huguet F., Huchet A.M., Gerard P. and Narcisse G. (1987): *Brain Res.*, 412: 125-130.

22.   Dooley D.J., Lickert M., Lupp A. and Osswald H. (1988): *Neurosci. Lett.*, 93: 318-323.

23.   Govoni S., Rius R.A., Battaini F., Bianchi A. and Trabucchi M. (1985): *Brain Res.*, 333: 374-377.

24.   Battaini F., Govoni S., Rius R.A. and Trabucchi M. (1988): *Neurosci. Lett.*, 61: 67-71.

EFFECT OF AGING ON BRAIN VOLTAGE-DEPENDENT CALCIUM CHANNELS

Fiorenzo Battaini, Marco Trabucchi,
Valerian Chikvaidze*, and Stefano Govoni**

Department Experimental Medicine and Biochemical
Sciences, II University of Roma, Rome, Italy

*Institute of Physiology, Georgian Academy of Science
Tbilisi, Georgia

**Pharmacobiology Department, University of Bari,
Bari, Italy

Control of interneuronal communication involves a variety of homeostatic mechanisms in which free calcium ions play a fundamental role (1). An altered metabolism of calcium in the brain that occurs with aging has been observed since the beginning of this century (2). A number of data have been accumulated mostly in the last decade, indicating that aging modifies a number of functions that are strictly related to, and controlled by, calcium ions (3). Release processes, enzymatic activities, and neuronal plasticity are all impaired as a consequence of age, so that a unifying model of brain aging stressing the importance of this ion has been proposed (4).

Recent data have demonstrated a direct involvement of calcium and calcium-dependent functions in processes related to neuronal plasticity and to neuronal death (5-6). This has stimulated the search for age-related modifications of the mechanisms controlling homeostasis of this ion. These studies are aimed on the one hand at explaining the age-related decline in a number of brain functions and, on the other hand, at finding new pharmacological tools of intervention for the aging brain.

The processes intervening in the control of calcium movements and its intracellular actions are accessible to biochemical analysis. The calcium channels regulated by voltage that allow ion influx from the extracellular milieu have been characterized in detail in neurons. These channels are subdivided according to activation properties (low or high threshold) (7) or to the classification considering general electrophysiological as well as pharmacological characteristics (8,9) (L,N,P, and T), see fig. 1. Tools of natural origin (toxins such as omega conotoxin and funnel web spider toxin) and substances of inorganic or organic origin interfere with various

231

T. Godfraind et al. (eds.), Calcium Antagonists, 231–240.
© 1993 *Kluwer Academic Publishers and Fondazione Giovanni Lorenzini.*

degrees of specificity (10). In fact, drugs like dihydropyridines (DHP; such as nitrendipine, isradipine), phenylalkylamines (PHA; verapamil, desmethoxyverapamil) and benzothyazepines (diltiazem) interact with L-type channels at specific sites in the channel structure (11). Toxins interact in a negative manner with N-type (omega conotoxin-sensitive) (12) and P-type (funnel web spider toxin-sensitive) (9) channels. T-type channels are inhibited with some degree of specificity by amiloride (12) and by diphenylalkylamine drugs such as flunarizine (13).

Tritiated dihydropyridines, or alkylamines and iodinated toxins have permitted identification of binding sites in the brain and have demonstrated the plasticity of these channels in a variety of physiopathological conditions (10,14). These channels have been tentatively suggested to be involved in different neuronal functions, such as normal excitability (T-type), hyperexcitability (L-type), and neurotransmitter release (N-type) (10).

P-type channels have been described in the cerebellum, where they may participate in the control of Purkinje cell function (9). Pharmacological studies show that calcium antagonists may control some behavioral and electrophysiological modifications related to brain aging. For example DHPs have been reported to improve associative learning in aged rabbits (15) and to ameliorate hippocampal impaired excitability in aged rats (16).

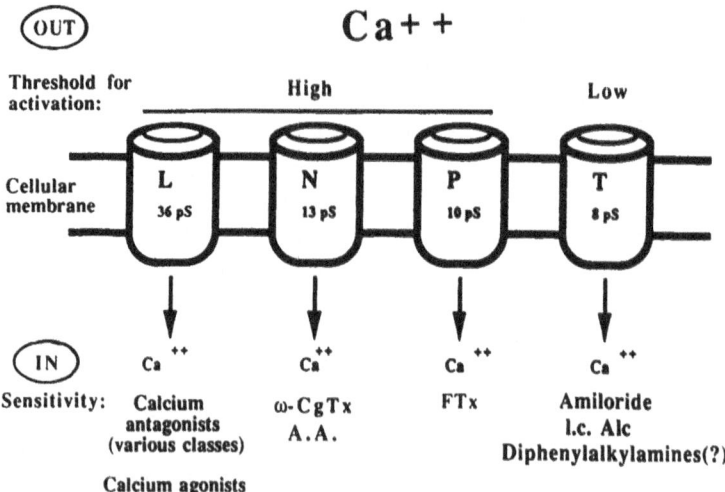

FIG. 1. Electrophysiological and pharmacological characteristics of the various subtypes of voltage dependent calcium channels. ω-CgTx: ω-conotoxin fraction GVIA (from the mollusc *Conus Geographus*); A.A.: Aminoglycoside antibiotics; FTx: funnel-web spider toxin (from the spider *Agelenopsis Aperta*); 1.c. Alc: long chain alcohols. Further details in (12).

This article will review and analyze recent work dealing with age-dependent changes in voltage-dependent calcium channels of the L- and N-type in the brain, utilizing labeled calcium antagonists and omega conotoxin respectively.

CHANGES IN L-TYPE CHANNELS WITH AGING

Studies were performed on the developmental expression of nitrendipine-labeled channels from birth to 24 months of age in Sprague Dawley rat cortex (17). Analysis of the data indicates that B_{max} values increase gradually up to 3 months of age. The values in 10- and 24-month old rats do not differ significantly from the values observed in 3-month rats. K_d values remain constant up to 10 months of age and almost double in absolute values in the 24-month rats, decreasing binding affinity by more than 80%.

Considering that calcium ions are needed *in vitro* to maximally express nitrendipine association with the channel (10) and that calcium ion concentrations are modified during aging (3), the calcium sensitivity of the binding of nitrendipine to L-type channels was analyzed in the presence of *in vitro* increasing calcium concentrations. Cortical membranes of rats of different ages were treated with chelating agents to remove endogenous calcium ions and then the calcium-dependent [³H]-nitrendipine-binding reconstitution process analyzed. The reconstitution curve is bell shaped at all ages investigated, peaking at 300, 400, and 500 μM when using membranes prepared respectively from rats of 3, 10, and 24 months of age. Although in this strain of rats we did not find the changes in B_{max} seen by other authors (see the results of Bangalore et al., on Fischer 344 rats, this volume) the data on calcium sensitivity as well as the observed increase in K_d values agree, indeed, with a reduced DHP binding efficiency with aging.

It is known that although DHP- and PHA-binding sites are separate entities on the channel molecule, they allosterically interact in a negative manner and calcium ions modulate such interaction (18). Study of the competition between various calcium antagonists and the calcium agonist Bay K 8644 on nitrendipine binding to cortical membranes of 3- and 24-month old rats reveals that while Bay K 8644 and flunarizine displace bound nitrendipine equally efficiently in the two age groups, verapamil displaces [³H]-nitrendipine by a maximum of 30% in young rat membranes with an IC_{50} of 60 ± 8 nM, and by 55% with a significantly lower IC_{50} value (28 ± 4 nM) in membranes prepared from 24-month old rats. These data suggest the possibility of either a higher affinity or an increased number of binding sites for verapamil in aged rats. To answer this question, verapamil-binding kinetic characteristics were evaluated during aging. No modification was observed in K_d values at all ages while

B_{max} values increased gradually with age, being 30 and 70% higher in 10- and 24-month old rats (19). The concomitant decrease of DHP-binding affinity and increase of [^{3}H]-verapamil B_{max} is in accordance with *in vitro* biochemical data indicating that verapamil modifies DHP binding affinity (18,20). The possibility cannot however be excluded that the increased binding might be due to changes in glial cells displaying significant levels of calcium antagonist binding (21).

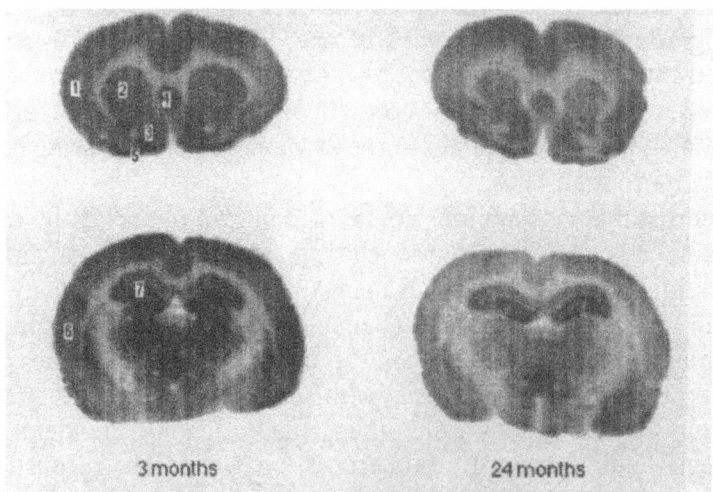

FIG. 2. Autoradiography of [^{3}H]-isradipine binding to Wistar rat brain. 12 μm brain sections on gelatin-coated slides were reacted with 0.5 nM [^{3}H]-isradipine as detailed in (22). The slides were exposed to [^{3}H]-hyperfilm for 7 days and analyzed with the Image 1.37 Program of Dr. Wayne Rusband, NIMH, Bethesda, USA. 1 and 6: cerebral cortex; 2: striatum; 3: nucleus accumbens; 4: septum; 5: olfactory tubercle; 7: hippocampus + dentate gyrus.

Recently the binding of DHPs and PHAs in brain thin sections from young and aged Wistar rats was analyzed in more detail. For this purpose, higher affinity ligands such as PN200-110 (isradipine) for DHP sites (fig. 2) and desmethoxyverapamil (D888) for PHA sites were used.

The analysis of the data (table 1) indicates that in this strain of rats, isradipine binding is reduced at cortical and hippocampal levels in accordance with observations made by others in Sprague Dawley (23) and Fisher 344 rats (see Bangalore et al., this volume). In addition, DHP binding displays age-dependent decreases in other areas such as the striatum, nucleus accumbens, septum, and olfactory tubercle.

TABLE 1.   Microdensitometric analysis of [³H]-isradipine
binding to selected rat brain areas

Area	Binding density (nCi/mg tissue)		% decrease in 24-month old rats
	3 months	24 months	
1:	1.79 ± 0.1	0.99 ± 0.3	45
2:	1.83 ± 0.3	0.93 ± 0.3	49
3:	1.90 ± 0.3	0.94 ± 0.2	51
4:	1.84 ± 0.2	1.10 ± 0.4	41
5:	2.10 ± 0.2	1.33 ± 0.3	37
6:	1.20 ± 0.3	0.86 ± 0.2	28
7:	1.64 ± 0.4	1.10 ± 0.2	31

The values in nCi/mg tissue were calculated using Amersham [³H]-microscales as standards. The values are means ± SD of 4 rats with 2-4 measurements for each area investigated. Areas as in legend fig. 2.

In contrast, preliminary analysis at the same levels of [³H]-D888 binding indicates no age-dependent modifications (data not shown) again stressing that the two binding sites have different destiny during aging.

AGING AND N-TYPE CHANNELS

[¹²⁵I]-ωCgTx binds with high affinity and specificity to N-type channels. Calcium antagonists do not interfere *in vitro* (up to 10 mM concentration) with the binding of [¹²⁵I]-ωCgTx to cortical membranes prepared from 3-month old rats. Table 2 shows the binding of [¹²⁵I]-ωCgTx to membranes prepared from cortex, striatum, and hippocampus of 3- and 24-month-old Sprague Dawley rats (24).

The B_{max} decreases with age in the striatum and cortex, while no modification is observed in the hippocampus. K_d values are not modified in all the areas investigated. These data are in accordance with previously published data on age-related variations on the same strain of rats (25).

TABLE 2. $[^{125}I]\omega CgTx$-binding kinetics in striatal, cortical, and hippocampal membranes prepared from 3- and 24-month-old rats

Area	Age (months)	B_{max} (fmol/mg prot)	Kd (pM)	% decrease in 24-month-old rats
Cortex	3	916	11.7	-
	24	634	11.2	31
Striatum	3	813	11.7	-
	24	577	10.9	29
Hippocampus	3	941	11.8	-
	24	930	12.3	1

Values are means of 4 experiments in which SD was less than 15%.

DISCUSSION

Recent work analyzing calcium homeostasis during aging in the rat cortex indicates that the fast calcium uptake is reduced while the slow phase is unmodified, and that the decline in free intracellular calcium to basal values following stimulus is slowed, resulting in increased levels of the free ion (26). This decreased buffering capacity in aged tissue is corrected by verapamil, suggesting an increased influx of calcium ions through verapamil-sensitive channels. These functional data appear to be in agreement with the reported biochemical findings showing that the number of cortical N-type channels is decreased with aging, thereby possibly contributing to the impairment of stimulated calcium influx. The increased or unmodified number of verapamil-labeled channels and the observation that nitrendipine binding is less sensitive to the action of calcium are in accordance with the sustained calcium concentrations measured following a stimulus, an effect counteracted by verapamil addition. At a hippocampal level, electrophysiological data indicate an age-dependent increase in free intracellular ion concentrations in postsynaptic elements (16). The addition of the calcium antagonist nimodipine corrects the decreased excitability (observed as a prolongation of the calcium-dependent afterhyperpolarization) in hippocampal slices prepared from aged rats (27). This would indicate, as in the case of verapamil in the cortex, an age-related increase in the expression of calcium antagonist-labeled channels in the hippocampus. In contrast, the biochemical data reported on [³H]-nimodipine and [³H]isradipine binding to tissues homogenates (27,23) and with [³H]PN 200-110 in tissue sections (fig. 2) indicate that dihydropyridine labeling is decreased with aging in the hippocampus. A possible explanation could be that *in vitro* DHPs bind to the channel in an "inactivated" form, so that a decreased binding with aging could be interpreted as an increased proportion of the channels in the "activated" form, thus allowing a higher calcium influx as observed in cortical structures. The modifications in L-type calcium channels with aging are more difficult to explain from a physiological perspective, mainly because functional data are not conclusive. It has been postulated that in conditions of neuronal hyperexcitability, for example during withdrawal from alcohol (28) or morphine (29), L-type channels play a major role in inducing the hyperactive state (measured biochemically as an increase in calcium fluxes) and the consequent increase in neurotransmitter release (30). Calcium antagonists, in fact, reduce withdrawal symptomatology and, under these conditions, inhibit transmitter release (28). Impaired neurotransmitter release with aging has been observed in both cortical and striatal structures for cholinergic transmission (31,32). The observed decrease in ω-CgTx binding in the cortex and striatum could be related to these observations, although it must be recalled that a defect in mechanisms beyond the channel could contribute to impaired neurotransmitter release. An example is the reduced ability of the calcium ionophore A 23187 to promote acetylcholine release from aged cortical synaptosomes (33). Additional mechanisms, equally important for control of calcium homeostasis in the brain, are affected by the aging process. In fact, extrusion (see above) or compartmentation into intracellular stores (calcium influx into endoplasmic reticulum and into mitochondria) are depressed in aged rodents (34,35). Moreover, changes in calcium release from intracellular compartments may participate in altering calcium homeostasis during aging (36) and therefore warrant further investigation. The different modification in L- and N-type channels with aging could also be related to their differential localization. L channels are preferentially located on neuronal cell bodies while N channels are preferentially located on nerve terminals (37). These data are strengthened by clinical observations obtained at the striatal level in pathological states. In fact, in Huntington's disease, the density of DHP-binding sites is decreased compared to matched controls, while in Parkinson's disease no changes are observed, underlying the postsynaptic localization of DHP-binding sites (38). The observation that during aging 3,4-diaminopyridines (that increase calcium influx indirectly though inhibition of K^+ efflux) are able to counterbalance the decreased neurotransmitter release and also, in certain conditions, memory deficits (39,40), underlines the concept of a differential balance between calcium homeostasis at presynaptic (decreased) in comparison to the postsynaptic (increased) level (3). It is important to underline that the detection of age-dependent modifications in calcium homeostasis, although functionally important, does not necessarily identify the primary defect, because of the complex interactions which regulate the movements of the ion (entry, extrusion, intracellular buffering). It is probably due to this complexity of interactions and to the different strains of rats utilized that the analysis of free intracellular calcium in the brain during aging does not give a single or consistent answer and consensus (41-44). Nevertheless, the observation that pharmacological treatments with calcium-interfering agents partially reverse the pre- or postsynaptic alterations in calcium homeostasis and some related behavioral deficits strengthen the concept that the aged organism may still be able to respond to appropriate stimuli.

REFERENCES

1. Kennedy M.B. (1989): *Trends in Neurosci.*, 12: 417-478.
2. Novi. A. (1912): *Arch. Ital. Biol.*, 58: 333-336.
3. Gibson G. and Peterson C. (1987): *Neurobiol. Aging*, 8: 329-344.
4. Katchaturian Z.S. (1987): *Neurobiol. Aging*, 8: 345-346.
5. Olton D.S., Golski S., Mishkin M., Gorman L.K., Olds, J.L. and Alkon D.L. (1991): *Brain Res. Rew.*, 16: 206-209.
6. Siesjo B.K. and Bengtsson F. (1989): *J. Cereb. Blood Flow Metab.*, 9: 127-140.
7. Carbone E. and Lux H.O. (1984): *Nature*, 310: 501-511.
8. Nowychy M.C., Fox A.P. and Tsien R.W. (1985): *Nature*, 316: 443-446.
9. Hillman D., Chen S., Aung T.T., Cherksey B., Sugimori M. and Llinas R.R. (1991): *Proc. Natl. Acad. Sci.*, 88: 7076-7080.
10. Ferrante J. and Triggle D.J. (1990): *Pharmacol. Rev.*, 43: 29-44.
11. Glossman H. and Striessnig J. (1990): *Rev. Physiol. Biochem. Pharmacol.*, 114: 1-105.
12. Hess P. (1990): *Ann. Rev. Neurosci.*, 13: 337-356.
13. Tytgat J., Vereccke J. and Cornelliet E. (1988): *Naunyn-Schmiedeberg's Arch. Pharm.*, 337: 690-692.
14. Govoni S., Battaini F., Magnoni M.S., Lucchi L., Rius R.A. and Trabucchi M. (1988): *Ann. N.Y. Acad. Sci.*, 522: 187-202.
15. Straube K.T., Deyo R.A., Moyer J.R. and Disterhoft J.F. (1990): *Neurobiol. of Aging*, II: 659-661.
16. Pitler T.A. and Landfield P.W. (1990): *Brain Res.*, 508: 1-6.
17. Govoni S., Rius R.A., Battaini F., Bianchi A. and Trabucchi M. (1985): *Brain Res.*, 33: 374-377.
18. Boles R.G., Yamamura H., Schoemaker H. and Roeske W.R. (1984): *J. Pharmacol. Exp. Ther.*, 229: 333-339.

19. Battaini F., Govoni S., Rius R.A. and Trabucchi M. (1985): *Neurosci. Lett.*, 61: 67-71.
20. Battaini F., Govoni S., Del Vesco R., Di Giovine S. and Trabucchi M. (1987): *Biochem. Biophys. Res. Comm.*, 144: 1135-1142.
21. Hertz LJ. (1989): *Neurosci. Res.*, 22: 209-215.
22. Ferry D.R., Goll A., Gadow C. and Glossman H. (1984): *Naunyn-Schmiedeberg's Arch. Pharmacol.*, 327: 183-187.
23. Huguet F., Huchet A.M., Gerard P. and Nacisse G. (1987): *Brain Res.*, 412: 125-130.
24. Moresco R.M., Govoni S., Battaini F., Trivulzio S. and Trabucchi M. (1990): *Neurobiol. Aging*, 11: 433-436.
25. Dooley D.J., Lickert M., Lupp M. and Osswald H. (1988): *Neurosci. Lett.*, 93: 318-323.
26. Giovannelli L. and Pepeu G.C. (1989): *J. Neurochem.*, 43: 392-398.
27. Landfield P.W. (1989): In: *Nimodipine and CNS Actions: New Vistas*, edited by J. Traber and W.H. Gispen pp. 227-238. Shattauer, Stuttgart.
28. Littleton J.M. and Little H.J. (1988): *Ann. New York Acad. Sci.*, 522: 199-202.
29. Bongianni F., Carla' V., Moroni F. and Pellegrini G. (1986): *Br. J. Pharmacol.*, 88: 561-567.
30. Ramkumar V. and El Fakahany E. (1984): *Eur. J. Pharmacol.*, 102: 371-374.
31. Pedata F., Slavikova J., Kotas A. and Pepeu G. (1983): *Neurobiol. Aging*, 4: 31-35.
32. Gibson G.E. and Peterson C. (1981): *J. Neurochem.*, 37: 978-984.
33. Meyer E.M., Crews F.T., Otero D.H. and Larsen K. (1986): *J. Neurochem.*, 47: 1244-1246.
34. Vitorica J. and Satrustegui J. (1986): *Brain Res.*, 378: 36-48.
35. Michaelis M.L., Johe K. and Kitos K.E. (1984): *Mech. Aging Dev.*, 25: 215-225.
36. Burnett, D.M., Daniell L.C. and Zahniser N.R. (1990): *Mol. Pharmacol.*, 37: 566-571.
37. Miller R.J. (1987): *Science*, 235: 46-52.
38. Watson D.L., Carpenter C.L., Marks S.S. and Greenberg D.A. (1988): *Ann. Neurol.*, 23: 303-305.
39. Peterson C. and Gibson G.E. (1983): *J. Biol. Chem.*, 258: 11482-11486.
40. Peterson C. and Gibson G.E. (1983): *Neurobiol. Aging*, 4: 25-30.
41. Martinez A., Vitorica J. and Satrustegui J. (1988): *Neurosci. Lett.*, 88: 336-342.
42. Farrar R.P., Mehdi Rezazadeh S., Morris J.L., Dildy J.E., Gnau K. and Leslie S.W. (1989): *Neurosci. Lett.*, 100: 319-325.

43.   Manger T., Bowers J. and Gibson G. (1987): *Soc. Neurosci. Abstr.*,
       13: 1238.
44.   Martinez-Serrano A., Blanco P. and Satrustegui J. (1992): *J. Biol.
       Chem.*, 267:4672-4679.

CALCIUM OVERLOAD BLOCKADE IN NEURONS AND CARDIOMYOCYTES

M. Borgers, L. Ver Donck, and H. Geerts

Life Sciences, Janssen Research Foundation
B-2340 Beerse, Belgium

An important role in cell death following an ischemic insult has been attributed to Ca^{2+}, although the mechanisms by which altered Ca^{2+} homeostasis induces cell injury are poorly understood. A good insight into the factors responsible for abnormal Ca^{2+} load in brain and heart cells is critical for the development of pharmacological agents that afford protection during ischemia. It is generally assumed that the depletion of ATP during ischemia triggers an impairment of ion homeostasis. Intracellular Ca^{2+} increases to a concentration which might exceed the equilibrium between influx and extrusion or sequestration capacities. This situation imposes a heavy burden on the already compromised energy household and eventually might be the cause of cell death. In addition to Ca^{2+} transients, Na^+ transients are considered to play a crucial role in the development of ischemic cell damage and Na^+ overload might precede excessive Ca^{2+}-influx. Factors liberated during ischemia and reperfusion (e.g., reactive O_2 species, lysophosphatides) are known to impair the inactivation kinetics of the Na^+ channel causing excessive Na^+-entry. The resulting Na^+-overload may then drive the Na^+/Ca^{2+}-exchanger in the direction of Ca^{2+}-overload. Moreover, altered Na^+ channels may directly contribute to Ca^{2+}-overload by carrying Ca^{2+} in addition to Na^+.

Flunarizine and R 56865, compounds with low affinity for the L-type Ca^{2+} channel, protect respectively the brain and heart when subjected to conditions that predispose these organs to Ca^{2+}-overload. Flunarizine's protective effect in various experimental conditions of brain ischemia has been amply documented and compared with other Ca^{2+}-antagonists (1-4). A comparative functional and cyto-morphological study investigating the cytoprotective effect in the heart of a typical L-type Ca^{2+}-antagonist, verapamil, with that of R 56865 has been performed (5). The latter provided considerable protection at a dose that did not influence heart rate and contractility, thus ruling out the possibility of energy sparing as the cardioprotective mechanism. This work confirmed previous studies on the action of R 56865 in ouabain-induced cardiac toxicity (6) and against functional defects elicited by ischemia or hypoxia (7,8). The mechanism of action of this new drug is summarized in this paper and in the work by

T. Godfraind et al. (eds.), Calcium Antagonists, 241–247.
© 1993 *Kluwer Academic Publishers and Fondazione Giovanni Lorenzini.*

Ver Donck et al. (this issue).

EFFECTS OF R 56865 IN ISOLATED NEURONS

The cytoprotective effects of flunarizine in neuronal cells against veratridine-induced toxicity (9), against long-term exposure to high K^+ (10), and after withdrawal of neurotrophic support (11) have been demonstrated. In cultures of neonatal hippocampal neurons, the $[Ca^{2+}]_i$ rise which is observed after application of 250 μM glutamate is antagonized by 10^{-8} M flunarizine (unpublished data). Such effects of R 56865 in neurons have not yet been reported. In the present study the effects of R 56865 were investigated in adult rat dorsal root ganglion cells isolated and cultured as described (12). They were allowed to recover for 24 hours, and experiments were performed up to 72 hours after isolation. Measurements of intracellular Ca^{2+} were performed as described before (13). The effects of R 56865 on K^+-mediated Ca^{2+} rises were studied in conditions of increasing membrane depolarization: 5.4-50 mM K^+. Fig. 1 illustrates the effect of 1 μM R 56865, on the neuronal Ca^{2+} homeostasis when extracellular K^+ is increased from 5.4 to 25 mM. As in cardiomyocytes, the rise in $[Ca^{2+}]_i$ is biphasic: a transient peak, followed by a steady-state Ca^{2+} level. When the neurons were pretreated for 5 minutes with R 56865, the peak $[Ca^{2+}]_i$ was inhibited for about 50%. This inhibition increased to 90% when $[K^+]_e$ was raised to 50 mM. At 0.1 μM R 56865, however, the effect was reduced to about 10%. Exposure to R 56865 after depolarization resulted in an immediate decrease in $[Ca^{2+}]_i$. This effect is independent of the degree of K^+-depolarization. About 70% of the cells were protected at 1 μM and 40% at 0.1 μM.

The data in fig. 1 suggests R 56865, at a concentration of 1 μM, inhibits the transient and steady-state rise in $[Ca^{2+}]_i$ induced by 25 and 50 mM K^+. Drug effects on two possible pathways of K^+-induced Ca^{2+} influx may be involved. The first possibility, an inhibition of the Na^+/Ca^{2+} exchange mechanism, whereby Ca^{2+} enters the cells in exchange for N^+, is highly unlikely. In myocytes R 56865 has been shown not to interfere with Na^+/Ca^{2+} exchange (14). In addition, we have shown that Na^+/Ca^{2+} exchange in dorsal root ganglion cells is of minor importance. The other pathway involves voltage-sensitive Ca^{2+} channels. Whether R 56865 affects L-type or T-type Ca^{2+} channels at this concentration remains to be proven. It must be stated, however, that at 0.1 μM, the inhibition largely disappears. In the same experimental conditions, flunarizine inhibits the transient peak after 15 mM K^+ (10). This might correspond to the documented effect of flunarizine on the T-type Ca^{2+} channel (15). A most interesting observation, however, is the capacity of R 56865 to decrease $[Ca^{2+}]_i$ levels when applied after $[K^+]_e$ depolarization. As this effect is present at all depolarization levels (including 15 mM $[K^+]_e$) and

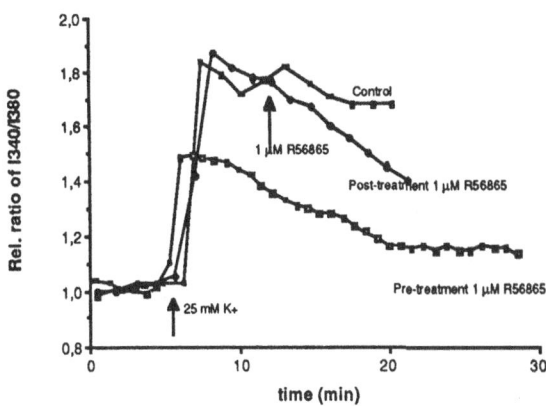

FIG. 1. Effect of 1 μM R 56865 (pre- and post-treatment) on the Ca^{2+} response to 25 mM K^+. The ratio of 340/380 nm intensity of the fura 2-signal (a measure of $[Ca^{2+}]_i$) for each individual cell is calculated relative to the first 5 minutes. Shown here are the mean of at least 8-10 cells. The response of untreated (control) cells to 25 mM K^+ is biphasic, consisting of a transient peak, followed by a steady-state level. Pretreatment with 1 μM R 56865 partially blocks the transient peak, but the steady-state level is returned to normal values. Posttreatment reduces the steady-state $[Ca^{2+}]_i$ level.

at concentrations as low as 0.1 μM, mechanisms influencing membrane-potential insensitive pathways may be involved. Amongst them, processes influencing Ca^{2+}-extrusion and/or sequestration are likely candidates. As neuronal cell death probably is associated with sustained high levels of $[Ca^{2+}]_i$, the capacity to decrease these levels even when the aggression has taken place, may provide an interesting therapeutic rational for neuroprotection.

EFFECTS OF R 56865 IN ISOLATED CARDIOMYOCYTES

Over the past decade, isolated cardiac cells have been used in a wide range of cardiovascular investigations. The outcome of electro-physiological, biochemical, morphological, and functional studies has greatly validated their use in this field of research (16). We have previously demonstrated that cardiomyocytes isolated from adult myocardium display irreversible shape change from rod-shaped to hypercontracted rounded cells when exposed to a variety of experimental conditions such as veratridine, reactive O_2 species (singlet O_2), ouabain, and lysophosphatidylcholine (LPC) (17). Shape change under these conditions entirely depended upon the presence of $[Ca^{2+}]_i$. Digital image processing of fura-2 fluorescence according to Geerts, et al. (10) was used to measure free $[Ca^{2+}]_i$. A median free $[Ca^{2+}]_i$ of 68 nM (n=43 cells) was measured in quiescent control cells (fig. 2). This corresponds well with concentrations reported in other single muscle cells and measured with several techniques (18). The majority of rat or rabbit cardiomyocytes exposed to veratridine or LPC display irreversible hypercontracture, coinciding with 10- to 20-fold increase in $[Ca^{2+}]_i$ (fig. 2). Thus, cellular shape change under such conditions also correlates with an excessive rise in $[Ca^{2+}]_i$, demonstrating a condition of intracellular Ca^{2+} overload.

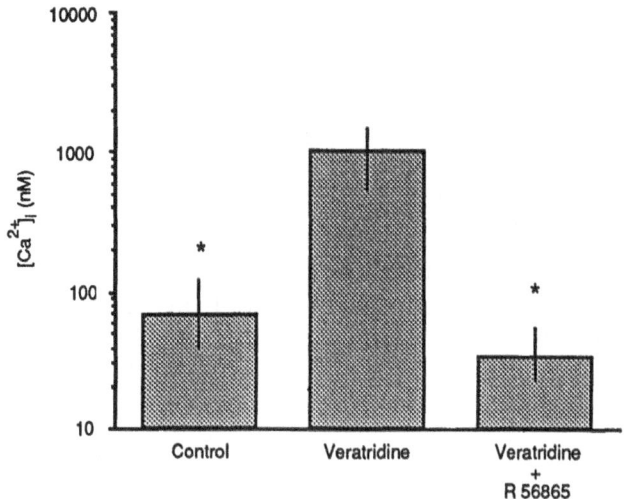

FIG. 2. $[Ca^{2+}]_i$ levels in single quiescent rat cardiomyocytes and effect of veratridine and R 56865 (5×10^{-7} M) (median values and 95% confidence limits, n=25-43 cells, *: P<.05 vs. veratridine, Kolmogorov-Smirnov test with Bonferoni in equality).

Since shape change (= Ca^{2+}-overload) of cardiomyocytes can be easily quantified by light microscopy, this property provides an interesting model to evaluate cellular Ca^{2+}-overload and Ca^{2+}-overload blockade (=cytoprotection) by cardioprotective drugs (17). The cytoprotective effects of the new drug R 56865 was investigated in this model of hypercontracture and Ca^{2+}-overload induced by the above-mentioned experimental conditions in isolated cardiomyocytes. Table 1 demonstrates that pretreatment with R 56865 largely prevents shape change of cardiomyocytes induced by veratridine, singlet O_2, ouabain, or LPC.

TABLE 1. Effect of R 56865 on shape change of rat and
          rabbit cardiomyocytes induced by experimental
          conditions (mean ± SEM, n=4-6)

Experimental condition	Species	Drug Concentration (M)	%Rod-shaped cells	
			Solvent	R 56865
Veratridine	Rat	1×10^{-7}	6.0 ± 1.7	61.5± 5.0
Singlet O_2	Rat	5×10^{-7}	3.6 ± 2.1	44.6± 2.0
Ouabain	Rat	2×10^{-7}	4.0 ± 0.7	41.2± 2.8
LPC	Rabbit	1×10^{-6}	27.5 ± 7.5	56.3±13.1

The IC_{50} values for inhibition of shape change by R 56865 in rat cardiomyocytes range from 1.2×10^{-7} to 3.5×10^{-7} M. A very interesting observation was that cells that were protected from contracture by R 56865 maintained very low $[Ca^{2+}]_i$ levels, comparable to control cells (fig. 2). It was remarkable that even hypercontracted cells after exposure to veratridine in the presence of the cytoprotective drug has a slightly (not significant) reduced $[Ca^{2+}]_i$ content (not shown).

Digital image processing of SBFI, a fluorescent probe selective for free $[Na^+]_i$ (19), demonstrated that veratridine induced a TTX-sensitive rise in $[Na^+]_i$ in rat cardiomyocytes (fig. 3). Application of R 56865 (2×10^{-7} M) largely attenuated the veratridine-induced rise in $[Na^+]_i$. This suggests that prevention of Na^+ load, that otherwise activates Na^+/Ca^{2+}-exchange resulting in Ca^{2+}-overload, contributes to the compounds' cytoprotective effect. A similar conclusion was reached by Haigney, et al. (20) studying the protective effect of R 56865 in hypoxic cardiac myocytes. From these observations, it can be concluded that the prevention of Ca^{2+}-overload by R 56865 may result from the attenuation of a preceding excessive Na^+ loading.

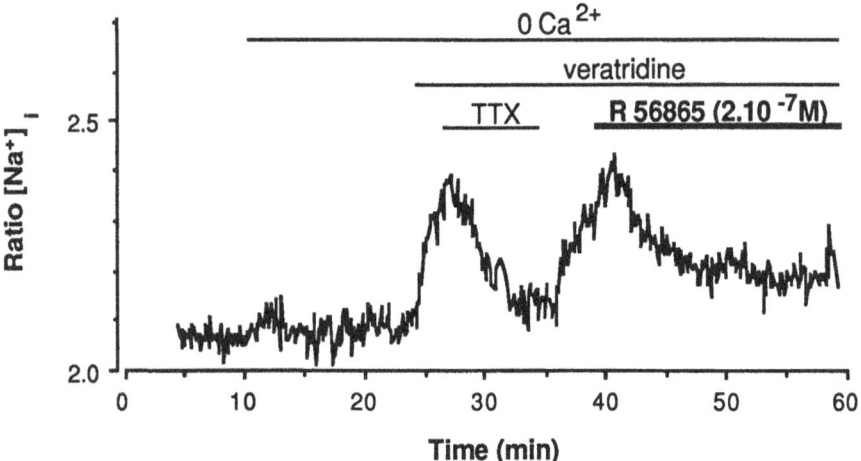

FIG. 3. [Na$^+$]$_i$ level (indicated by fluorescence ratio) in a representative single quiescent rat cardiomyocyte exposed to veratridine; effect of tetrodotoxin (TTX) (10^{-5} M) and R 56865 (2×10^{-7} M).

REFERENCES

1. Desphande J.K. and Wieloch T. (1986): *Anesthesiology*, 64: 215-224.
2. Silverstein F.S., Buchanan K., Hudson C. and Johnson M.V. (1986): *Stroke*, 17: 477-482.
3. Alps B.J., Calder C., Hass W.K. and Wilson A.D. (1988): *Brit. J. Pharmacol.*, 9: 877-883.
4. Borgers M. and Van Reempts J. (1989): *J. Neurosurg. Anesthesiol.*, 1: 368-374.
5. Vandeplassche G., Thoné F. and Borgers M. (1991): *Eur. J. Physiol.*, (in press).
6. Vollmer B., Meuter C. and Janssen P.A.J. (1987): *Eur. J. Pharmacol.*, 142: 137-140.
7. Garner J.A., Bernier M. and Hearse D.J. (1990): *J. Cardiovas. Pharmacol.*, 16: 468-479.
8. Koch P., Willfert B. and Peters Th. (1990): *Cardiovasc. Drug Rev.*, 8: 238-255.
9. Pauwels P.J., Van Assouw H.P., Leysen J.E. and Janssen P.A.J. (1989): *Molec. Pharmacol.*, 36: 525-531.
10. Geerts H., Nuydens R., Nuyens R. and Cornelissen F. (1990): In: *Ionic Currents and Ischemia,* edited by J. Vereecke, P.P. Van Bogaert and F. Verdonck pp. 310-313. University Press, London.
11. Rich K.M. and Hollowell J.P. (1990): *Science*, 248: 1419-1421.
12. Delree P., Leprince P., Schoenen J. and Moonen G. (1989): *J.*

Neurosc., 23: 198-206.

13.   Geerts H., Nuyens R., Nuydens R. and Ver Donck L. (1989): *Cardiovas. Res.*, 23: 797-806.

14.   Ver Donck L. and Borgers M. (1991): *Am. J. Physiol. (Heart Circ. Physiol.)*, (in press).

15.   Akaike N., Kostyuk P.G. and Osipchuk Y.N. (1989): *J. Physiol.*, 412: 181-195.

16.   Clark W.A., Decker R.S. and Borg T.K., editors (1988): *Biology of Isolated Adult Cardiac Myocytes.* Elsevier, New York.

17.   Borgers M., Ver Donck L. and Vandeplassche G. (1988): *Ann. N.Y. Acad. Sci.*, 522: 433-453.

18.   Dubell W.H. and Houser S.R. (1986): *Cell Calcium,* 8: 259-268.

19.   Minta A. and Tsien R.Y. (1989): *J. Biol. Chem.*, 264: 19449-19457.

20.   Haigney M.C.P., Ver Donck L., Stern M.D. and Silverman H.S. (1991): *Circulation,* (suppl.) (in press).

EFFECT OF NICARDIPINE ON ANTIOXIDANT ENZYMATIC ACTIVITIES AND LTC$_4$ RELEASE IN EXPERIMENTAL SAH

P. Gaetani, R. Rodriguez y Baena, D. Lombardi, F. Marzatico*,
R. Bellazzi**, S. Quaglini** and P. Paoletti

Department of Surgery, Neurosurgery, IRCCS Policlinico S. Matteo,
*Institute of Pharmacology and **Dipartimento di Informatica e
Sistemistica, University of Pavia, Pavia, Italy

INTRODUCTION

Modifications of biochemical and physiological parameters following subarachnoid hemorrhage (SAH) such as the activation of lipid peroxidation and liberation of free radicals (1,2), variations in scavenger enzyme activities (3), and the enhancement of arachidonic acid metabolism (4) could be considered indicators of brain damage in evolution after SAH. An impaired homeostasis of Ca^{2+} and its relationship with lipid-peroxidative processes has been largely studied with special regards for the pathogenesis of arterial vasospasm and its ischemic consequences.

Nicardipine has been largely investigated mainly in experimental animal models of cerebral ischemia while in SAH, an inhibitory effect of the compound on the release of leukotriene C$_4$ (LTC$_4$), which is considered the most important product in the cascade of arachidonate metabolism and a reduction of nonenzymatic lipid-peroxidative processes, has been reported in previous studies of our group (5,6).

In the brain, there are different protective systems against peroxidative and free radical-generating reactions: superoxide radicals are inactivated by superoxide-dismutase and the hydrogen peroxide produced in this reaction is inactivated by glutathione peroxidase (GSH-Px) (7,8). The aim of the present study was to verify whether nicardipine could provide the brain metabolic protection against Ca^{2+}-dependent lipid-peroxidative processes in relation to the release of leukotrienes and antioxidant enzymatic activities.

MATERIALS AND METHODS

Experimental Plan

We used a well-described experimental model of SAH in the rat (9,10):

T. Godfraind et al. (eds.), Calcium Antagonists, 249–255.

the hemorrhage is obtained by injection of 0.3 ml of autologous arterial blood into the cisterna magna. The following experimental groups were considered (8 animals each):

- sham-operated animals: injection of 0.3 of mock CSF into cisterna magna;

- SAH group: rats subjected to the standard procedure of SAH induction;

- SAH-treated group: rats treated, after SAH induction, with nicardipine.

Treatment was started immediately after surgical procedure (1.2 mg/kg, i.p.) and administered every 8 hours in rats sacrificed at 48 hours after SAH induction. Animals were sacrificed and brain samples were obtained from cerebral cortex and hippocampus at 1, 6, and 48 hours after SAH induction.

The following enzymatic antioxidant activities were assayed: CuZn-superoxide dismutase (SOD) and GSH-Px activities for the cytoplasmatic compartment, Mn-SOD for the mitochondrial compartment, according to previously reported methods (11,12). Data about the *ex vivo* release of LTC_4 were considered only for cerebral cortex and mutated from previous experiments (4,5).

Statistical Analysis

The ANOVA and the Tukey's test were used in order to assess any difference in each brain area at any of the times considered between sham-operated, SAH, and SAH-treated animals.

The statistical model of linear regression and Fisher's F test were used in order to analyze the relationship between the time-dependent trend of LTC_4 release and, as the other variable, the trend of antioxidant enzymatic activities.

RESULTS

Cerebral Cortex

In cerebral cortex CuZn-SOD activity was significantly reduced 1 hour after SAH induction; the enzymatic activity in treated animals appeared to be restored only in the late phases (table 1). The mitochondrial Mn-SOD activity was significantly affected at 1, 6, and 48 hours after the hemorrhage. Mn-SOD activity was restored only at 6 and 48 hours after nicardipine treatment (table 1). GSH-Px activity was significantly reduced only in the late phase (48 hours) after SAH. In nicardipine-treated rats this enzymatic activity returned up to control values with a significant recovery ($p < 0.5$) at 48 hours (table 1).

The *ex vivo* release of LTC$_4$ in brain cortex is significantly enhanced since 1 hour after SAH induction while in nicardipine-treated rats it is significantly reduced until control values (4,5).

TABLE 1.   Antioxidant enzymatic activities in cerebral cortex of rats subjected to experimental SAH procedure. Values are expressed in enzymatic U/mg of protein ± SEM

	Sham-op	SAH	SAH + Nic
CuZn-SOD			
- 1 hour	4.85 ± 0.43	3.69 ± 0.16**	3.91 ± 0.11
- 6 hours	4.51 ± 0.34	2.96 ± 0.06***	2.61 ± 0.15***
- 48 hours	4.84 ± 0.12	3.35 ± 0.15	3.95 ± 0.15¶¶
Mn-SOD			
- 1 hour	1.08 ± 0.05	0.60 ± 0.02*	0.54 ± 0.06
- 6 hours	1.13 ± 0.04	0.55 ± 0.04**	0.99 ± 0.07¶
- 48 hours	1.11 ± 0.07	0.56 ± 0.04*	0.83 ± 0.09¶
GSH-Px			
- 1 hour	53.8 ± 7.0	54.7 ± 3.5	61.3 ± 3.3
- 6 hours	57.8 ± 6.1	62.6 ± 4.9	63.6 ± 3.4
- 48 hours	60.4 ± 2.0	41.8 ± 1.8***	62.6 ± 2.3¶

Statistical analysis: ANOVA and Tukey's test: *$p < .05$, **$p < .02$ and ***$p < .01$ between sham-operated and SAH; ¶ $p < .05$ and ¶¶$p < .02$ between SAH and SAH + nicardipine-treated rats. Nic = nicardipine.

Hippocampus

In the hippocampus CuZn-SOD activity was significantly reduced at 1 and 6 hours after SAH (table 2). Mn-SOD is significantly reduced at 6 and 48 hours, confirming that the mitochondrial compartment is most susceptible to peroxidative damage in the hippocampus (table 2) as well as in the cerebral cortex.

However, the reduction of enzymatic activity is in contrast with data observed in the brain cortex where especially in the late phase the enzymatic activity was reduced. A characteristic effect of nicardipine treatment was evident at 48 hours with a significant recovery of the mitochondrial SOD activity.

GSH-Px activity in the hippocampus is significantly affected only at 48 hours after SAH induction; thereafter, the effect of nicardipine on GSH-Px activity is less significant at any time (table 2).

TABLE 2.   Antioxidant enzymatic activities evaluated
in hippocampus of rats subjected to experimental
SAH procedure. Values are expressed in enzymatic U/mg
of protein ± SEM

	Sham-op	SAH	SAH + Nic
CuZn-SOD			
- 1 hour	4.66 ± 0.35	4.03 ± 0.06	4.11 ± 0.11
- 6 hours	4.23 ± 0.06	3.38 ± 0.13**	3.91 ± 0.14
- 48 hours	4.49 ± 0.20	4.00 ± 0.24*	4.00 ± 0.13
Mn-SOD			
- 1 hour	0.74 ± 0.15	0.73 ± 0.07	0.79 ± 0.04
- 6 hours	0.70 ± 0.14	0.43 ± 0.06*	0.44 ± 0.04
- 48 hours	0.76 ± 0.06	0.23 ± 0.02**	0.58 ± 0.04
GSH-Px			
- 1 hour	53.7 ± 3.7	49.2 ± 1.2	57.1 ± 2.3
- 6 hours	52.2 ± 2.1	58.7 ± 1.9*	55.7 ± 3.6
- 48 hours	50.9 ± 2.0	40.1 ± 1.7**	45.5 ± 1.7

Statistical analysis: ANOVA and Tukey's test: $*p < .05$ and $**p < .02$ between sham-operated and SAH; $p < .05$ between SAH and SAH + nicardipine-treated rats.

DISCUSSION

The results of the present study suggest the existence of an integrated and complex interplay among lipid-peroxidative processes, Ca^{2+}, antioxidant enzymatic activities, and the effects of calcium antagonists (13). There is a significant difference between effects of nicardipine treatment on enzymatic antioxidant activities in the cerebral cortex and the hippocampus. The selective vulnerability of the hippocampus probably reflects different metabolic patterns. In this region we noted a significant response to nicardipine treatment at 6 hours.

According to a vascular theory, the selective vulnerability of the hippocampus during ischemia, was attributed to the characteristic distribution of the tortuous arterial branches; analogously, after SAH, considering also the early impairment of CBF, a delayed delivery of nicardipine may be related to the late effect on antioxidant activities (14). In this sense nicardipine treatment may attenuate brain hypoperfusion following SAH, restore a valid CBF and consequently favor hippocampal local microcirculation, and restore metabolic and enzymatic conditions. Furthermore, behind the vascular theory there are many differences in the morphological, metabolic, and biochemical characteristics of the hippocampus. The ischemic and the hemorrhagic damage may determine a depolarization of hippocampal cell membranes and this results in calcium overload in the mitochondrial compartment. Nicardipine may exert a significant effect also in this sense.

Lipid peroxidation and Ca^{2+} are strictly linked in the pathogenesis of the neuronal damage. In fact, Ca^{2+} can amplify the disruption of membrane function related to lipid peroxidation; moreover, Ca^{2+} and lipid peroxidation may synergistically collaborate through a peroxidation-induced increase of membrane permeability to extracellular Ca^{2+} (13). Nicardipine may limit lipid-peroxidative processes (6), reduce the release of eicosanoids (5) which participate in the peroxidation-induced neuronal damage, and, lastly, preserve a quite normal antioxidant enzymatic activity, as shown in the present study.

Looking at the time-dependent trend of different patterns considered, we observed that after SAH induction, there is a significant inverse correlation between the increasing trend of LTC$_4$ release and the decreasing trend of GSH-Px and of mitochondrial Mn-SOD activity, with a statistical significance of $p < .01$ (personal unpublished results).

These results confirm that both lipid-peroxidative processes and impairment of antioxidant activities are strictly linked and, in particular, that the mitochondrial compartment is specifically susceptible to lipid-peroxidative damage and/or the impairment of scavenger factors. In nicardipine-treated rats this inverse correlation is meanwhile highly significant for GSH-Px, showing a significant reduction of the release of LTC$_4$ related to the progressive recovery of GSH-Px activity (fig. 1).

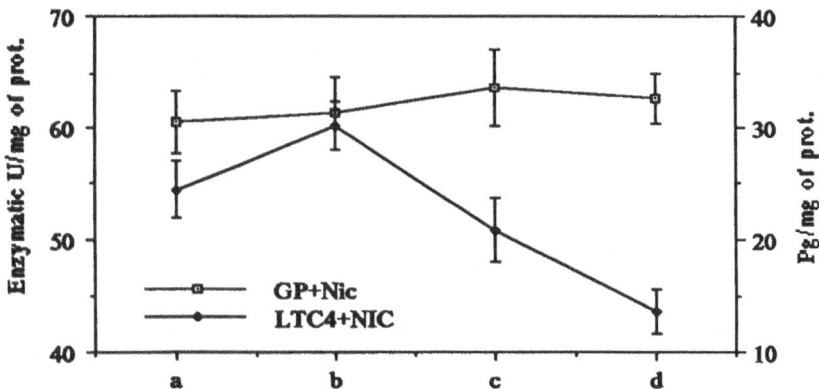

FIG. 1. Time-trend evaluation of *ex vivo* release of LTC$_4$ and glutathione peroxidase (GP) activity in brain cortex of rats subjected to experimental SAH induction and nicardipine treatment (see Material and Methods for details). Times of evaluation: (a) 30 minutes, (b) 1 hour, (c) 6 hours, and (d) 48 hours. Statistical analysis: linear regression and Fisher's F test: $p < .01$ at 6 and 48 hours.

The recovery of mitochondrial Mn-SOD activity only in the delayed phase at 48 hours after SAH induction is time related to the reduction of LTC_4 release (fig. 2). This kind of analysis did not show any kind of statistical correlation between LTC_4 release and CuZn-SOD activity which is the major antioxidant enzyme in the cytosolic compartment, nor in hemorrhagic, nor in nicardipine-treated animals. The brain-protective effect of nicardipine after SAH may depend on an integrated metabolic effect specifically exerted in the mitochondrial compartment and simultaneously involves the preservation of antioxidant enzymatic activities and the inhibition of lipid-peroxidative processes.

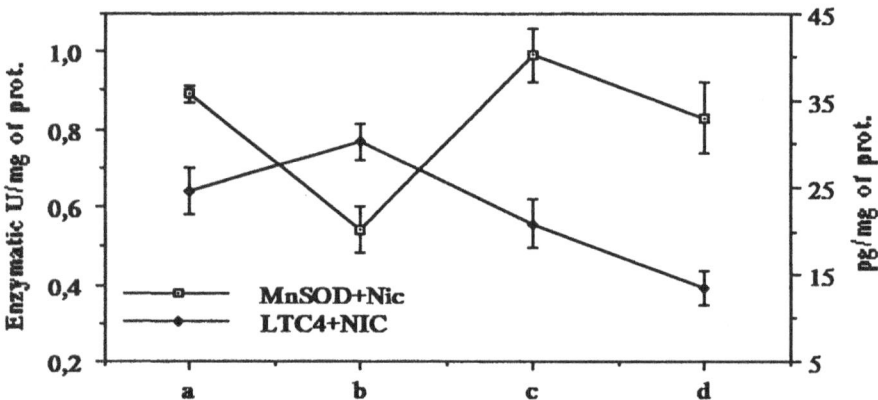

FIG. 2. Time-trend evaluation of *ex vivo* release of LTC_4 and Mn-SOD activity in brain cortex of rats subjected to experimental SAH induction and nicardipine treatment (see Material and Methods for details). Times of evaluation: (a) 30 minutes, (b) 1 hour, (c) 6 hours, and (d) 48 hours. Statistical analysis: linear regression and Fisher's F test: $p < .01$ at 6 and 48 hours.

REFERENCES

1.   Sano K., Asano T., Tanishima T. and Sasaki T. (1980): *Neurol. Res.*, 2: 253-272.
2.   Marzatico F., Gaetani P., Rodriguez y Baena R., Silvani V., Fulle I., Lombardi D., Ferlenga P. and Benzi G. (1989): *Mol. Chem. Neuropathol.*, 11: 99-107. 3.Sakaki S., Kuwabara H. and Ohta S. (1986): *Stroke*, 17: 196-202.
4.   Gaetani P., Marzatico F., Rodriguez y Baena R., Pacchiarini L., Viganò T., Grignani G., Crivellari M.T. and Benzi G. (1990): *Stroke*, 21: 328-332.
5.   Rodriguez y Baena R., Gaetani P., Marzatico F., Benzi G., Pacchiarini L. and Paoletti P. (1989): *J. Neurosurg.*, 71: 903-908.

6.   Marzatico F., Gaetani P., Spanu G., Buratti E. and Rodriguez y Baena R. (1991): *Acta. Neurochir.,* 108: 128-133.

7.   Chan P.H., Longar S. and Fishman R.A. (1987): *Ann. Neurol.,* 21: 540-547.

8.   Chan P.H., Chu L. and Fishman R.A. (1988): *Brain Res.,* 439: 388-390.

9.   Solomon R.A., Lobo A.J., Chen R.Y.Z., Bland L. and Chien S. (1985): *Stroke,* 16: 58-64.

10. Marzatico F., Gaetani P., Rodriguez y Baena R., Silvani V., Paoletti P. and Benzi G. (1988): *Stroke,* 19: 378-384.

11. Brannan T.S., Maker H.S., Weiss C. and Cohen G. (1980): *J. Neurochem.,* 35: 1013-1014.

12. Crapo J.D., McCord J.M. and Fridovich I. (1978): *Meth. Enzymol.,* 53: 382-393.

13. Braughler J.M. and Hall E.D. (1989): *Free Rad. Biol. Med.,* 6: 289-301.

14. Hadani M., Young W. and Flamm E.S. (1988): *Stroke,* 19: 1125-1132.

TREATMENT OF ACUTE ISCHEMIC STROKE WITH CALCIUM ANTAGONISTS

J. Gheuens, M. De Ryck, J. Van Reempts and T. Peters*

Janssen Research Foundation
Turnhouteweg 30, 2340 Beerse, Belgium and
*Raiffeisentr. 8, 4040 Neuss 21, Germany

Calcium exerts normal physiologic functions as a membrane stabilizer, metabolic regulator, and second messenger. However, it also can be involved in anoxic and toxic cell death (1). In animal models of global brain ischemia, which try to mimic the cerebral sequelae of cardiac arrest, loss of cellular calcium homeostasis has been related to: (a) selective neuronal necrosis, presumably in neurons with a higher density of calcium channels; and (b) delayed neuronal death, ascribed to cytosolic calcium overload. Similar pathologic processes have been assumed for neuronal tissue at risk in the "ischemic penumbra," i.e., an area of reversibly damaged brain tissue surrounding an irreversibly necrotic core in thrombotic stroke. The hypothesis of calcium-induced cell damage implies that pathological increases in cytosolic calcium concentration produce cell damage and death by causing enhanced lipolysis and proteolysis, energy wastage, activation of protein kinases, and altered gene expression. From this, a strategy for acute treatment of ischemic cerebrovascular accidents with calcium antagonists has been developed. However, at the present time this treatment strategy has not yet been fully confirmed by the results from clinical trials with calcium antagonists in stroke. It is not clear whether this discrepancy is due to incomplete or incorrect insight in pathogenetic mechanisms of ischemic stroke, or to methodological problems in the clinical trials.

EFFECTS OF CALCIUM ANTAGONISTS ON ISOLATED CELL PREPARATIONS

Calcium antagonists are divided into four subclasses according to the WHO classification (2). This classification was based upon important differences in their pharmacological profile in several models. In isolated myocytes, calcium antagonists with high affinity for L-type calcium channels were clearly less potent in protecting these cells from calcium-induced cell death than diphenylalkylamines, e.g., flunarizine (3). Similar effects were reported earlier for erythrocytes and endothelial cells

T. Godfraind et al. (eds.), Calcium Antagonists, 257–263.

(4,5). In synaptosomal preparations subjected to stimulated calcium influx, a comparable differentiation between subclasses of calcium blockers has been reported (6; Cohan, this issue). Calcium load in cultured neuronal cells could be antagonized with both class II and class IV blockers (7). Measurements of ion shifts and membrane potential in cerebral cells after ischemic or anoxic challenge revealed a twofold increase in latency of calcium influx and subsequent anoxic depolarization with flunarizine and with nimodipine (8). Recent voltage clamp experiments on the interaction of calcium antagonists with the different types of calcium channels on neuronal membranes confirmed important differences. Nifedipine preferentially inhibited the L-type calcium current. On the other hand, the T-type channel was blocked most potently by flunarizine (9). In addition, flunarizine also inhibited the high-threshold inactivating N-type calcium channel (10). These findings were recently confirmed in synaptosomal preparations (Cohan, this issue).

Other membrane-related pathways, which allow calcium overload in brain cells in ischemic conditions, include activation of sodium/calcium exchange mechanisms and entry of calcium through pathologically modified sodium channels. Flunarizine has been shown to prevent calcium-mediated neuronal cell death induced by sodium channel toxins, which prolong the opening of that channel, thus leading to excess sodium entry (11). Although the effects of different calcium antagonists vary, and the precise mode of action is not fully elucidated for some, the results described above provide a basis to explain the mechanisms by which calcium antagonists can reduce ischemic tissue damage.

EFFECTS OF CALCIUM ANTAGONISTS IN EXPERIMENTAL MODELS OF BRAIN ISCHEMIA AND STROKE

There exist several animal models of focal and global cerebral ischemia. Pathogenesis differs among these models, global ischemia being dominated by delayed neuronal decay in selectively vulnerable areas, and focal ischemia being characterized by progressive expansion of the lesion from the very labile peri-infarct area, the penumbra. However, the final pathway leading to cell death is similar in both conditions: loss of membrane integrity and consequent excessive rise of free intracellular calcium concentrations.

Global ischemia, and the influence thereon by calcium antagonists has been studied in detail in different species, mainly by histologic criteria. Neuronal necrosis after incomplete ischemia in rats and gerbils could be reduced by flunarizine, nicardipine, and nimodipine, albeit with variable success (12-18). There appeared to be no correlation between the cell protective effect of flunarizine and its influence upon cerebral blood flow (12,17).

Studies on the effects of calcium blockers on mortality and structural damage after cardiac arrest showed a variable degree of protection in dogs and primates (19-23). In rat models, post-treatment was successful with flunarizine (24,25) but not with nimodipine (24,26). A large number of *in vivo* studies focused on regional ischemia obtained by middle cerebral artery (MCA) occlusion. Class II calcium antagonists reduced infarct size after MCA occlusion in rats (27-29).

After thromboembolic obstruction of the MCA in dogs, flunarizine exerted a positive effect on the evolution of stroke (30). In rabbits, nimodipine and nicardipine as well as lidoflazine failed to improve neurological damage (31).

Protection of cerebral function constitutes the principal goal for any stroke therapy. Experimental models of stroke should therefore also address clinical priorities, i.e., protection of function. To this end, thrombotic infarcts were produced in the neocortex of rats by means of a photochemical technique. This photochemical method is virtually noninvasive, allows for reproducible infarct size and location, and by inducing endothelial damage/platelet aggregation interactions, simulates stroke-like ischemic events. The technique also makes it possible to confine the infarct to a selected brain structure and to adapt the behavioral analysis to predictable behavioral deficits. Flunarizine showed a significant reduction of cortical infarct size in this model of photothrombotic stroke in the rat (32,33). Thrombotic infarcts in the hindlimb area of the sensorimotor neocortex reliably produced long-lasting deficits in tactile and proprioceptive hindlimb placing reactions, contralateral to the side of infarction. This was most clearly revealed, when independent limb movements were required, indicating that this model allows for evaluation of deficits in independent limb use, a behavioral parameter possibly relevant to human stroke. Forelimb placing recovered, at least partially, within the first two weeks after infarction. By contrast, hindlimb placing deficits hardly recovered at all, and could be reliably demonstrated up to at least two months after infarction. This provided a robust deficit against which to evaluate the efficacy of drug treatment (33-35).

In a first series of experiments, administration of 1.25 mg/kg flunarizine IV 30 minutes after infarction resulted in marked sparing of sensorimotor function. In contrast to infarcted control rats, which remained deficient for at least 21 days, 75% of the flunarizine-treated rats showed normal hindlimb placing on the first postinfarct day, whereas the remainder recovered within five days. When infarcts were measured four hours after induction, at the time of maximal spread of ischemic damage in this model, flunarizine, administered as mentioned above, reduced infarct size by 31% (33).

Post-treatment with an intravenous bolus of flunarizine 5 minutes after neocortical infarction also protected sensorimotor function in a dose-dependent manner (35). The time-window of therapeutic opportunity was investigated by varying the post-treatment delay and dose of treatment with intravenous flunarizine. Post-treatment with 0.31 mg/kg flunarizine IV was totally ineffective, even when the bolus was administered as early as 5 minutes after infarction. At 0.63 mg/kg, flunarizine was effective only when administered within 30 minutes after infarction. At 1.25 mg/kg, scores were significantly better than control values with treatment delays up to and including 6 hours after infarction (De Ryck et al., in preparation). Thus, flunarizine antagonized neurological deficits when administered up to six hours after infarction, a therapeutic time-window that overlaps with the first four hours after infarction when the zone of ischemia has been shown to expand in this model. It should be noted that a 3 to 4 hour long therapeutic time-window has been proposed for the distal MCA occlusion model in rats, which produces a rather specific penumbral dysfunction in the neocortex (29,36,37). Thus, the penumbral window closes within 3 to 6 hours after a focal cerebrovascular insult in the rat model.

Although there is some variation in the results reported, there is a large amount of evidence that calcium entry blockers are able to attenuate structural deterioration of neuronal cells and that neurological function can be significantly improved in animal experiments of global and focal ischemia. With regard to acute treatment of stroke, the concept of an initial penumbra and a window of therapeutic opportunity lasting 3 to 6 hours should therefore be viewed as an incitation to early intervention.

GENERAL COMMENTS ON CLINICAL STUDIES OF STROKE WITH CALCIUM ANTAGONISTS

Despite the existence of a biological rationale and pharmacological results, clinical trials with different calcium antagonists (and other compounds) in acute ischemic stroke have been largely inconclusive, or the results have been difficult to reproduce. It is submitted that this may be at least in part due to methodological problems in the clinical trials.

Stroke clinical trials have suffered from heterogeneous patient samples. Many different clinical syndromes, etiologies, and possibly different pathogenic mechanisms are associated with stroke. This clinical reality is in sharp contrast with the standardization of the experimental stroke models used to assess the effect of calcium antagonists.

Errors in diagnosis add to the difficulties. If brain imaging is not used for screening the patients, the clinical misdiagnosis rate is 10-15%, making it virtually impossible to draw conclusions on therapeutic benefit (37,38).

The expectations of therapeutic effects should not be exaggerated. Including patients with major irreversible necrotic lesions and in clinical conditions that are beyond any treatment at all reflects an overly optimistic attitude towards expected therapeutic effects. Another problem pertains to the optimal way of assessing therapeutic efficacy in clinical stroke trials. There exist several scales that score neurological disability in a fundamentally different manner. For instance, the MCA scale [Orgogozo and Dartigues (39)], Scandinavian Stroke Study Scale (40), and Toronto Scale (41) differ from one another in being ordered and weighted, more linear, and nominally descriptive, respectively. In addition to a neurological scale, activities-of-daily-living (ADL) scales are incorporated into clinical trials. Of these, the Barthel index seems to be the most sensitive and reliable (42). One problem of ADL is that, in contrast to the neurological disability, it can not be scored upon entry of the acute stroke patient in the hospital, making comparison of evolution versus baseline values impossible.

Perhaps most importantly, the window of therapeutic opportunity which is documented in the experimental work is not always reflected in the time-to-treatment in clinical trials. The current pattern of clinical practice is based on a more or less fatalistic attitude towards stroke treatment, leading to selection bias and delay in referral of the patient to the hospital. Therefore, prolonged delays in the onset of treatment have been allowed in earlier clinical trials, under the pressure of having to recruit sufficient patients. There is a growing consensus that clinical trials in stroke must respect the window of therapeutic opportunity indicated by the experimental work, and restrict time-to-inclusion preferably to 6 hours or less.

CONCLUSION

The distinction between receptor- and voltage-operated calcium channels, as well as the further subdivision of the latter into the subtypes T, N, and L, along with the possible relationships of some of these calcium channels to other ion channels, continues to provide ever new challenges to the development of calcium antagonists aimed at the neuropathology involved in cerebral ischemia. The experimental stroke models have produced evidence that a significant benefit can be obtained from calcium antagonist therapy. However, the standardized conditions of these models are hardly ever encountered in the actual clinical trials. In view of the above considerations, it must be concluded that calcium antagonists have not yet been given a decisive challenge in a clinical trial in ischemic stroke.

REFERENCES

1.   Siesjö B.K. and Bengtsson F. (1989): *J. Cereb. Blood Flow Metab.*,

9: 127-140.

2.   Vanhoutte P.M. and Paoletti R. (1987): *Trends Pharmacol. Sci.*, 8: 4-5.

3.   Borgers M., Ver Donck L. and Vandeplassche G. (1988): *Ann. N.Y. Acad. Sci.*, 522: 433-453.

4   De Clerck F. and Hladovec J. (1984): In: *Calcium Entry Blockers in Cardiovascular and Cerebral Functions*, edited by T. Godfraind, A.G. Herman and D. Wellens pp. 81-90. Martinus Nijhoff Publishers, The Hague.

5.   Hladovec J. (1979): *Arzneimittelforsch*, 29: 1101-1103.

6   Wibo M., Delfosse I. and Godfraind T. (1983): *Arch. Int. Pharmacodyn. Ther.*, 263: 333-334.

7.   Geerts H., Nuydens R., Nuyens R., Cornelissen F. and Pauwels P. (1991): In: *Ionic Currents and Ischemia*, edited by J. Vereecke, P.P. Van Bogaert and F. Verdonck pp. 310-312. Leuven University Press, Leuven.

8.   Marrannes R., De Prins E., Fransen J. and Wauquier A. (1989): Satellite Symposium on *Ionic Regulation in Nervous Tissue*, XXXIth International Congress of Physiological Sciences, Helsinki, Finland, abstract.

9.   Akaike N., Kostyuk P.G. and Osipchuk Y.V. (1989): *J. Physiol.*, 412: 181-195.

10.   Tytgat J., Pauwels P.J., Vereecke J. and Carmeliet E. (1991): *Brain Res.*, 549: 112-117.

11.   Pauwels P.J., Leysen J.E. and Janssen P.A.J. (1991): *Life Sci.*, 8: 1881-1893.

12.   Deshpande J.K. and Wieloch T. (1986): *Anesthesiology*, 64: 215-224.

13.   Grotta J., Spydell J., Pettigrew C., Ostrow P. and Hunter D. (1986): *Stroke*, 17: 213-219.

14.   Van Reempts J., Haseldonckx M., Van Deuren B., Wouters L. and Borgers M. (1986): *Drug Dev. Res.*, 8: 387-395.

15.   Bunnell O.S., Louis T.M., Saldanha R.L. and Kopelman A E. (1987): *Med. Sci. Res.*, 15: 1513-1514.

16.   Alps B.J., Calder C., Hass W.K and Wilson A.D. (1988): *Br. J. Pharmacol.*, 93: 877-883.

17.   Beck T., Nuglisch J. and Sauer D. (1988): *Eur. J. Pharmacol.*, 158: 271-274.

18.   Krieglstein J., Sauer D., Nuglisch J., Karkoutly C., Beck T., Bielenberg G.W., Rossberg C. and Mennel H.D. (1989): In: *Cerebral Ischemia and Calcium*, edited by A. Hartmann and W. Kuschinsky pp. 223-231. Springer-Verlag, Berlin.

19.   Steen P.A., Newberg L.A., Milde J.H. and Michenfelder J.D. (1983): *J. Cereb. Blood. Flow Metab.*, 3: 38-43.

20.   Steen P.A., Gisvold S.E. and Milde J.H. (1985): *Anesthesiology*, 62:

406-414.

21. Newberg L.A., Steen P.A., Milde J.H. and Michenfelder J.D. (1984): *Stroke*, 15: 666-671.

22. Edmonds H.L., Wauquier A., Melis W., Van Den Broeck W.A.E., Van Loon J. and Janssen P.A.J. (1985): *Am. J. Emerg. Med.*, 3: 150-155.

23. Kumar K., Krause G., Koestner A., Hoehner T. and White B. (1987): *Exp. Neurol.*, 97: 115-127.

24. Wauquier A., Melis W. and Janssen P.A.J. (1989): *Neuropharmacol.*, 28: 837-846.

25. Lu H.R., Van Reempts J., Haseldonckx M., Borgers M. and Janssen P.A.J. (1990): *Am. J. Emerg. Med.*, 8: 1-6.

26. Calle P.A., Bogaert M.G., De Ridder L. and Buylaert W.A. (1990): *Arch. Pharmacol.*, 341: 586-591.

27. Germano I.M., Bartkowski H.M., Cassel M.E. and Pitts L.H. (1987): *J. Neurosurg.*, 67: 81-87.

28. McCulloch J., Graham D.I., Harper A.M. and Teasdale G.M. (1989): In: *Cerebral Ischemia and Calcium*, edited by A. Hartmann and W. Kuschinsky pp. 177-186. Springer-Verlag, Berlin.

29. Sauter A., Rudin M. and Wiederhold K.-H. (1989): In: *Cerebral Ischemia and Calcium*, edited by A. Hartmann and W. Kuschinsky pp. 282-291. Springer-Verlag, Berlin.

30. De Ley G., Weyne J., Demeester G., Stryckmans K., Goethals P. and Leusen I. (1989): *Stroke*, 20: 357-361.

31. Lyden P.D., Zivin J.A., Kochlar A. and Mazzarella V. (1988): *Stroke*, 30: 1020-1026.

32. Van Reempts J., Van Deuren B., Van de Ven M., Cornelissen F. and Borgers M. (1987): *Stroke*, 18: 1113-1119.

33. De Ryck M., Van Reempts J., Borgers M., Wauquier A. and Janssen P.A.J. (1989): *Stroke*, 20: 1383-1390.

34. De Ryck M. (1990): *Eur. Neurol.*, 30(suppl.2): 21-27.

35. De Ryck M., Duytschaever H., Pauwels P.J. and Janssen P.A.J. (1990): *Stroke*, 21: III-158-III-163.

36. Brint S., Jacewicz M., Kiessling M., Tanabe J. and Pulsinelli W. (1988): *J. Cereb. Blood Flow Metab.*, 8: 474-485.

37. Norris J.C.O. and Hachinski V.C. (1982): *Lancet*, 1: 328.

38. TRUST Study Group (1990): *Lancet*, 336: 1205-1209.

39. Orgogozo J.M. and Dartigues J.F. (1986): In: *Acute Brain Ischemia*, edited by N. Battistini et al. pp. 281-283. Raven Press, New York.

40. Scandinavian Stroke Study Group (1985): *Stroke*, 16: 885-890.

41. Norris J.C.O. (1982): *Stroke*, 13: 527-528.

42. Granger C.V., Greer D.S. and Liset E. (1975): *Stroke*, 6: 34-41.

PHARMACOLOGICAL APPROACH TO ACUTE ISCHEMIC STROKE

C. Fieschi, D. Toni, M. Frontoni, M. Fiorelli, M.L. Sacchetti
C. Argentino, and M. Gentile

Department of Neurological Sciences
University "La Sapienza"
Viale dell'Università, 30
00185 Rome, Italy

Negative results of trials on the pharmacological treatment of acute ischemic stroke have represented a constant, distressing experience for the clinicians involved. In fact, none of the drugs effective in preventing or limiting cerebral ischemia in animals has thus far overcome the proof of a clinical trial. Nevertheless, the existence of an area of ischemic penumbra surrounding infarction (1), liable to complete recovery if blood flow is promptly restored, has been revealed by animal models. Moreover, knowledge on the cascade of events leading to irreversible neuronal death, through the final common pathway of intracellular calcium increase, also comes from experimental data (2). These acquisitions on the pathophysiology of cerebral ischemia are fundamental cornerstones on which a rational therapeutic approach should be based. In this light, two main strategies of pharmacological treatment, neuroprotection and early revascularization, appear currently suitable.

NEUROPROTECTION

The effectiveness of the best-known neuroprotective agents lies in their capability to prevent influx of calcium ions, by acting on two fundamental sites, the receptor-operated and the voltage-operated calcium channels, respectively. Competitive and noncompetitive N-Methyl-D-Aspartate (NMDA) antagonists of glutamate (3) specifically interact with the first, whereas calcium antagonists selectively block the second ones (4). So far, only calcium antagonists have been clinically tested, and published trials concern only four studies on orally administered nimodipine (table 1). The final results of experiences with nicardipine, isradipine and flunarizine are awaited.

T. Godfraind et al. (eds.), Calcium Antagonists, 265–270.

TABLE 1.   Clinical trial of nimodipine in ischemic stroke

	Gelmers (1988)	Martinez (1990)	Bogousslavski (1990)	TRUST (1990)
Nimodipine treatment	30mg/6h x 28 days	30mg/6h x 28 days	30mg/6h x 14 days	40mg/8h x 21 days
Enrolled	186	164	60	1215
Excluded	-	41	8	-
Treated	93	81	24	608
Treatment delay	24h	48h	48h	48h
Admit neurologic grade			Mathew scale score ≥50 ≤75	
Outcome measures	Mathew scale	Mathew scale	Mathew scale	Mathew scale
Case fatality (No. of Pts.)*	8 N 19 P	6 N 10 P	- 1 P	92 N 75 P

*N: treated with nimodipine; P: receiving placebo.

The designs of Gelmers' and Martinez-Vila's trials are very similar (5,6), showing a small number of enrolled patients and a large delay (24-48 hours) between stroke onset and beginning of treatment. Both studies claim a decrease of death rate and a better outcome in treated patients. Nevertheless, case fatality ratios are not significantly different when all randomized patients are considered according to the intention-to-treat analysis. Moreover, nimodipine appears effective in limiting cardiac and pulmonary fatal events, but no significant difference exists in cerebral deaths between the two groups.

As regards the neurological outcome of survivors, the significant improvement reported by authors is obtained by subtracting the neurological score at the end of follow-up from that at baseline. This method would be rejected as clinically irrelevant by most of biostatisticians.

Over twelve hundred patients were randomized in the TRUST study (7). Despite the respectable sample size, inclusion criteria do not seem selective enough to test efficacy. Furthermore, CT scan was performed only in 25% of treated patients and the interval between stroke onset and randomization was over 48 hours. In Bogousslavsky's trial, nimodipine was administered to 24 patients with nonsevere acute ischemic stroke within 48 hours from onset (8). Lack of efficacy of the drug emerged from the study but, unfortunately, once again small sample size, selection criteria, and treatment delay make debatable an experience otherwise well conducted from a statistical point of view. The negative results of all published studies are therefore inconclusive.

The failure of nimodipine in the treatment of ischemic stroke is in contrast with its effectiveness in preventing ischemic complications of subarachnoid hemorrhage (SAH). Data from the British Aneurism Nimodipine Trial (BRANT) showed in fact a less severe neurological deficit in treated subjects with respect to controls (9). Since similar rates of symptomatic vasospasm were found in the two groups, these results could be attributed to neuroprotective effects of the drug. In this study nimodipine was administered before the occurrence of vasospasm, thus reproducing the experimental condition of stroke pretreatment with that Ca antagonist. Nimodipine might therefore be considered as a preventing measure in patients at high risk of ischemic stroke (repeated transient ischemic attacks, bilateral internal carotid stenosis, etc.).

Considering the neurotoxicity of glutamatergic receptor activation, which in turn leads to an excessive Ca influx, an important role could be played in neuroprotection by competitive and noncompetitive NMDA antagonists of that receptor. Phase III studies on these agents are still ongoing, so that their effectiveness is to be kept, at present, as potential. To date, in conclusion, there is no definitive and undebatable data on the efficacy of neuroprotective agents in clinical practice.

REVASCULARIZATION

Pharmacological possibilities of brain protection are not expected to completely resolve events leading to irreversible ischemia, when the primary event is an occlusion of major intracranial arteries downstream of the anastomotic circle of Willis. Early thrombolysis is likely the therapeutic approach of choice in this case.

Overestimation of hemorrhagic complications, due to lack of systematic diagnostic confirmation by CT scan, and aggressive technique made previous studies on local intra-arterial fibrinolysis somewhat unpopular (10,11,12). More recently, successes of early thrombolytic treatment of myocardial infarction (13) and availability of intravenously administered newer compounds, selectively acting on clot, such as rt-PA (14,15), have prompted numerous studies in the United States and Europe (tables 2 and 3).

TABLE 2.   Trials of efficacy and safety of rt-PA* in acute ischemic stroke

	NIH SSG**	NIH SSG**	rtPA SSG**	German-PA SSG**	Heidelberg MCA-Trial***
No. of pts.	74	20	104	12	26
Dose	<85 mg/kg	>85 mg/kg	0.12-0.75	70 mg	100 mg
Treatment delay	90 min	90-180 min	8 h	2-18 h	6 h
Angiography (Yes/No)	Y	N	Y	Y	Y
Anticoagulant	N	N	N	Y	N

* rt-PA: tissue-Plasminogen Activator obtained by recombinant DNA technology
** SSG: Stroke Study Group
*** MCA: Middle Cereberal Artery

TABLE 3.   Clinical trials in progress

Study	No. of pts.	Dose of rt-PA	Interval stroke onset/treatment	Angiography (Yes/No)
NINDS	280	0.90 mg/kg	<3h	N
Genentech	100	0.90 mg/kg	<6h	Y
Kopenhagen	?	15 + 85 mg	<6h	Y
HIMEJI	31	30 MU IV	<6h	Y
Okada	15	15 MU IV	<6h	Y
	11	30 MU IV	<6h	Y
Yamagouchi	49	20 MU IV	<6h	Y

A pathophysiological study of our group has further supported this approach (16). Eighty patients with first acute ischemic stroke were submitted to cerebral angiography within 6 hours from clinical onset. In this study, cerebral embolism appeared more common than previously reported. Angiograms showed, in fact, complete occlusion, possibly embolic in nature and liable to spontaneous or therapeutic resolution, in

66% of cases. Moreover, patients with intracerebral occlusion and poor collateral circulation had the worst clinical outcome. Therefore, the presence of a good collateral blood supply seems to be a prerequisite for thrombolytic intervention (17).

Thrombolysis should not take place in the absence of good collateral supplies or later than the first few hours from onset. Furthermore, the possibility of tissue damage from recanalization must also be taken into account, since cascade mechanisms similar to those of previous ischemic event can occur in the early phase of revascularization. For these reasons, neuroprotection is, in our opinion, complementary rather than an alternative to thrombolysis.

We think that neuroprotective agents should be administered to extend the temporal limits of ischemic penumbra, thus allowing a longer delay for revascularization. Once viable tissue is obtained by reperfusion, neuroprotection could then counteract the above mentioned toxic effects of early revascularization itself. By consequence, the treatment of ischemic stroke includes the possibility of coupling, in future clinical experiences, the two main branches of pharmacological approaches. Nonetheless, progress in this area should not be expected, even with the best drugs available, if the rules of times, sample size, and patients selection (18), appropriate to a well-designed clinical trial, are not respected.

REFERENCES

1. Astrup J., Siesjo B.K. and Symon L. (1981): *Stroke*, 12: 723-725.
2. Siesjo B.K. (1981): *J. Cereb. Blood Flow Metab.*, 1: 155-185.
3. Rothmans S.M. and Olney J.W. (1987): *TINS*, 10: 299-302.
4. Auer L.M. (1983): In: *Cerebrovascular Diseases*, edited by M. Reivich and H.I. Hurtin pp. 375-384. Raven Press, New York.
5. Gelmers H.J., Gorter K., De Weerdt C.J. and Wiezer H.J.A. (1988): *New Engl. J. Med.*, 318: 203-207.
6. Martinez-Vila E., Guillen F., Villanueva J.A., Matias-Guiu, Bigorre J., Gil P., Carbonell A. and Martinez-Lage J.M. (1990): *Stroke*, 21: 1023-1028.
7. TRUST Study Group (1990): *Lancet*, 336: 1205-1209.
8. Bogousslavsky J., Regli F., Zumstein V. and Kobberling W. (1990): *Eur. Neurol.*, 30: 23-26.
9. Pickard J.D., Murray G.D., Illingworth R., Shaw M.D.M., Teasdale G.M., Foy P.M., Humphrey P.R.D., Lang D.A., Nelson R., Richards P., Sinar, J., Bailey S. and Skene A. (1989): *Br. Med. J.*, 298: 636-642.
10. Del Zoppo G.J., Ferbert A., Otis S., Bruckmann H., Hacke W., Zyroff J., Harker L.A. and Zeumer H. (1988): *Stroke*, 19: 307-313.

11.   Haremberg J., Zimmermann R., Heuch C.C., Schmidt-Gayle U., Simon B. and Wahl P. (1980): In: *Fibrinolysis and Urokinase*, edited by V. Tilsner and H. Lenau pp. 391-404. Academic Press, London.

12.   Zeumer H., Freitag H.C., Grzyska U. and Nuenzig H.-P. (1989): *Neuroradiol.*, 31: 336-340.

13.   Wilcox R.G., Von der Lippe G., Olssen C.G., Jensen G., Skene A.M. and Hampton J.R. (1988): *Lancet*, 2: 525-530.

14.   Sloan, MA (1987): *Arch. Neurol.*, 44: 748-768.

15.   Zivin J.A., Lyden P.D., De Girolami U., Kochbar R., Mazzarella V., Hemenway C.C. and Johnston P. (1988): *Arch. Neurol.*, 45: 387.

16.   Fieschi C., Argentino C., Lenzi G.L., Sacchetti M.L., Toni D. and Bozzao L. (1989): *J. Neurol. Sci.*, 91: 311-322.

17.   Bozzao L., Fantozzi L.M., Bastianello S., Bozzao A., Argentina C., Lenzi G.L. and Fieschi C. (1989): *Stroke*, 20: 735-740.

18.   Fieschi C., Argentino C., Lenzi G.L., Fantozzi L.M., Sacchetti M.L., Pace A., Rasura M., Bastianello S., Bozzao L., Zanette E., Buttinelli C., Giubilei F. and Pantano P. (1988): *Ann. N.Y. Acad. Sci.*, 522: 662-666.

CALCIUM ANTAGONIST AND SUBARACHNOID HEMORRHAGE

Fredric B. Meyer, M.D.

Mayo Clinic
Department of Neurosurgery
200 First Street SW
Rochester, Minnesota 55905

There is a growing body of literature which indicates that calcium antagonists decrease the morbidity from vasospasm following subarachnoid hemorrhage by attenuating delayed ischemic deficits. At present, the dihydropyridine calcium antagonist nimodipine is approved for use by the Food and Drug Administration for treating patients suffering from aneurysmal hemorrhage. It is reasonable to anticipate that in the future additional calcium antagonists will be developed for treatment of neurological disorders. The mechanisms by which calcium antagonists exert a beneficial effect on subarachnoid hemorrhage remains controversial. For example, the initial concept for using calcium antagonists focused on the inhibition or reduction of vasospasm with resulting increases in cerebral blood flow. However, intensive angiographic studies, both in animal experiments and in patients, have failed to demonstrate consistent reversal of hemorrhage induced vasoconstriction. Thus, there is an apparent effect without an explanation. Alternative hypotheses for this beneficial effect have focused on vasodilatation of small surface-conducting arteries not visualized on angiography or on direct neuronal protection. Therefore, it seems reasonable to review the data within the framework of vascular versus neuronal effect. Accordingly, a brief overview of pertinent physiology will be discussed followed by a review of the major clinical studies which have tested the effects of calcium antagonists in subarachnoid hemorrhage.

POTENTIAL MECHANISMS

Modulation of Cerebral Blood Flow

The end result of vasospasm is a significant reduction in cerebral blood flow (CBF) to critical levels which reduces neuronal function and ultimately compromises membrane integrity. It is important to note that the response of neuronal tissue to reductions in blood flow is not an all-or-none phenomenon. In awake man, normal CBF is approximately 50

T. Godfraind et al. (eds.), Calcium Antagonists, 271–277.

ml/100 g brain tissue/min. When blood flow declines to approximately 15-18 ml/100 g brain tissue/min, there is loss of neuronal electrical function as observed in patients undergoing carotid endarterectomy. If CBF further declines to approximately 10-12 ml/100 g brain tissue/min, there is a loss of neuronal metabolic function which includes ionic homeostasis, brain pH regulation, and energy production. Although the tolerance of brain tissue to these profound reductions in CBF is not entirely known, it is assumed that after 1-2 hours, irreversible neuronal injury occurs. Between these two thresholds of electrical failure and membrane failure there is a small zone of reduced blood flow that will supply enough substrate for low-level metabolism allowing ionic homeostasis despite neuronal functional loss (1). With subsequent increases in CBF, functional recovery can occur. This concept of thresholds of ischemia offers a rationale for why patients suffering from a delayed ischemic deficit due to vasospasm can demonstrate significant recovery even if blood flow is marginally increased. Therefore, if calcium antagonists increase CBF through either reversal of angiographically observed vasospasm or dilatation of small parenchymal conducting arteries, the result can be dramatic.

A variety of *in vitro* studies have demonstrated that the dihydropyridine calcium antagonists can inhibit in a dose-dependent fashion vasoconstriction induced by a variety of agents such as potassium, serotonin, phenylephrine, and CSF from subarachnoid hemorrhage patients (2-4).

In vivo studies on the effects of calcium antagonists on CBF and brain metabolism are controversial. There is general agreement that in nonischemic models, agents like nimodipine will increase CBF without increasing oxygen consumption or glucose utilization (5). Although it is likely that an increase in CBF is caused by vasodilatation, there is also some evidence that calcium antagonists may increase microcirculatory flow by decreasing platelet aggregation. The mechanism by which calcium antagonists increase CBF in brain ischemic models may be through reversal of ischemic vasoconstriction which is thought to be due to a calcium mediated mechanism. In models of focal ischemia, although CBF is generally increased both in core and borderzone ischemic regions, the outcome on metabolic and histological parameters is controversial. Various laboratories have reported inconsistent results regarding brain pH, thresholds of ischemic edema and ionic influx, and infarction size (6,7).

Animal models testing the effects of calcium antagonists on subarachnoid hemorrhage have also yielded inconsistent results (8-12).

Neuronal Protection

The other potential mechanism by which calcium antagonists may be beneficial in subarachnoid hemorrhage is by retarding Ca^{2+} influx into ischemic neurons. As suggested above, degradative metabolic cascades transpire when CBF is reduced to approximately 10 ml/100 g/min. The original calcium hypothesis of brain injury postulated that with profound reductions in CBF, there was rapid ATP depletion. This in turn caused membrane depolarization with opening of voltage-dependent L channels (13). Accordingly, it was postulated that dihydropyridine calcium antagonists could be beneficial by blocking calcium influx through these L channels. More recent investigations have demonstrated the critical role that excitatory amino acids play in ischemic neuronal injury. This more recent "excitotoxic hypothesis" proposes that there is both an early and delayed from of ischemic injury. The early ischemic injury is attributable to an influx of NA^+, CL^-, and H_2O with osmolysis. The delayed deficit is thought to be secondary to Ca^{2+} influx. However, the route by which calcium enters neurons is thought to occur primarily through receptor-operated calcium channels such as the N-methyl-D-aspartate receptor. In this current hypothesis, Ca^{2+} influx through the voltage-dependent L channel is considered to be of secondary importance. Of further concern is the fact that intracellular Ca^{2+} could increase during ischemia through mechanisms independent of membrane channels, including release from endoplasmic reticulum or through the Na^+/Ca^{2+} antiport pump.

Given the fact that only the voltage-dependent L channel is modulated by organic calcium antagonists, it is not surprising that the evidence demonstrating that these antagonists can modulate neuronal Ca^{2+} concentrations is quite limited. Perhaps the most convincing evidence lies in the find that calcium antagonists have direct anticonvulsant properties in experimental seizures. Since calcium influx has been linked to the initiation of spontaneous neuronal discharge and epileptic activity, it has been theorized that the anticonvulsant effect of calcium antagonists is due to attenuation of transmembrane Ca^{2+} influx (14).

In summary, although some of these calcium antagonists possess the physical properties necessary for blood-brain barrier penetration and neuronal binding sites for them exist, the evidence that calcium antagonists can directly protect neurons or modulate function is inconclusive.

CLINICAL TRIALS

The calcium antagonist that has been most thoroughly evaluated as a treatment for subarachnoid hemorrhage is nimodipine. Nimodipine has the theoretical advantage of being cerebrovascular-smooth-muscle selective and of effectively penetrating the blood-brain barrier. Other calcium antagonists that have been investigated or proposed for the treatment of subarachnoid hemorrhage include nicardipine, flunarizine, and diltiazem. There are five major prospective clinical studies which merit comment (15).

In 1983, Allen and colleagues (16) published the first prospective study in which nimodipine or placebo was randomly administered to 116 patients within 96 hours of subarachnoid hemorrhage. All of these patients were good grade patients with limited neurological deficits following the aneurysm hemorrhage. Follow-up was assessed at 21 days. These investigators noted that although the overall outcome between the drug and placebo groups were not significantly different, nimodipine statistically reduced the severity of the neurological deficits caused by vasospasm. The number of patients who died or had severe ischemic deficits was 1.7% in the nimodipine group versus 13.3% in the placebo group (P <0.05).

Philippon and colleagues (17) performed a prospective study in 70 patients suffering from subarachnoid hemorrhage. They observed that nimodipine reduced the incidence of poor outcomes from vasospasm alone but did not influence the overall outcome between the drug and placebo groups. The permanent neurological deficits, including death from vasospasm, occurred in 6.4% of nimodipine-treated patients versus 25.6% in the placebo-controlled group.

In 1987, Neil-Dwyer and coinvestigators (18) randomized 50 patients suffering from subarachnoid hemorrhage noting that 26% of the nimodipine group had a good outcome at three months as compared to 52% of the placebo group. Unique in this trial was the cisternal administration of nimodipine intraoperatively. In addition, cerebral blood flow measurement with xenon[133] could not demonstrate any increases in CBF.

In 1988, the Canadian study (19) demonstrated that the oral administration of nimodipine increased both the incidence and severity of delayed ischemic deficits in patients in poor neurological condition following subarachnoid hemorrhage. The follow-up period was three months and included 72 nimodipine patients and 82 placebo controls. The incidence of angiographically proven vasospasm was similar between the two groups: 64% nimodipine versus 66% placebo. However, the number of patients with permanent neurological deficits was 29% in the nimodipine group as compared to 47% in the placebo controls. At three-month follow-up, 29% of the treatment group had made a good recovery as compared to only 9.8% of the placebo patients. It should be noted that the mortality rate in the nimodipine group was slightly higher but not statistically significant and that CT analysis of the three-month survivors demonstrated no significant differences in infarction size between the two treatment groups.

Finally, Pickard and colleagues (20) published a multicenter British trial in which 278 patients were administered nimodipine and 276 received a placebo. At the three-month follow-up, it was observed that in patients of all grades, nimodipine reduced the incidence of both infarction (61 patients, 22%) and poor outcome (55 patients, 20%) when compared with controls in which 92 (33%) and 91 (33%) patients suffered infarction and poor outcome, respectively. The authors note that there was no difference in the incidence of angiographic vasospasm between the two treatment groups.

There are a variety of prospective studies in which the results with using nimodipine were compared to historical controls (21,22). Although these clinical investigations were not placebo controlled, they agree with the above reported results. There are also several studies which have reported on the intravenous use of nimodipine in subarachnoid hemorrhage patients (23,24). These studies also suggest that nimodipine reduced the severity of neurological deficits attributable to vasospasm.

There are several investigations which have attempted to look at the effects of calcium antagonists on CBF, either by the use of transcranial doppler or xenon[133] inhalation (17,25). Overall, these clinical investigations have not demonstrated any significant increases in CBF attributable to the use of calcium antagonists.

SUMMARY

A critical review of the clinical data indicates that nimodipine does decrease the severity of neurological deficits and improves outcome following subarachnoid hemorrhage. There are currently several ongoing investigations testing the effects of nicardipine in this clinical setting (26,27). Mechanisms by which mortality and morbidity are reduced by nimodipine are still controversial. First, the frequency of vasospasm is not altered. Second, the consistent reversal of vasospasm once present has not been demonstrated either angiographically or by noninvasive CBF studies. These observations suggest that there is either modification of microcirculatory flow (i.e., dilatation of pial conducting vessels or decrease aggregation) or a direct neuronal protective effect. As suggested above, support for either mechanism is not resolute and further investigation is necessary. Given the similar findings from well-conducted prospective placebo-controlled studies, the risk/benefit ratio clearly supports the prophylactic use of nimodipine in patients of all clinical grades suffering from a subarachnoid hemorrhage. There is some evidence which indicates that starting nimodipine after the onset of delayed ischemic deficits may be of benefit. Finally, it can be predicted that in the future, additional calcium antagonists with more selective vascular or neuronal effects will be developed for use in subarachnoid patients.

REFERENCES

1. Astrup J., Siesjö B.K. and Simon L. (1981): *Stroke*, 12: 723-725.
2. Allen G.S. (1985): *Am. J. Cardiol.*, 55: 149B-153B.
3. Allen G.S. and Bahr A.L. (1979): *Neurosurgery*, 4: 43-47.
4. Brandt L., Andersson K.E., Ljunggren B., Saveland H. and Ryman T. (1988): *Acta. Neurochir.*, 45(suppl.): 11-20.
5. Haws C.H. and Heistad D.D. (1984): *Am. J. Physiol.*, 247: H170-H176.
6. Harper A.M., Craigen L. and Kazda S. (1981): *J. Cereb. Blood Flow Metab.*, 1: 349-356.
7. Meyer F.B., Sundt T.M. Jr., Anderson R.E. and Tally P. (1988): *Ann. NY Acad. Sci.*, 522: 502-515.
8. Bevan R.D., Bevan J.A. and Frazee J.G. (1988): *Stroke*, 19: 73-79.
9. Brandt L., Andersson K.E., Edvinsson L. and Ljunggren B. (1981): *J. Cereb. Blood Flow Metab.*, 1: 339-347.
10. Espinosa F., Weir B., Overton T., Castor W., Grace M. and Boisvert D. (1984): *J. Neurosurg.*, 60: 1167-1175.
11. Krueger C., Weir B., Nosko M., Cook D. and Norris S. (1985): *Neurosurgery*, 16: 137-140.
12. Nosko M., Weir B., Krueger C., Cook D., Norris S., Overton T. and Boisvert D. (1985): *Neurosurgery*, 16: 129-136.
13. Siesjö B.K. (1984): *J. Neurosurg.*, 60: 883-908.
14. Meyer F.B. (1989): *Brain Res.*, 14: 227-243.
15. Gilsbach J.M. (1988): *Acta. Neurochir.*, 45(suppl.): 41-50.
16. Allen G.S., Ahn H.S., Preziosi T.J., Battye R., Boone S.C., Chou S.N., Kelly D.L., Weir B.K., Crabbe R.A., Lavik P.J., Rosenbloom S.B., Dorsey F.C., Ingram C.R., Mellits D.E., Bertsch L.A., Boisvert D.P., Hundley M.B., Johnson R.K., Strom J.A. and Transou C.R. (1983): *N. Engl. J. Med.*, 308: 619-624.
17. Philippon J., Grob R., Dagreou F., Guggiari M., Rivierez M. and Viars P. (1986): *Acta. Neurochir.*, 82: 110-114.
18. Neil-Dwyer G., Mee E., Dorrance D., and Lowe D. (1987): *Eur. Heart J.*, 8: 41-47.
19. Petruk K.C., West M., Mohr G., Weir B.K.A., Benoit B.G., Gentili F., Disney L.B., Khan M.I., Grace M., Holness R.O., Karwon M.S., Ford R.M., Camercon G.S., Tucker W.S., Purves G.B., Miller J.D.R., Hunter K.M., Richard M.T., Durity F.A., Chan R., Clein L.J., Maroun F.B. and Godon A. (1988): *J. Neurosurg.*, 68: 505-517.

20. Pickard J.D., Murray G.D., Illingworth R., Shaw M.D., Teasdale G.M., Foy P.M., Humphrey P.R., Lang D.A., Nelson R., Richards P., Sinar J., Bailey S. and Skene A. (1989): *Br. Med. J.,* 298: 636-642.
21. Auer L.M. (1984): *Neurosurgery,* 15: 57-66.
22. Ljunggren B., Brandt L., Saveland H., Nilsson P.-E., Cronqvist S., Andersson K.-E., and Vinge E. (1984): *J. Neurosurg.,* 61: 864-873.
23. Jan M., Buchheit F., and Tremoulet M. (1988): *Neurosurgery,* 23: 154-157.
24. Ohman J. and Heiskanen O. (1988): *J. Neurosurg.,* 69: 683-686.
25. Harders A. and Gilsbach J. (1988): *Acta. Neurochir.,* 45(suppl.): 21-28.
26. Flamm E.S. (1989): *Am. Heart J.,* 117: 236-242.
27. Flamm E.S., Adams H.P., Beck D.W., Pinto R.S., Marler J.R., Walker M.D., Godersky J.C., Loftus C.M., Biller J., Boarini D.J., O'Dell C., Banwart K. and Kongable G. (1988): *J. Neurosurg.,* 68: 393-400.

PHARMACOLOGICAL BASIS FOR USE OF CALCIUM ANTAGONISTS IN DEMENTIA

Alexander Scriabine

Miles Inc., 400 Morgan Lane, West Haven, Connecticut 06516

Calcium overload is thought to be one of the processes contributing to neuronal death. Pathological enhancement of calcium transients can lead to disruption of cytoskeletal and membrane structures, adversely altering cellular function (1). Cytosolic free calcium increases in ischemic brain (2) and certain neurotoxins are known to elevate cytosolic calcium prior to the death of neurones (3,4).

Premature neuronal death is known to occur in patients with various types of dementias, including Alzheimer's disease. If calcium contributes to neuronal death, it can be expected that blockade of calcium entry into neurones by drugs will slow the rate of mental deterioration in demented patients. Since there are no specific animal tests for antidementia drugs, Ca^{2+} antagonists can be selected for clinical studies on the basis of their pharmacological properties assumed to be advantageous in the treatment of dementias (table 1).

TABLE 1.   Criteria for selection of Calcium antagonist for clinical evaluation in dementias

In Vitro   (Neurones or brain tissue)
    Binding affinity for specific receptors
    Blockade of Ca^{2+} currents
    Prevention of Ca^{2+} overload
    Antagonism of neurotoxins

In Vivo
    Ability to enter CNS
    Prevention of memory loss and behavioral deficiencies:
      a)   in aging animals
      b)   in animals subjected to prenatal or early postnatal hypoxia
      c)   in animals with selected brain lesions
    Lesser tendency to lower arterial pressure

T. Godfraind et al. (eds.), Calcium Antagonists, 279–284.
© 1993 *Kluwer Academic Publishers and Fondazione Giovanni Lorenzini.*

The only Ca^{2+} antagonist which was evaluated clinically in dementias is nimodipine (1,4-dihydro-2, 6-dimethyl-4-(3-nitrophenyl)-3,5-pyridinedicarboxylic acid 2-methoxyethyl 1-methylethyl ester). In initial pharmacological studies, nimodipine prevented postischemic cerebral hypoperfusion, dilated cerebral arteries, and antagonized cerebral vasoconstriction caused by 5-HT, norepinephrine or other cerebral neurotransmitters (5-7). It was also shown to have high affinity for dihydropyridine receptors in the rat or human brain (8,9) (table 2)

TABLE 2: Inhibition of isradipine binding to human brain by 1-4,dihydropyridines*

Drug	IC_{50}, nM $\pm$ SEM
Nimodipine	0.4 $\pm$ 0.1
Nitrendipine	1.02 $\pm$ 0.3
Nifedipine	3.0 $\pm$ 0.5

*From Quirion (9)

In freshly dispersed rabbit dorsal root ganglion neurones, nimodipine reduced Ca^{2+} currents through L-type channels at 2 nM (10). K^{+}-induced elevation of cytosolic calcium in rat hippocampal cells was antagonized by nimodipine at 10 nM (11). Ischemia-induced elevation in intracellular Ca^{2+} in cat brain was also antagonized by nimodipine *in vivo* (2).

Nimodipine (1 M) antagonized hypoxia- or methyl-mercury-induced neurotoxicity in PC12 cells (12), MDMA (methylene-dioxymetamphetamine)-induced toxicity in fetal rat serotonergic neurones (13), or AIDS virus coat protein (GP120)-induced Ca^{2+} uptake in rat retinal ganglion cells or hippocampal neurones (4). Excitatory amino acids (EAAs)-induced LDH release in mouse cortical cell cultures (14), and NMDA receptor-mediated lethality of rat retinal ganglion cells (3) were also reduced by nimodipine.

Nimodipine reduced behavioral deficiencies of rats with brain lesions in visual neocortex, septal area, or anterior hippocampus (15-17, table 3) and protected acetylcholinesterase-reactive cells in medial septal area of rats subjected to fimbria fornix lesions (18). These last effects were observed at a very low nimodipine dose (70 g/kg/day i.p. for 3 days).

TABLE 3.   Effects of nimodipine in rats with
                                     brain lesions

Location of Lesions	Nimodipine Dose/Route mg/kg	Parameters	Effects	Ref
Visual neocortex	1 or 3 prior to each/p.o	Visual discrimination task	Facilitation of recovery	15
Septal area	0.07/day for 4 days/i.p.	Hypermotion-ality	Antagonism	16
Anterior hippocampus	15/day for 14 days/p.o.	Performance in DRL* tasks	Increase in efficiency	17
Fimbriafornix	0.07/day for 8 days/i.p.	Staining of cells in septal area	Protection	18

*DRL: Differential Reinforcement of Low Rates of Response.

Beneficial effects of nimodipine by oral administration to aging animals included prevention of age-associated changes in walking pattern, behavioral deficits, cerebral microvascular pathology, as well as acceleration of acquisition of associative learning in rabbits, and improvement in the performance of aging monkeys in delayed-response task (19-23).

The initial clinical studies with nimodipine in vascular and old age and Alzheimer's dementias were published (24-26) (table 4). Improvements in the ability of patients to perform everyday activity, in Wechsler memory, global deterioration, and Buschke long-term retrieval scales were reported. In most patients, nimodipine was used at 30 mg t.i.d. In patients with Alzheimer's disease, nimodipine was effective at 30 but not at 60 mg t.i.d. (26). Nimodipine is likely to delay neuronal death and, consequently, deterioration of patients with dementias.

TABLE 4.   Initial clinical studies with
              nimodipine (30 mg t.i.d., p.o.) in dementias

Type of Dementia	Patients #/Mean Age Yrs	Duration of Therapy # Weeks	Improvement in	Ref
Vascular	65/80	24	Short memory. Ability to perform daily activity	24
Old-age	178/75	12	Wechsler memory scale. Global deterioration scale	25
Alzheimer's	195/70	12	Long-term retrieval and storage tests	26

Pharmacology of nimodipine was also summarized in recent extensive reviews (27,28).

SUMMARY

Prevention of neuronal Ca^{2+} is likely to delay neuronal death caused by ischemia or neurotoxins. Since L-type voltage-dependent Ca^{2+} channels exist in neurones, blockade of these channels is likely to be beneficial not only in the therapy of cerebral ischemia but also of dementias.

Nimodipine, a dihydropyridine, which enters CNS more readily than nifedipine, and has high affinity for its binding sites in brain, blocks Ca^{2+} currents in neurones at concentrations as low as 2 nM, antagonizes various neurotoxins *in vitro* and reduces behavioral deficiencies caused by brain lesions or advanced age. The initial clinical studies suggested that nimodipine, 30 mg t.i.d., has beneficial effects in the treatment of vascular, old-age, and Alzheimer's dementias.

REFERENCES

1.   Siesjö B.K. (1990): *Eur. J. Neurol.,* 30: (suppl.2): 3-9.
2.   Uematsu D., Greenberg J.H., Hickey W.F. and Reivich M. (1989): *Stroke,* 20:1531-1537.
3.   Sucher N.J., Lei S.Z. and Lipton S.A. (1991): *Brain Res.,* 297: 297-302.

4.   Dreyer E.B., Kaiser P.K., Offermann J.T. and Lipton S.A. (1990): *Science,* 248: 364-367.

5.   Kazda S., Garthoff B., Krause H.P. and Schlossmann K. (1982): *Arzneimittelforsch,* 32: 331-338.

6.   Hoffmeister F., Benz U., Heise A., Krause H.P. and Neuser V. (1982): *Arzneimittelforsch,* 32: 347-360.

7.   Towart R. (1981): *Circ. Res.,* 48: 650-657.

8.   Bellemann P., Schade A. and Towart R. (1983): *Proc. Natl. Acad. Sci. USA,* 80: 2356-2360.

9.   Quirion R. (1986): *Eur. J. Pharmacol.,* 117: 139-142.

10. McCarthy R.T. and TanPiengco P.E. (1991): *J. Neurosci.,* (submitted).

11. Chisholm J.: personal communication.

12. Fahey J.M., Hedayat B., Cook J.R., Isaacson R.L. and Van Buskirk R.G. (1989): In: *Nimodipine and Central Nervous System Function: New Vistas,* edited by J. Traber and W.H. Gispen pp. 117-40. Schattauer, Stuttgart.

13. Azmitia E.C. (1989): In: *Nimodipine and Central Nervous System Function: New Vistas,* edited by J. Traber and W.H. Gispen pp. 141-159. Schattauer, Stuttgart.

14. Weiss J.H. and Choi D.W. (1991): In: *Nimodipine, Pharmacological and Clinical Results in Cerebral Ischemia,* edited by A. Scriabine, G.M. Teasdale, D.Tettenborn and W. Young pp. 57-63. Springer, Berlin.

15. LeVere T.E., Brugler T., Sandin M. and Gray-Silva S. (1989): *Behav. Neurosci,* 103: 561-565.

16. Poplawsky A. and Isaacson R.L. (1990): *Behav. Neural. Biol.,* 53: 133-139.

17. Finger S., Green L., Tarnoff M.E., Mortman K.D. and Andersen A. (1990): *Exp. Neurol.,* 109: 279-285.

18. Isaacson R.L., Shen Y. and Mandel A. (1990): *Rest. Neurol. Neurosci.,* 2: 1-7.

19. Schuurman T., Klein H., Beneke M. and Traber J. (1987): *Neurosci. Res. Commun.,* 1: 9-15.

20. Luiten P.G.M., deJong G.I., Malder A.B., Horvath E., Schuurman T. and Traber J. (1989): In: *Nimodipine and Central Nervous System Function: New Vistas,* edited by J. Traber and W.H. Gispen pp. 239-256. Schattauer, Stuttgart.

21. Nyakas C., Markel E., Schuurman T. and Luiten P.G.M. (1991): *Eur. J. Neurosci.,* 3: 168-174.

22. Deyo R.A., Straube K.T. and Disterhoft J.F. (1989): *Science,* 243: 809-811.

23. Sandin M., Jasmin S. and Levere T.E. (1990): *Neurobiol Aging,* 11: 573-575.

24.   Tobares N., Pedromingo A. and Bigorra J. (1989): In: *Diagnosis and Treatment of Senile Dementia*, edited by M. Bergener and B. Reisberg pp. 360-365. Springer, Berlin.

25.   Ban T.A., Morey L., Aguglia E., Azzarelli O., Balsano F., Marigliano V., Caglieris N., Sterlicchio M., Capurso A., Tomasi N.A., Crepaldi, G., Volpe D., Palmieri G., Ambrosi G., Polli E., Cortellaro M., Zanussi C. and Froldi M. (1990): *Prog. Neuropsychopharmacol. Biol. Psychiat.*, 14: 525-551.

26.   Tollefson G.D. (1990): *Biol. Psychiat.*, 27: 1133-1142.

27.   Scriabine A., Schuurman T. and Traber J. (1989): *FASEB J.*, 3: 1799-1806.

28.   Bär P.R., Traber J., Schuurman T. and Gispen W.H. (1991): *J. Neural. Transm.* (suppl.)31: 55-71.

CALCIUM CHANNELS IN INTESTINAL SMOOTH MUSCLE

D.J. Beech and T.B. Bolton

Department of Pharmacology and Clinical Pharmacology,
St. George's Hospital Medical School,
Cranmer Terrace, London SW17 ORE, United Kingdom

Tension in intestinal smooth muscle is critically dependent on electrical activity. In the small intestine this activity comprises frequent spontaneous action potentials often superimposed on slow waves of depolarization. It is usually calcium (Ca)-influx through voltage-gated Ca channels that generates the depolarization of the action potential and so these ionic channels have been the subject of many investigations and much is now known of their biophysics and pharmacology (1). Also, Ca channels in intestinal smooth muscle are subject to modulation by neurotransmitters and local hormones. Recent observations on these agonist effects are described.

RESULTS AND DISCUSSION

Biophysics and Pharmacology

Intestinal Ca-current is shown in fig. 1. Inward current started to activate positive of -40 mV and was large and rapid at 0 mV. Single exponentials describing current activation had time constants of 2.5 ms at -30 mV and 1 ms at -10 mV (25°C): tail-currents at -50 mV declined by 85% in 1 ms. The peak of the current-voltage relationship (not shown) was at 10 mV and at 80mV current reversed polarity. Inactivation occurred even at voltages where Ca-current was not activated: there was 10% inactivation at -50 mV and 50% at -35 mV (not shown). This is perhaps mostly voltage-dependent inactivation, Ca-induced inactivation being superimposed once Ca-influx occurs.

In most intestinal smooth muscles the Ca-current is dihydropyridine-sensitive with an activation threshold at about -40 mV. Cell-attached patch studies show the channels have a large conductance of 25-30 pS with 100-110 mM $BaCl_2$ as the charge carrier [e.g.,(2)]. The channels are often classified as "L-type" (3). Small and intermediate Ca channels have also been reported (2,4), but the absence of distinct whole-cell current components in most intestinal smooth muscles suggests either paucity of the smaller conductance channels or, if there are substantial numbers of

T. Godfraind et al. (eds.), Calcium Antagonists, 285–290.
© 1993 *Kluwer Academic Publishers and Fondazione Giovanni Lorenzini.*

each type, that their voltage-dependencies usually overlap sufficiently to prevent a dissection of the whole-cell current.

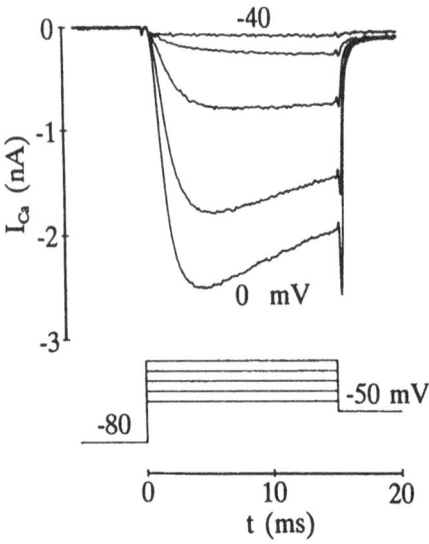

FIG. 1. Ca-current recorded from a single smooth muscle cell isolated enzymatically and mechanically from the guinea-pig ileum longitudinal muscle layer [as in (7)]. Voltage-clamp and intracellular dialysis were achieved by the whole-cell mode of the patch-clamp method. The bath solution was NaCl-based with 2 mM $CaCl_2$ and the pipette solution CsCl-based (to block K-currents) with 30 mM EGTA and 3 mM ATP. Leak current (measured with 100 μM lanthanum in the bath) was subtracted and 80% of the 2.5 MΩ series resistance was compensated.

We have investigated Ca-current block by pinaverium bromide (5). This drug has been used to treat intestinal hypermobility disorders and is thought to act by a Ca antagonist action (6). Fig. 2 shows block in a guinea-pig ileum smooth muscle cell and initiation of recovery on wash-out. The IC_{50} was 1.5 μM in the rabbit (5), similar to that for flunarizine, verapamil and diltiazem (1). Block did not show dependence on voltage or frequency of "use" of the channels and so pinaverium may bind equally both resting and inactivated but not open channel states, contrasting with the actions of other Ca antagonists (1).

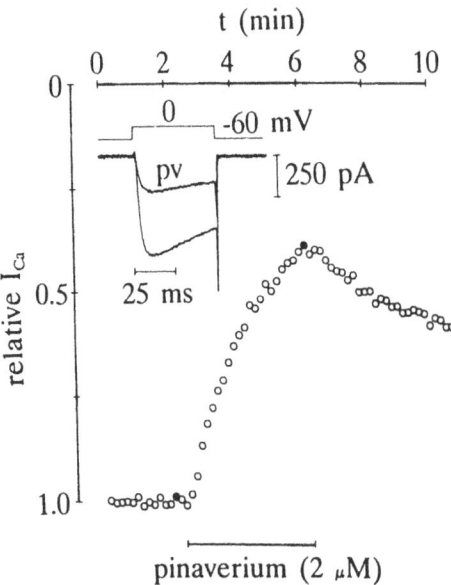

FIG. 2. Ca-current block by pinaverium bromide. The method was as for fig. 1 except leak current was assessed in 100 μM cadmium. Relative amplitude of the peak Ca-current (after subtraction of a small run-down) is plotted against time. Pinaverium was bath-applied and the inset shows actual currents for the filled circles.

Agonist-Induced Suppression of Calcium-current

Acetylcholine (ACh) and histamine suppress Ca-current in guinea-pig intestinal smooth muscle cells. Fig. 3 shows reversible suppression caused by histamine. Fig. 4 shows the response to ACh, which was often complex because a large cationic (inward) current occurred at the holding potential concomitant with Ca-current suppression. (The reversal potential for the cationic current was near 0 mV and so the Ca-current was observed essentially in isolation at this voltage.)

Because histamine and ACh release Ca from intracellular stores [e.g., (7)] a plausible explanation for the suppression is that receptor activation mobilized stored intracellular Ca by generating inositol 1,4,5-trisphosphate (IP$_3$), raised [Ca]$_i$ and thus inactivated the Ca channels. Indeed, Ca-induced inactivation of the Ca-current can be observed by double-pulse voltage protocols [not shown, but see (2)] and fig. 5 shows that inclusion of 30 mM EGTA, a Ca chelator, in the recording pipette inhibited suppression by ACh or histamine. Furthermore, in smooth muscle cells from rabbit jejunum, flash photolysis of caged IP$_3$ loaded into cells from the pipette caused a profound suppression of the Ca-current (8), an effect that was blocked when there was 10 mM EGTA in the pipette. However,

increases in [Ca]$_i$ caused by ACh or histamine are transient (7) and yet
the Ca-current suppression was sustained (fig. 3). Therefore, it is
probable that Ca-current remained suppressed even when [Ca]$_i$ was no
longer high. A similar observation is reported for suppression of Ca
channel activity by thyrotropin-releasing hormone in pituitary tumor GH$_3$
cells (9). The implication is that the [Ca]$_i$-transient acts as a "trigger" for
another process that maintains suppression or that suppression occurs
without regard for the [Ca]$_i$-transient but via a process that requires a
"normal" level of intracellular Ca. These possibilities are being
investigated.

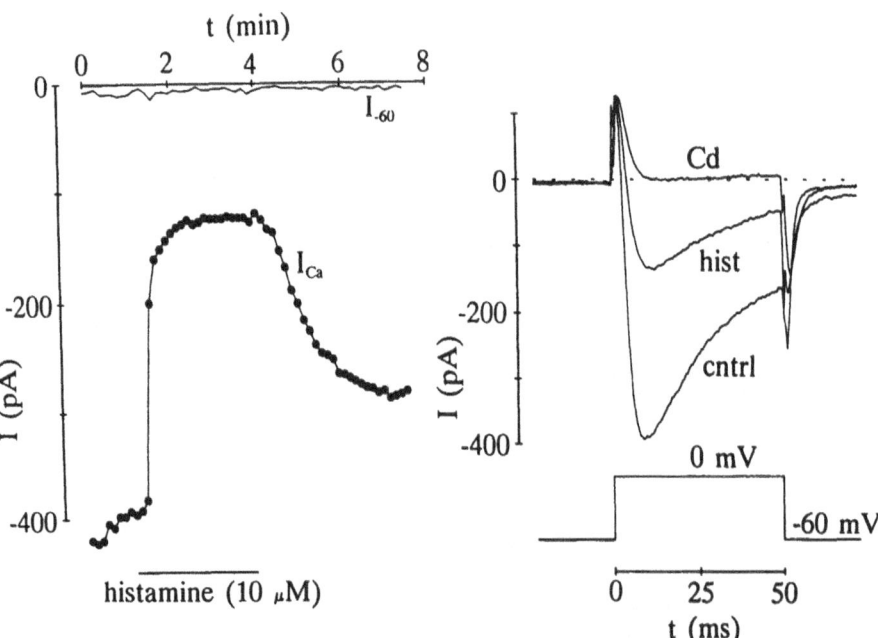

FIG. 3. Suppression of Ca-current by histamine. Current was recorded
as in fig. 2 except the pipette solution contained 100 time less EGTA and,
in addition, 15 mM creatine phosphate and 50 U/ml creatine
phosphokinase. On the left, peak Ca-current amplitude (filled connected
circles) and holding current (continuous line) are plotted against time;
histamine was bath-applied. On the right, actual (not leak-subtracted)
currents before, during histamine application and in 100 μM cadmium.

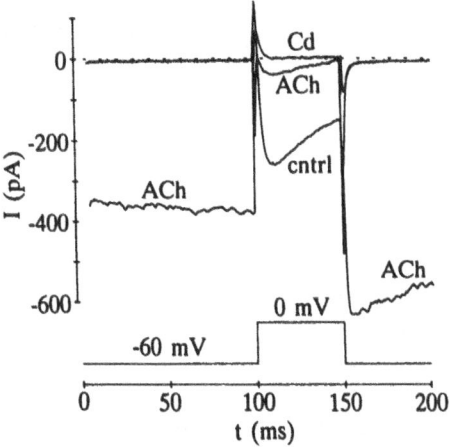

FIG. 4. Suppression of Ca-current by acetylcholine (ACh). Current was recorded as in fig. 3. Samples of 200 ms-long current trace taken every 8 s are shown for the control, in the presence of 10 μM ACh in the bath and in 100 μM cadmium. Current at the -60 mV holding potential is mostly ACh-activated cationic current and current during the 0 mV test step mostly voltage-gated Ca-current.

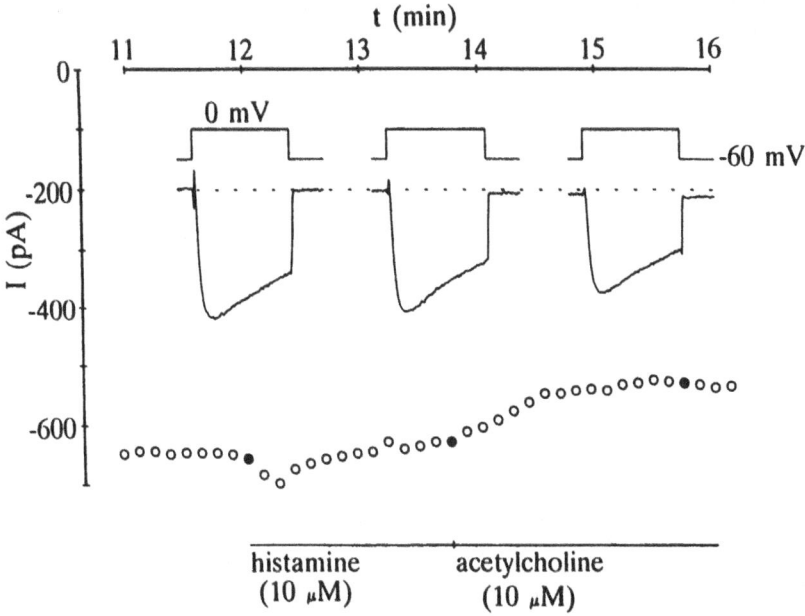

FIG. 5. High intracellular EGTA inhibits agonist actions. Current was recorded with 30 mM EGTA in the pipette. Peak Ca-current amplitude is plotted against time after break-through to the whole-cell. The filled circles mark which current traces are shown (insets) and the test steps were 50 ms long. Histamine and ACh were bath-applied.

This work was supported by grants from the UK Medical Research Council and the Wellcome Trust. We thank Latema (Paris) for the pinaverium bromide.

REFERENCES

1.   Kitamura K., Inoue Y., Inoue R., Ohya Y., Terada K., Okabe K. and Kuriyama H. (1989): *Gen. Physiol. Biophys.*, 8: 289-312.
2.   Yoshino M., Someya T., Nishio A., Yazawa K., Usuki T. and Yabu H. (1989): *Pflügers Archiv.*, 414: 401-409.
3.   Bean B.P. (1989): *Ann. Rev. Physiol.*, 51: 367-384.
4.   Vivaudou M.B., Singer J.J. and Walsh J.V. (1991): *Pflügers Archiv.*, 418: 144-152.
5.   Beech D.J., MacKenzie I., Bolton T.B. and Christen M.O. (1990): *Br. J. Pharmacol.*, 99: 374-378.
6.   Christen M.O. (1990): *Gen. Pharmacol.*, 21(6): 821-825.
7.   Pacaud P. and Bolton T.B. (1991): *J. Physiol.*, 441: 477-499.
8.   Komori S. and Bolton T.B. (1991): *Pflügers Archiv.*, 418: 437-446.
9.   Kramer R.H., Kaczmarek L.K. and Levitan E.S. (1991): *Neuron*, 6: 557-563.

PHARMACOLOGICAL ASPECTS OF CALCIUM CHANNELS IN NONVASCULAR SMOOTH MUSCLE

David J. Triggle

School of Pharmacy, State University of New York at Buffalo,
Buffalo, NY 14260

Smooth muscle is a heterogeneous tissue. However, a key component of excitation-contraction coupling in both vascular and nonvascular smooth muscle is an elevation of intracellular calcium subsequent to excitation. The processes by which intracellular calcium is elevated from the resting level of approximately 10^{-8}M to the stimulated level of approximately 10^{-6}M are quite diverse in smooth muscle and depend on tissue stimulus, component of response, and pathology (1,2).

In principle, calcium mobilization to satisfy the demands of E-C coupling can occur through several distinct pathways either alone or in combination. Pathways of particular importance to nonvascular smooth muscle include the mobilization of extracellular calcium through plasma membrane calcium channels, receptor-operated channels and voltage-gated channels (fig. 1). These channels have primary control by chemical and electrical signals respectively, but the voltage-gated channels are the better characterized. It should, however, be noted that many smooth muscles will use multiple calcium mobilization mechanisms.

VOLTAGE-GATED CALCIUM CHANNELS

Voltage-gated calcium channels may be considered as pharmacological receptors. As such, they possess the following general properties:
1.   Drug binding sites for activators and antagonists with specific structure-activity relationships.
2.   Drug binding sites linked to channel permeation and gating machinery.
3.   Association with guanine nucleotide binding proteins.
4.   Regulation by homologous and heterologous factors.
5.   Alteration of expression and function during disease states, both experimental and clinical.

These considerations apply to at least four major classes of calcium channel, each characterized by discrete electrophysiologic and pharmacologic characteristics (table 1). Within each of the major classes

T. Godfraind et al. (eds.), Calcium Antagonists, 291–297.

there likely exists several distinct subclasses (3-5). The extent to which the structural classification now being provided by molecular biology studies correlates with pharmacologic differentiation remains, however, to be determined (6,7).

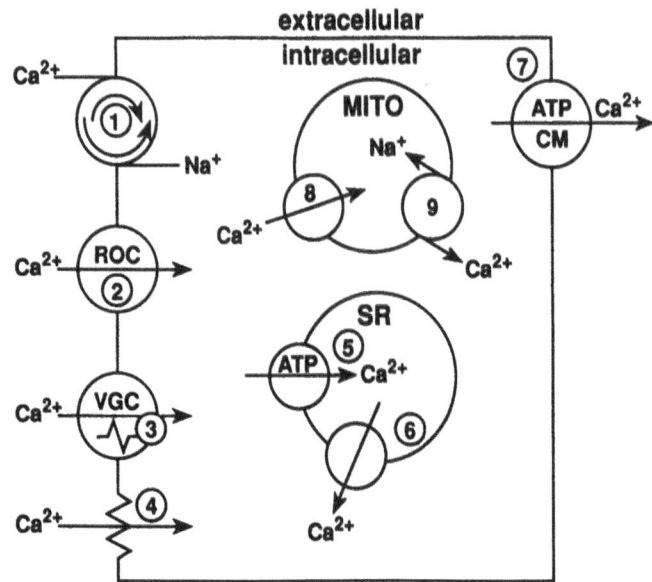

Figure 1. Representation of cellular Ca^{2+} regulation depicting: 1) plasmalemmal Na^+/Ca^{2+} exchange; 2) receptor-operated Ca^{2+} channels (ROC); 3) potential-dependent Ca^{2+} channels (PDC); 4) "leak pathway; 5) sarcoplasmic reticulum Ca^{2+}-ATPase; 6) Ca^{2+} release channel; 7) plasmalemmal Ca^{2+}-ATPase (calmodulin-dependent); 8) uniporter potential-driven Ca^{2+} uptake in mitochondria (MITO); 9) Na^+/Ca^{2+} exchanger for Ca^{2+} efflux in mitochondria.

The most important drugs are those active at the L-type channel and which have assumed major importance as primary cardiovascular drugs (6). The first generation of drugs is represented by verapamil, nifedipine, and diltiazem. This chemically heterogeneous group of drugs occupies discrete binding sites on the $alpha_1$ subunit which are linked allosterically one to the other and to functional machinery of the channel (8). The 1,4-dihydropyridine site is home to both antagonist and activator structures.

The primary actions of these antagonists are in the cardiovascular system. However, clinically and experimentally the calcium channel antagonists have been examined, with some success, for disorders ranging from achalasia to enuresis to irritable bladder and intestinal hypermotility. It is clear, however, that under most circumstances, the actions of calcium antagonists on nonvascular smooth muscle are not major. This lack of activity is not, however, due to an absence of L-type calcium channels since K^+ depolarization or Bay K 8644 can involve antagonist-sensitive tension responses (9,10).

SELECTIVITY OF DRUG ACTION

The observed selectivity of action of the calcium antagonists may arise from a number of factors:

1. Pharmacokinetic factors, including tissue distribution
2. Mode of calcium mobilization: intracellular and extracellular routes
3. Class and subclass of calcium channel
4. State-dependent interactions
5. Regulation by homologous and heterologous factors
6. Regulation by pathological (disease) state

The roles of these several factors have been discussed previously (11,12). Of particular interest are state-dependent interactions whereby stimulus intensity and frequency may both quantitatively and qualitatively alter drug activity.

These voltage-dependent interactions are likely observed in a number of situations. Fig. 2 depicts the activities of a series of achiral 1,4-dihydropyridines on the phasic and tonic components of the tension response in guinea pig ileal smooth muscle to K^+ depolarization or muscarinic receptor activation. Major differences in sensitivity of pharmacologic response occur relative to the binding properties in the same smooth muscle. For the phasic and tonic responses to K^+ depolarization the quantitative structure-activity relationship (QSAR) are as follows:

Phasic
$$\log 1/IC_{50} = 0.36\pi + 1.09\sigma_m - 0.13L_m - 1.93B_{1p} + 8.52$$

Tonic
$$\log 1/IC_{50} = 0.76\pi + 2.47\sigma_m - 0.47L_m - 3.41B_{1p} + 11.4$$

For the tonic response, which shows an essentiality 1:1 correlation to binding, the QSAR is identical to that for binding of the same compounds:

$$\log 1/IC_{50} = 0.68\pi + 2.50\sigma_m - 0.47L_m - 3.26B_{1p} + 11.0$$

We have proposed elsewhere (13) that these different structure-activity relationships reflect 1,4-dihydropyridine interactions with the resting state of the channel (phasic response) and with the activated/inactivated states

TABLE 1.   Classification of Voltage-Gated Calcium Channels.

Property	L	T	N	P
Conductance, pS:	25	8	12–20	10–12
Activation threshold	high	low	high	moderate
Inactivation rate:	slow	fast	moderate	rapid
Permeation:	$Ba^{2+} > Ca^{2+}$	$Ba^{2+} = Ca^{2+}$	$Ba^{2+} > Ca^{2+}$	$Ba^{2+} > Ca^{2+}$
Function:	E-Coupling cardiovascular system, smooth muscle, endocrine cells and some neurons	Cardiac SA node: neuronal spiking repetitive spike activity in neurons and endocrine cells	Neuronal only: neurotransmitter release	Neuronal only?: neurotransmitter release
Pharmacologic sensitivity:				
1,4-Dihydropyridines (Activators/antagonists)				
Phenylalkylamines				
Benzothiazepines	Sensitive	Insensitive	Insensitive	Insensitive
ω-conotoxin	Sensitive? (some)	Insensitive	Sensitive	Insensitive
Octanol, amiloride	Insensitive?	Sensitive	Insensitive	?
Funnel web spider toxin	Insensitive	Insensitive	Insensitive	Insensitive

of the channel (tonic response).

In addition to such state-dependent or relative differences in antagonist activity, absolute differences may also exist between different smooth muscle preparations. Such may be the situation in the comprehensive analysis by Yousif and Triggle (14) of the activities of a series of ten different calcium antagonists in a number of different smooth muscle preparations stimulated by a common K^+ depolarization. Significant, up to ten-fold, activity differences were observed independent of the native or potency of the antagonist (fig. 3). This suggests that the series of calcium antagonists may be probing differences in channel binding site or kinetic properties.

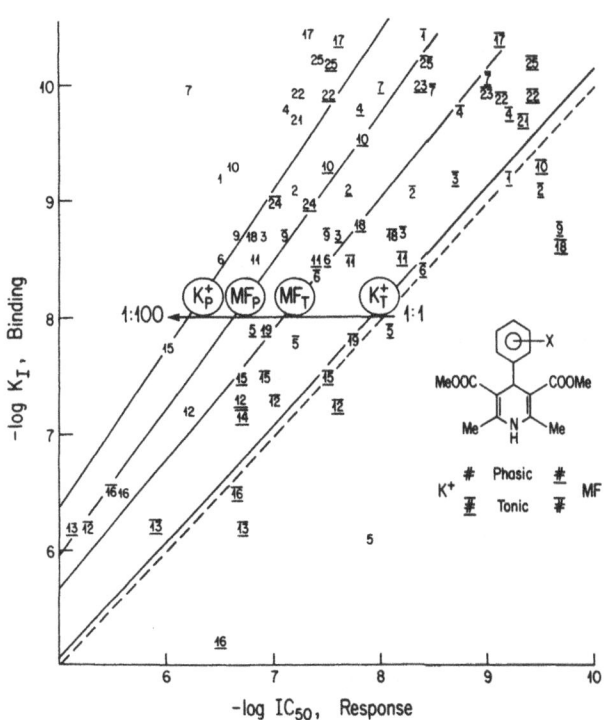

FIG. 2. Correlations between the abilities of a series of 1,4-dihydropyridines to inhibit specific [³H] nitrendipine binding to guinea pig ileal longitudinal smooth muscle membranes and to inhibit the phasic and tonic components of tension response to K^+ depolarization and methylfurmethide (muscarinic) receptor activation in the same tissue. The 1,4-dihydropyridines are 2,6-dimethyl-3,5-dicarbomethoxy-4-substituted phenyl-1,4-dihydropyridine where the substituents are 1, 2-CN; 2, 2-NO2; 3, 2-Me; 4, 2-Cl; 5, 2-OMe; 6, 2-F; 7, 3-NO2; 8, 3-O-Me; 9, 3-CN; 10, 3-Cl; 11, 3-F; 12, 3-Me; 13, 4-Cl; 14, 4-Me; 15, 4-F; 16, 4-NO2; 17, F5; 18, 2-F,6-Cl; 19, H; 21, nitrendipine; 22, nimodipine; 23, (-)3628; 24, (+)3629; 25, nisoldipine.

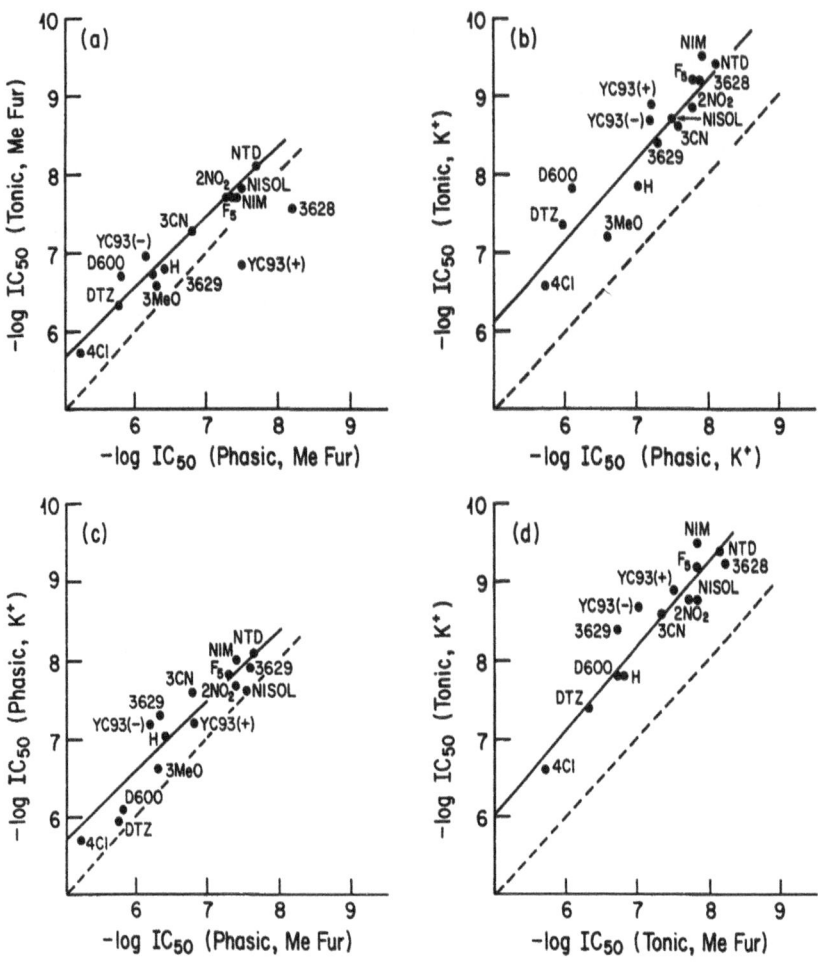

FIG. 3. Comparison of sensitivities of responses to K^+ depolarization in a series of smooth muscles to a series of calcium channel antagonists. a. Guinea pig bladder and trachea, tonic components of response. b. Guinea pig parenchyma and trachea, tonic components of response. c. Guinea pig parenchyma and pulmonary artery, tonic components of response. d. Guinea pig trachea and rat mesenteric arteries, tonic components of response. The solid line represents the correlation determined by linear regression and the dashed line represents unit slope. Reproduced with the permission of *Can. J. Physiol. Pharmacol.* from (14).

REFERENCES

1. Bolton T.B. (1979): *Physiol. Rev.*, 43: 606-718.
2. Hurwitz L. (1986): *Ann. Rev. Pharmacol. Toxicol.*, 26: 225-258.
3. Bean B.P. (1989): *Ann. Rev. Physiol.*, 51: 367-384.
4. Tsien R.W. and Tsien R.Y. (1990): *Ann. Rev. Cell Biol.*, 6: 715-760.
5. Miller R.J. (1987): *Science,* 235: 46-52.
6. Triggle D.J. (1990): In: *Cardiovascular Pharmacology,* edited by M. Antonaccio. Raven Press, New York.
7. Glossmann H. and Striessnig J. (1989): *Rev. Physiol. Biochem. Pharmacol.*, 114: 1-105.
8. Janis R.A., Silver P.J. and Triggle D.J. (1987): *Adv. Drug Res.*, 16: 307-591.
9. Wei X.-Y., Luchowski E.M., Rutledge A., Su C.M. and Triggle D.J. (1986): *J. Pharmacol. Exp. Therap.*, 70:
10. Yousif F.B., Bolger G.T., Ruzycky A. and Triggle D.J. (1987): *Can. J. Physiol. Pharmacol.*, 63: 453-462.
11. Triggle D.J., Zheng W., Hawthorn M., Kwon Y.-W., Wei X.-Y., Joslyn A., Ferrante J. and Triggle A.M. (1989): *Ann. N.Y. Acad. Sci.*, 560: 215-229.
12. Triggle D.J. (1992): *Cleveland Clinic J. Med.*, (in press).
13. Triggle D.J., Hawthorn M. and Zheng W. (1988): *J. Cardiovasc. Pharmacol.*, 12(suppl.4): S91-S93.
14. Yousif F.B. and Triggle D.J. (1985): *Can. J. Physiol. Pharmacol.*, 64: 273-283.

PHARMACOLOGY OF CALCIUM ANTAGONISTS ACTIVE ON GASTROINTESTINAL CALCIUM CHANNELS

M.O. Christen

SOLVAY PHARMA
Laboratoires de Thérapeutique Moderne (L.T.M.)
42, rue Rouget-de-Lisle, B.P. 22
92151 Suresnes Cedex, France

Current drug therapy for functional gastrointestinal (GI) motility disorders [e.g., irritable bowel syndrome (IBS), esophageal spasm] and other disorders (e.g., hypersecretory states, ulcers) of the GI tract has not been effective in many cases. Among the agents that are being used to treat motility disorders of the upper GI tract (e.g., esophageal spasm and achalasia) are metoclopramide, clebopride, domperidone, cholinomimetics, and anticholinesterases. Anticholinergics, antispasmodics, opiates,

TABLE 1.   Pharmacological Modulation of
            Irritable Bowel Syndrome

Peripherally Acting Agents

Opiates
Antispasmodics and anticholinergics
Cholecystokinin (CCK) antagonists
Serotonin antagonists
Somatostatin agonists
Substance P antagonists
Corticotropin-releasing factor (CRF) antagonists
Calcium antagonists

Centrally Acting Agents

Antidepressants
Anxiolytics
Antidopaminergic agents

---

and other drugs are also being used with variable success in efforts to

299

T. Godfraind et al. (eds.), Calcium Antagonists, 299–313.
© 1993 *Kluwer Academic Publishers and Fondazione Giovanni Lorenzini*.

manage IBS and other GI motility disorders (see table 1).

CALCIUM AND GI PHYSIOLOGY/PATHOPHYSIOLOGY

A new approach to such problems is based on the facts that Ca^{2+} is involved in excitation-contraction, stimulus-secretion, and excitation-release coupling and, therefore, directly or indirectly in all mechanisms controlling GI motility, secretion, and hormonal action. Some physiological and pathophysiological implications based on this idea are listed in table 2.

TABLE 2.   Roles of Calcium in GI
                             Physiology/Pathophysiology

Motility: Role in excitation-contraction coupling at different levels

Esophagus
Stomach
Intestine

Blood Flow: Ulcers

Hepatocellular injuries: Ca^{2+} overload

Others

Although psychological factors, as well as food ingestion, may provoke the symptoms of GI disorders, therapy with Ca^{2+}-antagonists can be envisaged by considering that activation of Ca^{2+} channels probably represents the "final common path" of all mechanisms governing GI functions. Ca^{2+}-antagonists inhibit Ca^{2+} entry into smooth muscle cells by selectively interacting with specific receptor sites that are associated with potential-dependent Ca^{2+} channels (PDCs) (1,2), thereby opposing the increase in concentration of internal ionized Ca^{2+} that is required to initiate tension in the contractile proteins of smooth muscle (3,4) and preventing the Ca^{2+} entry needed for hormone and gastric acid secretion and certain other digestive functions. Hence, it is not surprising that typical Ca^{2+}-antagonists (e.g., nifedipine, verapamil, and diltiazem) can influence GI activity in experimental animals and in patients suffering from GI disorders.

Some relevant studies supporting these contentions will be discussed briefly, and an attempt will be made to indicate the possible advantages and pitfalls associated with the use of Ca^{2+}-antagonists as therapy for GI disorders. Three classes of conventional Ca^{2+}-antagonists, i.e., the 1,4-dihydropyridines (DHPs; e.g., nifedipine); the phenylalkylamines (e.g., verapamil, gallopamil); and the benzothiazepines (e.g., diltiazem), will be discussed first, and a discussion of the atypical Ca^{2+}-antagonist pinaverium bromide will follow. At the outset, it might be noted that typical Ca^{2+}-antagonists influence GI activity both in basic *in vitro* and *in vivo* experiments and in the clinical setting.

Studies with Nifedipine
and Other 1,4-Dihydropyridines

In vitro studies.

In vitro experiments with monkey, dog, and human tissues have shown that nifedipine and nilvadipine reduce, but do not block, the contractile response of colonic muscle to acetylcholine (ACh), substance P, KCl, and electrical field stimulation (5,6). Some difference in sensitivity for DHP-type Ca^{2+}-antagonists might exist between vascular (rat aorta) and intestinal (guinea pig ileum) smooth muscle preparations (7). This contention derives support from experiments which have shown that the maximal binding capacity (B_{max}), or density, of [^{3}H]PN 200 110 binding is greater in membranes prepared from guinea pig ileal smooth muscle than in those prepared from rat aorta. Also, the potencies of nifedipine and nisoldipine in inhibiting K^+-depolarization-induced contractions differed between rat aortic and guinea pig ileal smooth muscle preparations (7). Such differences could be related to tissue-selectivity.

A more recent study has shown that two series of tiamdipine analogues inhibited specific [^{3}H]PN 200 110 binding in microsomal membranes of guinea pig ileal longitudinal muscle and rat heart ventricles in a concentration-dependent manner. However, no marked differences were observed in radioligand binding activity (K_i values) between the two tissues (8). Another recent study (9) has indicated that lacidipine is more vascular-selective than standard DHP derivatives. In rabbit ear artery, both lacidipine and nitrendipine caused marked inhibition of Ca^{2+}-induced contraction, but this occurred with much slower onset and longer duration with lacidipine. Perhaps the most interesting data obtained in this study showed that all Ca^{2+}-antagonists tested, except amlodipine, are significantly more potent as inhibitors of contractions of rat aorta than of rat colon or bladder. This apparent "lack" of tissue-selectivity of amlodipine provides a clue for synthesizing colon-(GI-)selective DHP derivatives. No selectivity of action existed among the various Ca^{2+}-antagonists tested with respect to their actions on colon versus bladder (two nonvascular smooth muscle preparations).

In vivo studies.

In conscious dogs, intravenous infusions not only of nifedipine but also of verapamil and diltiazem, exert profound inhibitory effects on intestinal motility (10). Nifedipine and nilvadipine reduced activity in the proximal colon of the dog (6). Oral or sublingual administration of nifedipine also decreased lower esophageal sphincter (LES) pressure and partially inhibited the LES response to tetragastrin or bethanechol in dogs (11,12).

Clinical studies.

Upper GI tract. In normal human volunteers, LES pressure was reduced by nifedipine (10-40 mg), but only by about 40% at the highest dose administered (11,13-15), and nifedipine generally had a much less pronounced effect on contraction amplitude in the body of the esophagus (11,14-18). Nifedipine (20 mg, sublingually) reduced LES pressure by about 30-35% in achalasia patients. Thus, in both healthy and achalasia humans, oral nifedipine reduces LES pressure. Nifedipine also showed some promise (reduced esophageal pressure) in patients with "nutcracker esophagus," very high LES pressures, or diffuse esophageal spasm (11,13,15,19), but symptoms were not improved (19) and adverse cardiovascular side-effects occurred in some cases . It was concluded that nifedipine could be useful in treating elderly achalasia patients possessing risk factors that would render pneumatic dilatation or surgery inappropriate (15).

Most recently, a placebo-controlled, double-blind crossover study was performed to compare the effects of oral nifedipine (up to 20 mg) and verapamil (160 mg) on LES pressure, esophageal contraction amplitude, and clinical symptomatology in patients with achalasia (20). Both nifedipine and verapamil significantly decreased mean LES pressure, but only nifedipine caused a significant decrease in the amplitude of esophageal contractions. No statistically significant improvement in overall clinical symptomatology was observed with these drugs, but some individual improvements in dysphagia and chest pain were noted in these achalasia patients. In another very recent trial, the effect of sublingual nifedipine (mean dose, 55 mg/day; i.e., 10-20 mg, 30 minutes before meals) and pneumatic dilatation were compared in patients suffering from mild or moderate esophageal achalasia (21). Significant decreases in LES pressure were produced by both treatments, and excellent or good clinical results were observed in 75% of the dilated patients and in 77% of the patients treated with nifedipine. One of the 14 patients treated with nifedipine could not tolerate the drug. It was concluded that long-term treatment with sublingual nifedipine and pneumatic dilatation are equally effective in treating mild-to-moderate esophageal achalasia.

In contrast to these positive results, Robertson et al. (22) found that nifedipine (10 mg, 15 minutes before meals) produced no detectable decrease in LES pressure in achalasia patients. However, the methods used by these latter workers appear questionable (see 21); e.g., the dose of nifedipine (10 mg) that was administered was too low and it was given at intervals before meals that were too short (15 min).

Lower GI tract. Nifedipine (30 mg, p.o.) did not affect fasting or postprandial duodenal or jejunal motility, and did not alter gastric emptying in normal subjects (15,23), but similar doses inhibited colonic spike activity induced by eating and decreased the abnormal colonic motor response to distension in patients afflicted with IBS (18,24,25). Sublingually-administered nifedipine (26) and intravenously-administered nicardipine (27) reduced colonic motor responses to a standard meal in IBS patients. Most recently, paired controlled studies were conducted in normal volunteers and patients with IBS to examine the effect of standard nicardipine (20 mg, p.o.) and sustained-release nicardipine (30 mg, t.i.d.) on the responses of the anorectum to rectal distension and a meal (28). In normal volunteers, standard nicardipine did not affect rectal responses to distension but did significantly reduce the postprandial motility index. In IBS patients, standard nicardipine caused a significant reduction in distention-induced rectal motor activity and increased the rectal sensory thresholds for desire to defecate and discomfort. Both formulations of nicardipine significantly reduced the postprandial motility index and symptoms, supporting the contention that Ca^{2+}-antagonists may be useful in managing IBS (28).

Studies with Verapamil and Gallopamil

In vitro studies.
In strips of guinea pig stomach, verapamil prevented the contractile actions of K^+, ACh and some prostaglandins (29). Gallopamil (10 μM) increased the rate of Ca^{2+} efflux from guinea pig taenia coli into Ca^{2+}-free medium by an action that differed from those of papaverine and benactyzine (30).

In vivo studies.
Intravenously-administered verapamil decreased both LES pressure and the amplitude of contractions of the esophageal body in anaesthetized opossums (31) and in conscious baboons (32). Both verapamil and nifedipine relaxed K^+-induced contractions of rat exteriorized intestinal tract, the most pronounced relaxant effect occurring in duodenum; however, effective doses of both agents produced significant cardiovascular effects (18,33).

Clinical studies.

Intravenously-administered verapamil decreased LES pressure both in normal volunteers and in achalasia patients (34) and reduced the acid secretion provoked by intravenous gastrin (35). Some promising results have been obtained in certain IBS patients who have been treated with verapamil (18,36).

Studies With Diltiazem

Diltiazem has not been as thoroughly studied as nifedipine or verapamil with respect to its influence on basic GI functions or pathology.

Clinical Studies.

In the earlier clinical studies that were conducted, conflicting results were reported for the effect of diltiazem in patients with GI motility disorders such as "nutcracker" esophagus (19,37,38), achalasia (38) and IBS (39). However, one open study has shown that diltiazem significantly decreased the symptoms of chest pain and dysphagia associated with "nutcracker" esophagus (40). Most recently, a randomized double-blind trial indicated that orally-administered diltiazem (60-90 mg, q.i.d.) appeared to improve noncardiac chest pain associated with high-amplitude, peristaltic esophageal contractions ("nutcracker" esophagus), as compared with placebo (41). This improvement was associated with a significant decrease in contraction pressure. As microvascular angina (coronary artery disease) probably coexists with nutcracker esophagus in a large number of patients suffering from chest pain, diltiazem could have a dual therapeutic action in these cases.

The Problem with Using Typical Calcium Antagonists

Collectively, the results obtained with typical Ca^{2+}-antagonists are encouraging, but adverse cardiovascular side-effects represent major disadvantages when using these drugs. Hence, the possible beneficial effects of these agents would likely be offset by their production of such side-effects. Other drugs possessing greater selectivity for the GI tract are required.

Studies with Pinaverium Bromide

This problem has been overcome by using the atypical Ca^{2+}-antagonist pinaverium bromide (Dicetel®; see fig. 1). Pinaverium bromide acts mainly by blocking PDCs (42-46), and displays selectivity for the GI tract. Experimental evidence supporting these contentions has been obtained in both *in vitro* and *in vivo* studies.

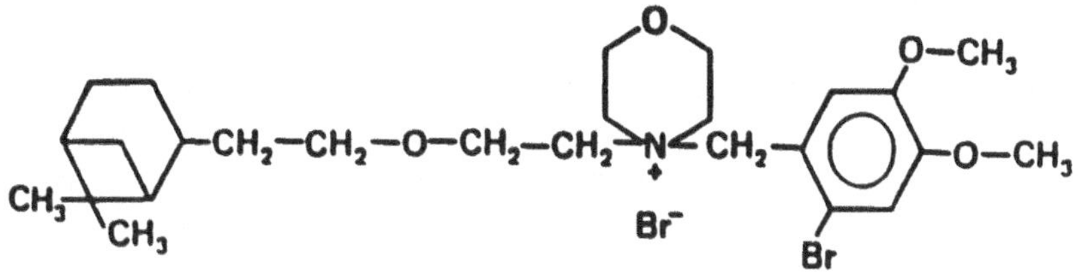

FIG 1. Pinaverium bromide: Dicetel®

In vitro studies.

In vitro results have shown that at concentrations of 0.4-4.0 μM pinaverium bromide antagonized $BaCl_2$-induced contractions of guinea pig isolated ileum and rat isolated duodenum (47), and at 1 μM it inhibited electrical activity in preparations of guinea pig ileum and taenia coli (42). Other experiments showed that micromolar concentrations of pinaverium bromide also inhibited the effects of electrical field stimulation, $BaCl_2$, ACh or synthetic MET-enkephalin analogue FK33-824 on *in vitro* intraluminal pressure responses of rat colonic segments (44), and that norepinephrine-induced contractions of rabbit ear artery are much less responsive to pinaverium bromide than those induced by K^+-induced depolarization (42).

Results obtained with the patch-clamp technique have shown that pinaverium bromide inhibits Ca^{2+} inward current in smooth muscle cells of rabbit jejunum with $IC_{50} \approx 1.5$ μM (48,49), as compared with IC_{50} values of 0.02 μM for nicardipine, 0.1 μM for nifedipine, 1.3 μM for gallopamil and 1.4 μM for diltiazem and flunarizine (49-52). This action of pinaverium bromide differed from those of verapamil, gallopamil, and diltiazem (and resembled that of nifedipine) in that it did not show use-dependence (i.e., not increased by repetitive stimulation) and differed from that of DHPs in that it was not conspicuously voltage-dependent (48,49). Very little shift of the "availability curve" was produced by pinaverium bromide with conditions under which DHPs, verapamil, and diltiazem cause distinct negative shifts, indicating that it has similar affinities for the closed available state and the inactivated state of the Ca^{2+} channel. DHPs combine with the inactivated state of the Ca^{2+} channel; verapamil and gallopamil act on the channel when it is opened by activity; and pinaverium bromide might act by combining with all three states of the channel with similar affinity (48).

In vivo studies.

The *in vivo* pharmacokinetic properties of pinaverium bromide differs markedly from those of conventional Ca^{2+}-antagonists after oral administration (53). The highly polar quaternary ammonium group of pinaverium bromide limits its passage across cell membranes, and therefore its *in vivo* selectivity for the GI tract, in addition to its tissue-selectivity, may be due to its low absorption and marked hepatobiliary excretion (46,53).

Clinical studies.

Double-blind studies have shown that oral administration of pinaverium bromide (50 mg, t.i.d.) is more effective than placebo and at least as effective as trimebutine and N-butyl hyoscine as therapy for IBS (46). Also, oral doses (150 mg single dose, or 200 mg b.i.d. for 3 days) relaxed Oddi's sphincter in patients with biliary dyskinesia (46). These findings gain added significance upon considering that pinaverium bromide has no vasodilatatory or anti-arrhythmic action in the clinical setting (46).

CONCLUDING COMMENTS

GI Selectivity

Collectively, the results obtained with typical Ca^{2+}-antagonists are encouraging, but the possible beneficial effects of these agents would likely be seriously limited by their dominant cardiovascular effects. Typical Ca^{2+}-antagonists that are widely employed in treating cardiovascular disorders are less effective on the GI tract, and this is why they have limited usefulness as therapy for GI disorders. Ca^{2+}-antagonists with augmented GI-selectivity are required. This problem related to GI-selectivity has been surmounted to some degree by introducing the atypical Ca^{2+}-antagonist pinaverium bromide (Dicetel®), the only drug currently registered as therapy for functional motility disorders (45,46). The unique properties of pinaverium bromide with respect to state- and use-dependency, together with its atypical pharmacokinetic profile, might contribute to its apparent selectivity for the GI tract. Pinaverium bromide exerts a spasmolytic action via its effect on Ca^{2+} channels, and it has no *in vivo* anticholinergic effect (54,55). As mentioned above, the GI-selectivity of pinaverium bromide appears to be related to its tissue-selectivity and its unique pharmacokinetic profile.

Although radioligand binding experiments have revealed no significant differences between the Ca^{2+} channels of GI smooth muscle and those of other types of smooth muscle (56,57), there are reasons for believing that a search for agents possessing GI selectivity would be successful; e.g., although all Ca^{2+} antagonists block the Ca^{2+} entry that occurs via L-subtype PDCs during smooth muscle activation (2), they do differ considerably from one another in chemical structure and with respect to their detailed molecular mechanism of action (58), indicating that L-channels of GI smooth muscle might possess certain distinct receptor sites. Factors such as pharmacokinetic characteristics of the drug, relative agonist use of PDCs, Ca^{2+} channel differences, pathological state of the tissue, mode of Ca^{2+} mobilization in the tissue, effective drug concentration at its site of action, state- and use-dependent interactions, and physicochemical and relative agonist/antagonist properties of the drug represent other determinants of tissue-selectivity that might vary among Ca^{2+}-antagonists (58,59).

Certain Ca^{2+}-antagonists do appear to possess GI-selectivity if their actions on GI versus cardiac tissues are compared; e.g., DHP-type Ca^{2+}-antagonists are more active in inhibiting ACh-induced contractions of guinea pig ileal smooth muscle (60) than in eliciting negative inotropic activity in cat papillary muscle (61), and pinaverium bromide is much more potent in inhibiting Ca^{2+}-induced contractions of K^+-depolarized rabbit colonic segments than in opposing the positive inotropic effect of Ca^{2+} on rabbit myocardium (62). Tissue-selectivity might also be related to relative potency of action. In this regard, the recent results of Xiong et al. (63) are of interest. They showed that the diltiazem derivative TA-3090 is about 10 times more potent than diltiazem itself in inhibiting Ba^{2+}-inward current recorded from smooth muscle cells of rabbit mesenteric artery. Also binding studies showed that [^{3}H]TA-3090 appears to have about 4-times higher affinity for rat myocardial benzothiazepine binding sites than does diltiazem (64).

Another property of Ca^{2+}-antagonists that could be related to GI-selectivity would be their interactions with sites other than Ca^{2+} channels. In this regard, the results of Terada et al (51) indicate that Ca^{2+} antagonists may block membrane channels other than the Ca^{2+} channels in intestinal smooth muscle cells; variations in this property among smooth muscle cells of different tissues could be a determinant of tissue-selectivity. Other mechanisms of action of drugs possessing Ca^{2+}-antagonistic activity could also contribute to GI-selectivity.

Further understanding of the differences that exist between the properties of ion channels of smooth muscle cells of the GI tract and those of vascular and other tissues will be required for developing GI-selective Ca^{2+}-antagonists.

Other Possible Uses for Ca^{2+}-Antagonists in Gastroenterology

Recent studies with animal models indicate that the possible therapeutic benefit of Ca^{2+}-antagonists in gastroenterology is not limited to motility disorders. Ca^{2+} entry is crucially implicated in the regulation of digestive secretions, and Ca^{2+}-antagonists are active on experimentally-induced gastric ulcers (65) and exert cytoprotective effects in other GI injuries. Thus, Ca^{2+}-antagonists may provide useful therapy for GI hypersecretory states and ulcerogenesis. In this regard, it has been reported that verapamil inhibits histamine-induced gastric acid secretion in isolated guinea pig stomach (35) and in isolated guinea pig parietal cells (66) and affects gastric acid secretion in humans (18). More recent reports indicate that verapamil, nifedipine, and diltiazem can reduce gastric acid secretion in various animal models and block stress- or bethanechol-induced gastric ulceration (67-75). However, further experiments showed that such Ca^{2+}-antagonists (e.g., verapamil) actually worsened ethanol-induced gastric mucosal damage (74,76). The most consistent and striking results obtained by Glavin (77) involved nitrendipine, which was found to abolish stress-induced gastric mucosal lesions at all doses tested (8,16, and 32 mg/kg, i.p.). Further tests with rats have revealed that nitrendipine (in these same doses) blocked bethanechol-stimulated gastric acid output, but not the acid secretion induced by dimaprit or by pentagastrin, indicating the nitrendipine interacts with cholinergic (vagal) input to the GI tract (65). Such findings support the contention that typical Ca^{2+}-antagonists could be useful in treating gastric or intestinal hypermotility states, ulcers, and certain esophageal motility disorders, but this may not be a valid assumption since the doses of these agents that are required to produce such actions would likely produce adverse cardiovascular effects. Further studies may reveal that certain other Ca^{2+}-antagonists will influence favorably intestinal or gastric secretion or ulcerogenesis in the clinical setting while inducing tolerable side-effects.

With more specific regard to the cytoprotective effects of Ca^{2+}-antagonists, it might be noted that a deregulation of Ca^{2+} homeostasis (i.e., a pathological increase in cytosolic Ca^{2+} concentration) is significantly involved in the development of hepatocellular injury. The increase in intracellular Ca^{2+} concentration that occurs following cellular damage (78,79) activates a variety of potentially destructive mechanism, including stimulation of Ca^{2+}-dependent degradative enzymes (phospholipases, proteases, endonucleases) potentiation of the destructive actions of oxygen-free radicals, and decreased mitochondrial ATP synthesis (79). Reducing the Ca^{2+} influx that follows cellular injury with Ca^{2+}-antagonists can protect hepatocytes against cellular injury. Such cytoprotective effects of Ca^{2+}-antagonists would involve the inhibition of Ca^{2+} release from mitochrondria (80,81), their antioxidant properties (82), and their interactions with inflammatory mediatory (83). In the few *in vivo* studies that have been conducted in this area, it has been shown that nicardipine significantly reduces the levels of liver enzymes in rats with acute hepatic injury induced by *d*-galactosamine (84), and that verapamil and nifedipine exert protective effects on hepatocellular injury in rats following CCl_4, chloroform, dimethylnitrosamine, thioacetamide, and paracetamol (85). Nifedipine (25 mg/kg) completely prevented centrilobular necrosis and provided almost total protection following dimethylnitrosamine and partial protection against chloroform toxicity; verapamil (25 mg/kg) provided almost complete protection against dimethylnitrosamine and some protection against CCl_4 toxicity. Diltiazem can also protect against the hepatotoxic effects of paracetamol in mice (79). A study of liver transplantation in the rat has revealed that addition of a very low concentration of nisoldipine (1.4 μM) to the fluid used for liver grafts significantly increased survival time of the transplanted rats (86). Taken together, these results indicated that Ca^{2+}-antagonists might be useful in controlling hepatocellular injury and certain other types of cellular injuries, but the use of conventional Ca^{2+}-antagonists for clinical cytoprotection will have to be considered in light of the possible deleterious cardiovascular (and other) side-effects of these agents.

REFERENCES

1.   Fleckenstein A. (1977): *Annu. Rev. Pharmac. Toxicol.*, 17: 149-166.
2.   Fleckenstein A. (1983): *Calcium Antagonism in Heart and Smooth Muscle.* John Wiley & Sons, New York.
3.   Bolton T. (1979): *Physiol. Rev.*, 43: 606-718.
4.   Bolger G.T., Gengo P.T., Luchowski E.M., Siegel H., Triggle D.J and Janis R.A. (1982): *Biochem. Biophys. Res. Commun.*, 104: 1604-1609.
5.   Zar M.A. and Gooptu D. (1983): *Br. J. Clin. Pharmacol.*, 16: 339-340.
6.   Barone F.C., White R.F., Ormsbee H.S. and Wasserman M.A. (1986): *J. Pharmacol. Exp. Therap.*, 237: 99-105.
7.   Godfraind T., Morel N., Salomone S. and Wibo M. (1989): In: *Calcium Antagonism in Gastrointestinal Motility*, edited by M.O. Christen, T. Godfraind and R.W. McCallum pp. 117-139. Elsevier, Paris.
8.   Galletti F., Zheng W., Gopalakrishnan M., Rutledge A. and Triggle D.J. (1991): *Eur. J. Pharmacol.*, 195: 125-129.
9.   Micheli D., Ratti E., Toson G. and Gaviraghi G. (1991): *J. Cardiovasc. Pharmacol.*, 17 (suppl.4): S1-S8.

10. De Ponti F., D'Angelo L., Frigo G.M. and Crema A. (1989): *Eur. J. Pharmacol.*, 168: 133-144.

11. Weiser H.F., Lepsien G., Golenhofen K., Shattenmann G. and Siewert R. (1978): In: *Gastrointestinal Motility in Health and Disease,* edited by H.L. Duthie pp. 565-572. MTP Press, Lancaster.

12. Hongo M. (1981): *Jpn. J. Smooth Muscle Res.,* 17: 47-51.

13. Blackwell J.N., Holt S. and Heading R.C. (1981): *Digestion,* 21: 50-56.

14. Hondo M., Traube M., McAllister R.G. and McCallum R.W. (1984): *Gastroenterology,* 86: 8-12.

15. McCallum R.W. (1989): In: *Calcium Antagonism in Gastrointestinal Motility,* edited by M.O. Christen, T. Godfraind and R.W. McCallum pp. 55-68. Elsevier, Paris.

16. Bortolotti M. and Labo G. (1981): *Gastroenterology,* 80: 39-44.

17. Gelfond M., Rozen P. and Gilat T. (1982): *Gastroenterology,* 83: 963-969.

18. Traube M. and McCallum R.W. (1984): *Amer. J. Gastroenterol.,* 79: 892-896.

19. Richter J., Dalton C., Bradley L. and Casteel D. (1987): *Gastroenterology,* 93: 21-28.

20. Triadafilopoulos G., Aaronson M., Sackel S. and Burakoff R. (1991): *Dig. Dis. Sci.,* 36: 260-267.

21. Coccia G., Bortolotti M., Michetti P. and Dodero M. (1991): *Gut,* 32: 604-606.

22. Robertson C.S., Hardy J.G. and Atkinson M. (1989): *Gut,* 30: 768-773.

23. Santander R., Mena I. and Valenzuela J.E. (1986): *Gastroenterology,* 90: 1614.

24. Blume M., Schuster M. and Tucker H. (1983): *Gastroenterology,* 84: 1109.

25. Narducci F., Pelli M.A., Vedovelli A. and Morelli A. (1983): *Gastroenterology,* 84: 1256

26. Narducci F., Bassotti G., Gaburri M., Farroni F. and Morelli A. (1985): *Amer. J. Gastroenterol.,* 80: 317-319.

27. Piror A., Harris S.R. and Whorwell P.J. (1987): *Gut,* 28: 1609-1612.

28. Sun W.M., Edwards C.A., Prior A., Rao S.S.C. and Read N.W. (1990): *Dig. Dis. Sci.,* 35: 885-890.

29. Ishizawa M. and Miyazaki E. (1978): *Prostaglandins,* 16: 591.

30. Shiba T., Uruno T, Kubota K and Takagi K. (1981): *Jpn. J. Pharmacol.,* 31: 551-561.

31. Goyal R.K. and Rattan S. (1980): *Am. J. Physiol.,* 238: G40-G44.

32. Richter J.E., Sinar D.R., Cordova C.M. and Casteel D.O. (1982): *Gastroenterology,* 82: 882-886.

33.   Maggi C.A., Grimaldi G. and Meli A. (1983): *Jpn. J. Pharmacol.*, 27: 9-14.
34.   Becker B.S. and Burakoff R. (1983): *Am. J. Gastroenterol.*, 78: 773-775.
35.   Kirkegaard P., Christiansen J., Petersen B. and Skov Olsen P. (1982): *Scand. J. Gastroenterol.*, 17: 533-538.
36.   Byrne S. (1987): *J. Clin. Psych.*, 48: 388.
37.   Richter J.E., Spurling T.J., Cordova C.M. and Castell D.O. (1987): *Dig. Dis. Sci.*, 29: 649-656.
38.   Silverstein B.D., Kramer C.M. and Pope C.E. (1982): *Gastroenterology*, 82: 1181.
39.   Perez-Mateo M., Sillero C., Cuesta A., Vasquez N. and Berbegal J. (1986): *Int. J. Clin. Pharm. Res.*, 6: 425-427.
40.   Richter J.E., Dalton C.B., Bradley L.A. and Castell D.O. (1984): *Gastroenterology*, 93: 21-28.
41.   Cattau E.L., Castell D.O., Johnson D.A., Spurling T.J., Hirszel R., Chobanian S.J. and Richter J.E. (1991): *Am. J. Gastroenterol.*, 86: 272-276.
42.   Droogmans G., Himpens B. and Casteels R. (1983): *Naunyn-Schmiedeberg's Arch. Pharmacol.*, 323-72-77.
43.   Mironneau J., Lalune C., Mironneau C., Savineau J.-P. and Lavie J.-L. (1984): *Eur. J. Pharmacol.*, 98: 99-107.
44.   Baumgartner A., Drack E., Halter F. and Scheurer U. (1985): *Br. J. Pharmacol.*, 86: 89-94.
45.   Christen M.O. and Tassignon J.P. (1989): *Drug. Dev. Res.*, 18: 101-112.
46.   Christen M.O. (1990): *Gen. Pharmacol.*, 21: 821-825.
47.   Bretaudeau J. and Foussard-Blanpin O. (1980): *J. Pharmacol.* (Paris), 11: 233-243.
48.   Bolton T.B., MacKenzie I. and Beech D.J. (1989): In: *Calcium Antagonism in Gastrointestinal Motility*, edited by M.O. Christen, T. Godfraind and R.W. McCallum pp. 109-115. Elsevier, Paris.
49.   Beech D.J., MacKenzie I., Bolton T.B. and Christen M.O. (1990): *Br. J. Pharamcol.*, 99: 374-378.
50.   Terada K., Kitamura K. and Hirosi K. (1987): *Pflugers Arch.*, 408: 552-557.
51.   Terada K., Kitamura K. and Kuriyama H. (1987): *Pflugers Arch.*, 408: 558-564.
52.   Terada K., Nakao K., Okabe K., Kitamura K. and Kuriyama H. (1987): *Br. J. Pharmacol.*, 92: 615-625.
53.   Jacquot C., Trouvin J.H. and Christen M.O. (1989): In: *Calcium Antagonism in Gastrointestinal Motility*, edited by M.O. Christen, T. Godfraind and R.W. McCallum pp. 141-151. Elsevier, Paris.
54.   Bretaudeau J., Foussard-Blanpin O., Baronnet R. and Despraires

R. (1975): *Thérapie,* 10: 919-930.

55. Roux J.-J., Salmon D. and Reny A. (1980): *Gaz. Méd.,* 87: 1033-1036.

56. Yousif F.B., Bolger G.T., Ruzycky A. and Triggle D.J. (1985): *Can. J. Physiol. Pharmacol.,* 63: 453-462.

57. Janis R.A., Silver P. and Triggle D.J. (1987): *Adv. Drug Res.,* 16: 309-591.

58. Triggle D.J., Ferrante J., Hawthorn M., Kwon Y.W. and Zheng W. (1989): *Calcium Antagonism in Gastrointestinal Motility,* edited by M.O. Christen, T. Godfraind and R.W. McCallum pp. 89-107. Elsevier, Paris.

59. van Breeman C., Aaronson P. and Loutzenhiser R. (1979): *Pharmacol. Rev.,* 30: 167-208.

60. Janis R.A. and Triggle D.J. (1983): *J. Med. Chem.,* 26: 775-785.

61. Rodenkirchen R., Bayer R., Steiner R., Bossert F., Meyer H. and Moller E. (1978): *Naunyn-Schmiedeberg's Arch. Pharamcol.,* 310: 69-78.

62. Szekeres L. and Papp J.G. (1989): *Calcium Antagonism in Gastrointestinal Motility,* edited by M.O. Christen, T. Godfraind and R.W. McCallum pp. 161-167. Elsevier, Paris.

63. Xiong Z., Sakai T., Inoue Y., Kitamura K. and Kuriyama H. (1990): *Naunyn-Schmiedeberg's Arch. Pharmacol.,* 341: 373-380.

64. Zobrist R.H. and Mecca T.E. (1990: *J. Pharmacol. Exp. Ther.,* 253: 461-465.

65. Glavin G.B. (1990): *Ann. NY Acad. Sci.,* 597: 293-297.

66. Sewing K.F. and Hannemann H. (1983): *Pharmacology,* 27: 1-8.

67. Canfield P., Coruzzi G., Curwain B. and Rundell T. (1985): *Eur. J. Pharmacol.,* 116: 89-95.

68. Ogle C.W., Cho C.H., Tong M.C. and Koo M.W.L. (1985): *Eur. J. Pharmacol.,* 112: 399-404.

69. Wait R., Leahy A., Vee J. and Pollock T. (1985): *J. Surg. Res.,* 38: 424-428.

70. Koo M., Cho C. and Ogle C. (1986a): *J. Pharm. Pharmac.,* 38: 845-848.

71. Brage R., Cortijo J., Esplugues J., Mati-Bonmati E. and Rodriguez C. (1986): *Br. J. Pharmacol.,* 89: 627-633.

72. Herling A.W., Wirth K. and Bickel M. (1986): *Archs. Pharmacol.,* 332(suppl.): 242.

73. Ghoneim M.T., El-Dakhakhny M.M., Arab A.M. and Barakat M.K. (1986): *Archs. Pharmacol.,* 334(suppl.): 117.

74. Glavin G. (1988): *J. Pharm. Pharmac.,* 40: 514-515.

75. Hertz F. and Cloarec A. (1989): *Gen. Pharmacol.,* 20: 635-640.

76. Koo M., Cho C. and Ogle C. (1986): *Eur. J. Pharmacol.,* 120: 355-358.

77.   Glavin G. (1989): *Dig. Dis. Sci.* 34: 1477.
78.   McLean A.E.M., McLean M. and Judah J.D. (1965): *Int. Rev. Exp. Pathol.*, 4: 127-157.
79.   Deakin C.D., Fagan E.A. and Williams R. (1991): *J. Hepatol.*, 12: 251-255.
80.   Kloner R.A. and Braunwald E. (1987): *Am. J. Cardiol.*, 59: 84B-94B.
81.   Buss W.C., Savage D.D., Stepanek J., Little S.A. and McGuffee L.J. (1988): *Eur. J. Pharmacol.*, 152: 247-253.
82.   Stein H.J., Oosthuizen M.M. and Hinder R.A. (1989): *Gastroenterology*, 96: A663.
83.   Weichmann B.M., Muccitelli R.M., Tucker S.S. and Wasserman M.A. (1988): *J. Pharmac. Exp. Ther.*, 225: 310-315.
84.   Garay G.L., Annesley P. and Burnett M. (1984): *Gastroenterology*, 86: 1319.
85.   Landon E.J., Naukam R.J. and Sastry B.V.R. (1986): *Biochem. Pharmacol.*, 35: 697-705.
86.   Takei Y., Marzi I., Kauffman F.C., Currin R.T., Lemasters J.J. and Thurman R.G. (1990): *Transplantation*, 50: 14-20.

ROLE OF CALCIUM CHANNELS IN PHARMACOLOGICAL MODULATION OF GASTROINTESTINAL MOTILITY

Jan D. Huizinga and Louis W.C. Liu

Intestinal Disease Research Unit
Departments of Biomedical Sciences
and Engineering Physics
McMaster University
Hamilton, Ontario, Canada L8N 3ZS

Synchronized circumferential contractions and propagating phasic contractile activity are important components of gastrointestinal motility involved in propulsion of food and food residues. These contractions are initiated and regulated by electrical activity (slow-wave-type action potentials or slow waves) of the smooth muscle cell membranes (1). Periodically electrical pacemaker potentials are generated by specialized pacemaker cells [a network of interstitial cells of Cajal (ICC) and smooth muscle cells] and propagated into the rest of the smooth muscle layer. Slow-wave-type action potentials generated in the smooth muscle layers consist of an upstroke phase and a plateau phase with or without superimposed spikes. Calcium channels play a role in both the initiation of the slow-wave and in the generation of the plateau phase, the latter being associated with generation and regulation of contraction. Pharmacological manipulation of calcium channel activity may therefore influence the slow wave frequency and propagation, determining the characteristics of propulsion and synchronization, as well as the duration and amplitude of the slow wave plateau phase, determining the force of the individual phasic contractions. One objective of this chapter is to discuss involvement of calcium channels in the generation of the slow-wave-type action potentials in the colon with the canine colon as primary model. Another objective is to discuss the role of calcium channels in the pharmacological control of colonic motility.

CALCIUM CHANNELS MEDIATING FORCE DEVELOPMENT IN COLONIC SMOOTH MUSCLE CELLS

Physiology

Colonic circular smooth muscle of the dog is electrically dominated by pacemaker activity generated by a network of ICC located at its

T. Godfraind et al. (eds.), Calcium Antagonists, 315–323.
© 1993 Kluwer Academic Publishers and Fondazione Giovanni Lorenzini.

submucosal border. This activity (a 6-cycles-per-minute depolarizing pulse) is transmitted to the circular muscle cells where it activates L-type calcium channels that lead to the creation of a slow-wave-type action potential. It is the plateau phase of the action potential, generated through L-type calcium channels, that is associated with generation of contraction. The amplitude and duration of the plateau phase depends on the level of excitation. Modification of this plateau potential by excitatory or inhibitory drugs is directly related to modification of force of contraction. At high levels of excitation, activity of L-type calcium channels creates spikes superimposed on the plateau phase of the slow waves further increasing force of contraction.

The contribution of the intrinsic properties of smooth muscle cells to the regulation of L-type calcium channels has to be studied in preparations devoid of pacemaker cells. In such preparations, the slow waves disappear and only spikelike action potentials occur created by voltage activation of L-type calcium channels (2). We learned that the action potential activity depends on the level of excitation of L-type calcium channels which to a large extent is determined by the K conductance of the cells. Varying the level of stimulation of L-type calcium channels with Bay K 8644, it was seen that this drug only affects action potential activity and hence contraction, at certain levels of K conductance regulated by $BaCl_2$ (fig. 1). It is likely that the spike and the plateau are mediated by different subtypes of L-type calcium channels. Because of the voltage dependency

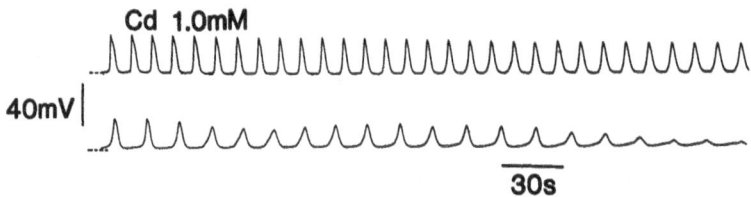

FIG. 1. A non-L type calcium channel is underlying pacemaker activity. Pacemaker activity generated by the network of interstitial cells of Cajal at the submucosal surface of the canine colon. The activity is recorded in the presence of the L-type calcium channel blocker, nifedipine. Cd^{2+} is added 30 seconds after the start of the top tracing. It inhibits the pacemaker potential, this effect is not associated with a change in the resting membrane potential.

of the L-type calcium channels, the value of the resting membrane potential is also a determining factor. Thus, the L-type calcium conductance, the K conductance, and the membrane potential all have to be within a certain region for the plateau phase and the spikes, and hence contraction, to be generated. Change in any of the three parameters will effect the electrical activity and can move the muscle cells out of the active region.

L-type calcium channel blockers affect the slow-wave-type action potential in gastrointestinal smooth muscle in a very characteristic manner. They inhibit the plateau potential without affecting the frequency of the activity (3-5). In many intestinal tissues, force of contraction is enhanced through development of spiking activity superimposed on the slow waves. Calcium channel blockers inhibit spike activity at concentrations lower than those needed to inhibit the slow-wave plateau.

L-type calcium channel blockers

Available blockers of L-type calcium channels are a chemically heterogeneous group of substances with, as prototypes, nifedipine, diltiazem, and verapamil (or D600) (6). These agents are effective in both cardiac and intestinal muscle suggesting that the channel structure is very similar in both types of muscle. The cardiovascular effects of these drugs may limit their applicability to certain intestinal motor disorders, although this cannot be predicted with certainty (7). Pinaverium bromide is a smooth muscle relaxant that has properties as a calcium channel antagonist without systemic cardiovascular effects when given orally. Beech et al. (8) studied the action of pinaverium bromide (0.1-10 μM) on calcium channel activity in single smooth muscle cells of the rabbit jejunum. Pinaverium blocked inward current through voltage-sensitive calcium channels thus establishing itself as a calcium entry blocker. The IC_{50} for pinaverium on calcium currents is about 10 times less than that for nitrendipine but similar to that of verapamil. Pinaverium bromide is a quaternary ammonium compound that is poorly absorbed by the gastrointestinal (GI) tract (only 5 to 10% enters the circulation) and rapidly metabolized and efficiently excreted through the hepatobiliary system (9). Elimination is thus essentially by the fecal route. Autoradiographic studies showed that most of the orally administered drug remained in the gastrointestinal tract. Thus, the specificity of pinaverium for GI smooth muscle may be due to the characteristics of its resorption kinetics. However, intravenous injection of pinaverium bromide did not have marked effects on cardiac function (10) indicating that pinaverium bromide may have specificity for the intestinal calcium channel.

Studies on the effects of L-type calcium channel blockers on intestinal motility in conscious dogs were recently performed by measuring electrical (11) as well as contractile (11,12) activities. Profound inhibitory effects were observed, the order of potency being nifedipine > verapamil > diltiazem = pinaverium. Significant cardiovascular effects were reported with nifedipine (an increase in heart rate and a decrease in systolic blood pressure) emphasizing the notion that the classical L-type calcium channel blockers also act upon cardiac channels. In addition, other nonintestinal smooth muscle actions have to be taken into account, although respiratory smooth muscle seems to be significantly and uniformly less sensitive to the blockers than other smooth muscle (6).

L-type calcium channel blockers have obvious potential in colonic motor disorders where contractile activity of excessive force is established. Nifedipine has been shown to reduce contractions and electrical spiking activity induced by eating of a meal in irritable bowel syndrome (IBS) patients (13). Pinaverium bromide has also been shown to reduce colonic motor activity in individuals following a test meal or after stimulation with neostigmine (9). Pinaverium bromide has been shown to reduce symptoms in IBS (14); this occurred without effects on amplitude or frequency of propulsive contractions. It is therefore clear that the mechanism of action of pinaverium bromide *in vivo* is not fully elucidated.

To increase our understanding of the pathophysiology of colonic motor disorders associated with IBS, and to make rational decisions about the use of calcium channel blockers, we have to characterize and define electrical and motor patterns underlying the motor abnormalities. Then we have to study the effects of calcium channel blockers on the motor patterns and on the abnormal colonic transit. Such studies may not only reveal benefits to patients, it will increase our understanding of motor abnormalities because the mechanism of action of the drugs is known at a cellular and molecular level. *In vivo* studies will allow us to understand the mechanism of action on a tissue level. Because of the complexity of colonic motor activities, 24-hour studies seem essential (15). Since IBS patients are a heterogeneous group with respect to pathophysiology, a clinical trial to assess usefulness of a drug for IBS patients in general seems destined to fail. It seems more relevant to try to find out why a drug worked in one but not in another patient by carefully monitoring the effect of the drug on measured parameters of function. Such studies could lead to calcium channel blockers being targeted to specific subclasses of IBS patients.

DO INTESTINAL PACEMAKER CELLS HAVE UNIQUE CALCIUM CHANNELS?

Cell physiology

A network of ICC connected by gap junctions (in some tissues such as the canine colon in gap junction contact with a few layers of smooth muscle cells) is involved in the generation of pacemaker activity for the gut smooth muscle layers. Particularly in the canine colon circular muscle the evidence is strong (1,2,16,17). Comparing the characteristics of the circular muscle layer with and without this network of ICC, it became clear that only the ICC network and possibly the associated smooth muscle cells generate an omnipresent oscillation, that is insensitive to L-type calcium channels and hyperpolarization. This oscillation acts as pacemaker activity to circular smooth muscle to generate the slow-wave-type action potential. The pacemaker potential is sensitive to Ni^{2+} and Cd^{2+} and omission of extracellular calcium (fig. 2) (18). Furthermore, the oscillation is sensitive to temperature and changes in intracellular cAMP (19). Since circular muscle preparations without the ICC network are electrically quiescent, the hypothesis seems warranted that cells in the ICC network possess a calcium conductance that is not primarily regulated by voltage changes, but by metabolic activity (19). Extensive gap junction contacts between ICC and a high density of mitochondria seem consistent with such a postulate. Activity generated in the ICC network will actively propagate into the circular muscle layer where the full slow-wave-type action potential is generated. The frequency of the slow-wave-type action potentials, local frequency gradients, and neuronal activity will determine contraction patterns and propagation characteristics; hence, pharmacological modulation of the pacemaker calcium channel in the ICC network will have marked influence on these parameters.

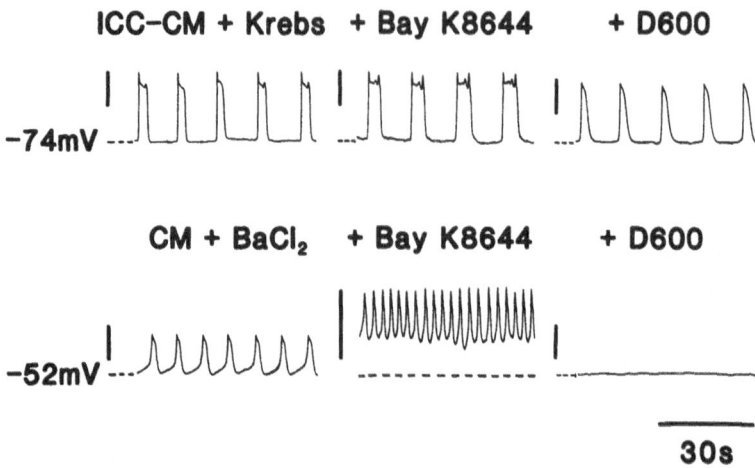

FIG. 2. Activation of L-type calcium channels. In ICC-CM, i.e., circular muscle with the submucosal network of interstitial cells of Cajal (ICC) attached, slow-wave-type action potentials are generated. Bay K 8644 increases the plateau phase and spiking activity. The slow-wave-type action potentials have a D600 insensitive component.

In CM, i.e., circular muscle without the ICC pacemaker cells, slow-wave-type action potentials are not generated. However, Ba^{2+} can evoke oscillatory activity. This activity is markedly influenced by Bay K 8644 but only at appropriate K conductances. L-type calcium blockers abolish this activity. Note that the oscillation frequeney is very "unstable" without the presence of pacemaker activity.

ICC and propulsive contractile activity

In vivo recordings from the dog colon revealed a cyclic occurrence of bursts of contractions propagating in anal direction at a speed of 3-20 cm/min (20). Similarly, Fioramonti et al. (21), recorded propagation of spike bursts at 14 cm/min. An *in vitro* study in the cat colon showed that slow waves were phase locked in longitudinal direction and were propagating at a velocity around 18 cm/min (22). This study clearly indicated that the longitudinal propagation occurred not through the circular muscle layer nor longitudinal muscle layer, but through the submucosal network of ICC. Indeed, in the dog colon the circular muscle layer is divided into circumferentially oriented lamella with connective tissue in between. There is no evidence of electrical communication between lamella. However, there is a continuous network of ICC covering the submucosal surface and hence electrically connecting the different circumferential bundles of smooth muscle. It is therefore clear that the network of ICC is crucial for normal propulsive contractile activity; pharmacological interference with these cells may be the most direct way of correcting abnormalities in colonic transit. Our studies point to a difference in calcium channel composition between circular muscle cells and ICC; further research may reveal more differences. Pharmacological studies aimed to interfere with normal or abnormal propagation cannot be limited to studies on muscle cells, but have to be directed specifically to the network of ICC.

Colonic transit and propulsive contractile activity leading to defecation is controlled in part by neural activity. The ICC are ideally positioned to receive neural input. In all regions of the GI tract, the ICC are densely innervated with very close contacts between nerve varicosities and interstitial cell membranes (15,23). In contrast, smooth muscle cells are much less innervated. In addition, there are no special structural contacts between nerves and smooth muscle cells.

The recognition that the pacemaker activity in the GI tract is generated by specific cell systems including ICC, that ICC are essential for propulsive contractile activity, and that ICC may have a unique ion channel composition and may also have unique innervation, should intensify research into the physiological and pharmacological role of ICC. This should provide new strategies for the development of drugs related to motor disturbances not primarily with respect to force of contraction but with respect to abnormal development of patterns of electrical and motor activity directly related to colonic transit (24).

In summary, in intestinal smooth muscle there are different types of calcium channels associated with different functions, probably situated in different cell types. The most significant functions are the regulation of frequency and force of phasic contractile activity. The calcium channels can be influenced through manipulation of intracellular messenger systems and K-channel activity, as well as directly through specific blockers or activators. The direct approach becomes interesting with the development of drugs specific for intestinal calcium channels.

ACKNOWLEDGEMENTS

The Medical Research Council of Canada provided a scholarship to JDH, a studentship to LWCL and operating grants.

REFERENCES

1.   Huizinga D., (1991): *Can. J. Physiol. Pharmacol.*, (in press).

2.   Liu L.W.C., Daniel E.E. and Huizinga J.D. (1991): *Can. J. Physiol. Pharmacol.*, accepted for publication.

3.   Szurszewski J.H. (1987): In: *Physiology of the Gastrointestinal Tract, Volume 1*, edited by L.R. Johnson pp. 383-422. Raven Press, New York.

4.   Barajas-López C. and Huizinga J.D. (1988): *Pflügers Arch.*, 12: 203-210.

5.   Barajas-López C. and Huizinga J.D. (1989): *Am. J. Physiol.*, 256: G570-G580.

6.   Triggle D.J. (1990): *Can. J. Physiol. Pharmacol.*, 68: 1474-1481.

7.   Castel D.O. (1985): *Am. J. Cardiol.*, 55: 210B-213B.

8.   Beech D.J., MacKenzie I., Bolton T.B. and Christen M.O. (1990): *Br. J. Pharmacol.*, 99: 374-378. 9.   Christen M.O. and Tassignon J.P. (1989): In: *Calcium Antagonism in Gastrointestinal Motility*, edited by M.O. Christen, T. Godfraind and R.W. McCallum pp. 169-181. Elsevier, Paris.

10. Guerot C., Khemache A., Sebbah J. and Noel B. (1988): *Curr. Med. Res. Opin.*, 11: 73-79.

11. De Ponti F., D'Angelo L., Frigo G.M. and Crema A. (1989): *Eur. J. Pharmacol.*, 168: 133-144.

12. Itoh Z., Iwanaga Y. and Mizumoto A. (1989): In: *Calcium Antagonism in Gastrointestinal Motility*, edited by M.O. Christen, T. Godfraind and R.W. McCallum pp. 153-159. Elsevier, Paris.

13. Narducci F., Bassotti G., Gaburri M., Farroni F. and Morelli A. (1985): *Am. J. Gastroenterol.*, 80: 317-319.

14. Galeone M., Stock F., Moise G., Cacioli D., Toti G.L. and Megevand J. (1986): *Curr. Ther. Res.*, 39: 613-624.

15. Huizinga J.D. (1986): *Clin. Gastroenterol.*, 15: 879-901.

16. Berezin I., Huizinga J.D. and Daniel E.E. (1988): *J. Comp. Neurol.*, 273: 42-51.

17. Barajas-López C., Berezin I., Daniel E.E. and Huizinga J.D. (1989): *Am. J. Physiol.*, 257:C830-C835. 18.   Huizinga J.D., Farraway L. and Den Hertog A. (1991): *J. Physiol. (Lond).*, 442: 15-29.

19. Huizinga J.D., Farraway L. and Den Hertog A. (1991): *J. Physiol. (Lond).*, 442: 31-45.

20. Sarna S.K., Condon R. and Cowles V. (1984): *Am. J. Physiol.*, 246: G355-G360.

21. Fioramonti J., Garcia Villar R., Bueno L. and Ruckebusch Y. (1980): *Dig. Dis. Sci.*, 25: 641-646.

22. Conklin J.L. and Du C. (1990): *Am. J. Physiol.*, 258: G894-G903.

23. Thuneberg L. (1989): In: *Handbook of Physiology, the Gastrointestinal System*, edited by G.S. Schultz, J.D. Wood and B.B. Raunert pp. 349-386. American Physiological Society, Bethesda,

Maryland.
24. Schang J.C., Hemond M., Hebert M.and Pilote, M. (1986): *Dig. Dis. Sci.,* 31: 1331-1337.

A NEW CLASS OF CALCIUM ANTAGONIST SELECTIVE FOR THE GI TRACT

M.O. Christen

SOLVAY PHARMA
Laboratoires de Thérapeutique Moderne (L.T.M.)
42, rue Rouget-de-Lisle, B.P. 22
92151 Suresnes Cedex, France

This article is aimed at providing evidence in support of the existence of a new class of Ca^{2+} antagonist, the prototype being pinaverium bromide, which exhibits selectivity for the gastrointestinal (GI) tract. It should be recalled that Ca^{2+} is implicated at different levels in the functions of the GI tract. As examples, Ca^{2+} plays roles in excitation-contraction coupling in smooth muscle of the esophagus, stomach, and intestine, it is involved in the regulation of blood flow (alterations of which could be related to the formation of ulcers), and Ca^{2+} overload can lead to hepatocellular injuries. Aspects related to GI motility will be emphasized herein.

Certain drugs (e.g., anticholinergics, opiates, phenothiazines) are currently used to treat functional GI motility disorders [e.g., irritable bowel syndrome (IBS), esophageal spasm, nausea, nonulcer dyspepsia, chronic intestinal pseudo-obstruction]. However, their effects have not been adequate in many cases, particularly in the treatment of IBS, indicating the need for a new approach. Spasmolytic doses of competitive anticholinergic spasmolytic agents (e.g., N-butylscopolamine) are recommended for controlling increased tone and/or motility of the GI tract (e.g., for IBS therapy). However, the use of such agents in treating IBS has several major shortcomings (1). Musculotropic spasmolytic agents are also being employed, but have not provided significant benefit in many cases. Although IBS can be pharmacologically modulated by using certain centrally-active drugs (e.g., antidepressants, anxiolytics, antidopaminergic agents), their efficacy also remains questionable.

A NEW APPROACH IS NEEDED -- POSSIBLE USE OF CALCIUM ANTAGONISTS

A new approach has been envisaged; i.e., one involving the search for Ca^{2+} antagonists that exhibit selectivity for the GI tract. Although stress, other psychological factors, and the ingestion of food may play significant

T. Godfraind et al. (eds.), Calcium Antagonists, 325–335.
© 1993 Kluwer Academic Publishers and Fondazione Giovanni Lorenzini.

roles in triggering the symptoms of functional GI motility disorders such as IBS (2), therapy with Ca^{2+} antagonists can be readily conceived upon considering that all smooth muscles examined to date possess Ca^{2+}-selective transmembrane channels which open when the cell membrane is depolarized, and that activation of these channels represents the "final common path" of all mechanisms regulating GI motility.

At least four types of potential-dependent Ca^{2+} channels (PDCs) have been characterized: long-acting (termed "L"); transient (termed "T"); neither T nor L, or neuronal (termed "N"); and so-called "P" channels. Only the L and T types appear to exist in smooth muscle (3). L-channels, whose properties of high conductance and relative resistance to inactivation render them likely to contribute a greater net Ca^{2+} influx than T-channels, are most significantly involved in the actions of typical Ca^{2+} antagonists (4) and in excitation-contraction coupling in smooth muscle cells. Ca^{2+} can also enter smooth muscle cells via receptor-operated channels (ROCs), via a "leak pathway" and/or via reversed Na^+/Ca^{2+} exchange (5,6). Typical Ca^{2+} antagonists generally block Ca^{2+} efflux through PDCs at relatively low concentrations. They can block some ROCs (usually at higher concentrations), but they do not appear to block the leak pathway (7).

Typical Ca^{2+} antagonists inhibit the contractions (i.e., excitation-contraction coupling) elicited by Ca^{2+} entry in depolarized smooth muscle mainly by interacting with the α_1-subunits of PDCs of the L-subtype (8-11). The chemical structures of some of these typical Ca^{2+} antagonists, as well as that of pinaverium bromide, are shown in fig. 1.

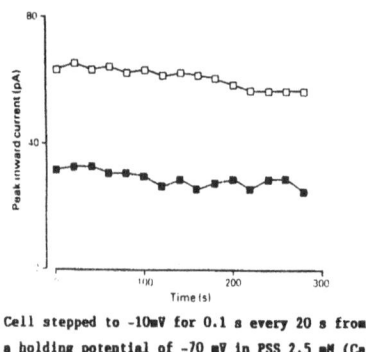

Cell stepped to -10mV for 0.1 s every 20 s from
a holding potential of -70 mV in PSS 2.5 mM (Ca)

FIG. 1. Typical Ca^{2+} antagonists and pinaverium bromide.

The problem of major interest with respect to developing therapy for functional GI disorders involves GI-selectivity of action. Typical Ca^{2+} antagonists (e.g., nifedipine, verapamil, diltiazem) can inhibit GI smooth muscle contraction to various extents (8,12) and they do alter GI activity in experimental animals and in patients suffering from functional GI disorders, but since they act primarily on the cardiovascular system their beneficial actions in patients are seriously limited by cardiovascular (and other) adverse side effects (13).

A CALCIUM ANTAGONIST WITH GI-SELECTIVITY IS NEEDED

Ca^{2+} antagonists that block L-subtype PDCs exhibit marked heterogeneity in molecular structure and in their detailed molecular mechanisms of action (14), indicating that L-channels apparently possess numerous distinct receptor sites (9). Hence, the L-channels of GI smooth muscle could possess certain discrete receptor sites. Also, the proportionate number of L-type channels in relation to other modes of tissue activation may contribute to tissue-selectivity (3).

Pharmacokinetic characteristics of the drug, state- and use-dependent interactions, pathological state of the tissue, mode of Ca^{2+} mobilization in the tissue, physicochemical and relative agonist/antagonist properties of the drug and a drug's effective concentration, and physicochemical properties represent additional important determinants of tissue-selectivity (9,11,15; table 1) which can be used to identify GI-selective Ca^{2+} antagonists.

TABLE 1.   Factors contributing to
                    tissue-selectivity of Ca^{2+} antagonists

---

Characteristics of the Drug
Ca^{2+} channel selectivity
Specificity of Ca^{2+} channel/other receptor (selectivity window)
Physicochemical properties (pK_a, partition coefficient)
Voltage and/or frequency dependency
Site of action within the channel; rate of interaction with the receptor
        sites
Ratio of Ca^{2+} antagonist/Ca^{2+} antagonist activity
Susceptibility to tissue metabolism

Properties of the Tissue
Relative contribution of various Ca^{2+} pools to final response
Level of resting membrane potential
Nature and duration of the stimulus
Gating control of Ca^{2+} channels (rate of activation/inactivation, state of
        phosphorylation)
Density and types of channels
Pharmacokinetic determinants (anatomical localization, importance of
        blood supply)

Abbreviations: pK_a, -log K_a, where K_a is the dissociation constant; also
termed ionization constant. (Reproduced with permission from reference
15)

THE CALCIUM ANTAGONIST PINAVERIUM BROMIDE (DICETEL®) HAS THE DESIRED PROPERTIES

The quaternary ammonium structure of the atypical Ca^{2+} antagonist
pinaverium bromide (Dicetel®) is shown in fig. 1. Pinaverium bromide
possesses GI-selectivity (16-18), and acts mainly by interfering with Ca^{2+}
influx through PDCs (16-21). *In vitro* results indicate that at 0.4-4.0 μM
it antagonizes $BaCl_2$-induced contractions of guinea pig isolated ileum and
rat isolated duodenum (22), and at 1.0 μM it inhibits electrical activity in
preparations of guinea pig ileum and taenia coli (19). Micromolar
concentrations of pinaverium bromide also inhibit the effects of electrical
field stimulation, $BaCl_2$, acetylcholine, or synthetic MET-enkephalin
analogue FK33-824 on *in vitro* intraluminal pressure responses of rat
colonic segments (21).
    The spasmolytic action of pinaverium bromide appears to be mediated
by an effect on Ca^{2+} channels, and it has no significant *in vivo*

anticholinergic effect (23,24). Selectivity of pinaverium bromide at the tissue level is indicated by the finding that it is more potent in inhibiting Ca^{2+}-induced contractions of K^+-depolarized rabbit colonic segments than in opposing the positive inotropic effect of Ca^{2+} on rabbit myocardium (25). Pinaverium bromide also influences GI motility *in vivo* in the conscious dog (26); spike potentials and the amplitude and frequency of slow waves were reduced, the duration of the myoelectric complex was increased, and contraction was significantly inhibited.

The existence of three states of the Ca^{2+} channel, resting, open and inactivated, which can have different affinities for a given drug [the "modulated receptor hypotheses"; (27-29)] provides a means for examining tissue selectivity (9,30-32). State-dependent interactions refer to ligand access or affinity as being dependent upon the equilibria between these three states of the Ca^{2+} channel (27,32). As these states likely represent different conformation of the channel, a given ligand may exhibit quantitatively and qualitatively different structure-activity relationships at each state. Also, since the transitions among these states are voltage-dependent, the apparent affinities of an antagonist will depend upon the tissue-dependent equilibrium that exists among the states (9). Voltage-dependency may be caused by the effect of potential on binding affinity (33). If one considers ligand-binding data, the binding of 1,4-dihydropyridines (DHPs) such as (+)PN 200-110 to Ca^{2+} channels is voltage-dependent, the highest affinity occurring when the membrane is depolarized (33). On this basis, it is considered that DHPs bind with highest affinity to the inactivated state of the Ca^{2+} channel, i.e., the state that is favored by the depolarized condition of the membrane preparation (9). Verapamil and diltiazem might also interact preferentially with the inactivated state of the L channel and/or with the open state (3). However, such state-dependency appears to depend to some extent upon the tissue preparation examined. Hering et al. (34) found little selectivity of nifedipine for the inactivated state of the L channel in single cells of rabbit ear artery, indicating that this DHP exhibits high affinity for the resting state of the channel, results that agreed with those of Loirand et al. (35) who studied smooth muscle of rat portal vein. In this regard, Bolton et al. (6), though concluding that studies with single smooth muscle cells show that DHPs bind more readily to the inactivated state of the channel, did mention that the "modulated receptor hypothesis" does not seem adequate for explaining certain features of DHP action.

Ca^{2+} channel blockade may also show use-dependency. A drug is considered to exhibit use-dependency (frequency-dependence) if it produces greater inhibition when the cell is stimulated at high frequency, or if it elicits greater inhibition with each successive stimulation at a given frequency until a steady state is achieved (36). After excitation, it is proposed that the channel will proceed from its active to its inactive state, and then to its resting state, but for a brief period after excitation the channel can exist in either its active or inactive state, indicating that a drug which shows use-dependency binds to the active or inactive state (33). For example, with higher stimulation rates (or increased "use" of the channels), verapamil produces greater inhibition of the channels, indicating that the affinity of this Ca^{2+} antagonist is increased when the channels are opened by activity or when they exist in an active or inactive state (33,37). Thus, the use-dependency of verapamil may be explained by its action of blocking the channel from the inside after entering the channel during the open state (33). Since the blocking of L channels by both verapamil and diltiazem is clearly augmented by repetitive stimulation, it appears evident that the channels must be opened before they can be blocked (see also 11). In contrast, DHPs shows much less use-dependency than verapamil (38,39). As mentioned above, nifedipine exhibits high affinity for resting channels in single cells of rabbit ear artery (34). Also, studies in isolated cells of rat mesenteric artery indicate that nitrendipine binds to the resting state of the L channel (29). Such state- and use-dependency of Ca^{2+} antagonist action may be important determinants of tissue selectivity.

The significance of state- and use-dependency in relation to the GI-selectivity of Ca^{2+} antagonists receives support from patch-clamp experiments. Pinaverium bromide inhibited Ca^{2+} inward current in smooth muscle cells that differed qualitatively from those of verapamil, gallopamil, and diltiazem and resembles that of the DHPs in not showing use-dependency (i.e., not increased by repetitive stimulation), and that it differed from that of DHPs in not being noticeably voltage-dependent (40). Pinaverium bromide inhibited Ca^{2+} inward current with $IC_{50} \approx 1.5$ μM (40,41), as compared with IC_{50} values of 0.02 μM for nicardipine, 0.1 μM for nifedipine, 1.3 μM for gallopamil and 1.4 μM for diltiazem and flunarizine (36,42). This lack of use-dependency of the action of pinaverium bromide is illustrated in fig. 2.

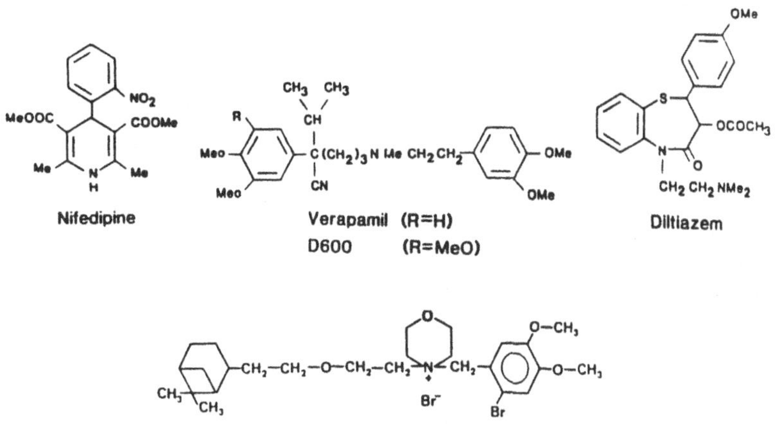

FIG.2. Block of Ca^{2+} current by pinaverium bromide. Lack of use-dependency. (Reproduced with permission from reference 40)

Very little shift of the "availability curve" was produced by pinaverium bromide using conditions under which DHPs, verapamil, and diltiazem cause distinct negative shifts, indicating that pinaverium bromide has similar affinities for the closed available state and the inactivated state of the Ca^{2+} channel. Thus, it may be considered that DHPs generally combine with the inactivated state of the Ca^{2+} channel, that verapamil and gallopamil act on the channel when it is opened by activity, and that pinaverium bromide may act by combining with all three states of the channel with similar affinity (41).

UNIQUE PHARMACOKINETIC PROPERTIES OF PINAVERIUM BROMIDE

The unique pharmacokinetic properties of pinaverium bromide also contribute to its *in vivo* selectivity for the GI tract. A comparison of the pharmacokinetic parameters of pinaverium bromide and some other Ca^{2+} antagonists is shown in table 2. After oral administration, the pharmacokinetic properties of pinaverium bromide, differ markedly from those of conventional Ca^{2+} antagonists (43). Studies with both experimental animals and humans indicate that the highly polar quaternary ammonium group of pinaverium bromide limits its passage across cell membranes (43). In humans, the absolute bioavailability of pinaverium bromide is less than 1%, and the drug is extensively metabolized and excreted.

TABLE 2. Pharmacokinetic parameters of Ca^{2+} antagonists
in the human

Ca^{2+} Antagonist	V_d (l/kg)	Protein binding (%)	f (%)	F (%)
Pinaverium bromide	1.5	97.0	5-10	<0.5
Nifedipine	1.2	98.0	>90	65-70
Verapamil	4.0	90.0	>90	10-22
Diltiazem	5.3	78.0	>90	10-40
Nicardipine	1.2	98.5	>90	15-40
Bepridil	7.5	65.0	>90	60

Abbreviations: V_d, volume of distribution; f (%), percent resorption; F (%), percent bioavailability. (Reproduced with permission from reference 43).

CLINICAL STUDIES WITH PINAVERIUM BROMIDE

Pinaverium bromide is currently registered for treating GI motility disorders (16-18). Double-blind studies have indicated that oral administration of pinaverium bromide (50 mg, t.i.d.) is more effective than placebo and at least as effective as trimebutine and N-butyl hyoscine as therapy for IBS (18). Also, oral doses of pinaverium bromide relaxed Oddi's sphincter in patients with biliary dyskinesia (18). These findings gain added significance when considered together with the observations that pinaverium bromide has no vasodilatatory or antiarrhythmic action in the clinical setting (18), and that conventional Ca^{2+} antagonists produce only limited beneficial effects in IBS patients.

CONCLUDING REMARKS

Collectively, these results indicate that pinaverium bromide represents the prototype of a new class of Ca^{2+} antagonist (6,40,41), its major therapeutic advantage being that it does not elicit cardiovascular side effects at doses that effectively relieve spasm, pain, transit disturbances, and other symptoms related to GI motility disorders. In contrast, the doses of typical Ca^{2+} antagonists (e.g. nifedipine, verapamil) that are required to attain beneficial therapeutic effects in GI disorders would likely elicit undesirable cardiovascular side effects.

The finding that the pharmacokinetic properties of pinaverium bromide (after oral administration) differ markedly from those of typical Ca^{2+} antagonists is useful in explaining its apparent *in vivo* GI-selectivity. Ca^{2+} antagonists with augmented GI-selectivity are essential for limiting cardiovascular (and other) side effects. This selectivity of pinaverium bromide may be due mainly to its low absorption and exceptional hepatobiliary excretion (18). The significance of experiments concerning state- and use-dependency in explaining the tissue-selectivity of Ca^{2+}-antagonist action is evident from the results of recent experiments conducted with pinaverium bromide.

It is conceivable that future studies may reveal that Ca^{2+} antagonists exert beneficial effects on other disorders of the GI tract and related organs/glands, since Ca^{2+} entry is involved not only in smooth muscle contraction, but also in stimulus-secretion coupling (e.g., gastric acid secretion and gastrin release) in the GI tract (44-46). In this regard, preliminary experiments with whole animal models have already indicated that oral administration of such compounds might have an antiulcer effect at the level of the stomach. On this basis (47,48), it might be expected that Ca^{2+} antagonists could exert cytoprotective effects with respect to gastric ulceration and hepatocellular injury.

REFERENCES

1. Maggi C.A. and Meli A. (1983): *Arch. Int. Pharmacodyn.* 262: 221-231.

2. Devroede G. (1989): In: *Calcium Antagonism in Gastrointestinal Motility,* edited by M.O. Christen, T. Godfraind and R.W. McCallum pp. 7-19. Elsevier, Paris.

3. Spedding M., Fraser S., Clarke B. and Patmore L. (1990): *J. Neural. Transm.,* 31(suppl.): 5-16.

4. Nowycky M.C., Fox, A.P. and Tsien R.W. (1985): *Nature,* 316: 440-443.

5. van Breemen C., Aaronson P. and Loutzenhiser R. (1979): *Pharmacol. Rev.,* 30:167-208.

6. Bolton T.B., MacKenzie I. and Aaronson P.I. (1988): *J. Cardiovas. Pharmacol.,* 12(suppl.6): S3-S7.

7. Cauvin C., Loutzenhiser R. and van Breemen C. (1983): *Annu. Rev. Pharmacol. Toxicol.,* 23: 373-396.

8. Fleckenstein A. (183): *Calcium Antagonism in Heart and Smooth Muscle.* John Wiley & Sons, New York.

9. Triggle D.J., Ferrante J., Hawthorn M., Kwon Y.W. and Zheng W. (1989): In: *Calcium Antagonism in Gastrointestinal Motility,* edited by M.O. Christen, T. Godfraind and R.W. McCallum pp. 89-107. Elsevier, Paris.

10. Glossmann H. (1990): *J. Neural. Transm.*, 31(suppl.): 1-3.
11. Godfraind T. (1990): *Communication at the 11th Panhellenic Annual Congress of Gastroenterology*, Salonica, November 1-2, 1990, pp. 1-3.
12. Godfraind T., Miller R. and Wilbo M. (1986): *Pharmacol. Rev.*, 38: 321-416.
13. Traube M. and McCallum R.W. (1984): *Amer. J. Gastroenterol.*, 79: 892-896.
14. Janis R.A., Silver P. and Triggle D.J. (1987): *Adv. Drug Res.*, 16: 309-591.
15. Godfraind T., Morel N., Salomone S. and Wibo M. (1989): In: *Calcium Antagonism in Gastrointestinal Motility,* edited by M.O. Christen, T. Godfraind and R.W. McCallum pp. 117-139. Elsevier, Paris.
16. Christen M.O. and Tassignon J.P. (1989): *Drug Dev. Res.*, 18: 101-112.
17. Christen M.O. and Tassignon J.P. (1989): In: *Calcium Antagonism in Gastrointestinal Motility,* edited by M.O. Christen, T. Godfraind and R.W. McCallum pp. 169-181. Elsevier, Paris.
18. Christen M.O. (1990): *Gen. Pharmacol.*, 21: 821-825.
19. Droogmans G., Himpens B. and Casteels R. (1983): *Naunyn-Schmiedeberg's Arch. Pharmacol.*, 323: 72-77.
20. Mironneau J., Lalune C., Mironneau C., Savineau J.-P. and Lavie J.-L. (1984): *Eur. J. Pharmacol.*, 98: 99-107.
21. Baumgartner A., Drack E., Halter F. and Scheurer U. (1985): *Brit. J. Pharmacol.*, 86: 89-94.
22. Bretaudeau J. and Foussard-Blanpin O. (1980): *J. Pharmacol.*, (Paris), 11: 233-243.
23. Bretaudeau J., Foussard-Blanpin O., Baronnet R. and Despraires R. (1975): *Thérapie*, 10: 919-930.
24. Roux J.-J., Salmon D. and Reny A. (9180): *Gaz. Méd.*, 87: 1033-1036.
25. Szekeres L. and Papp J.G. (1989): In: *Calcium Antagonism in Gastrointestinal Motility,* edited by M.O. Christen, T. Godfraind and R.W. McCallum pp. 161-167. Elsevier, Paris.
26. Itoh Z., Iwanaga Y. and Mizumoto A. (1989): In: *Calcium Antagonism in Gastrointestinal Motility,* edited by M.O. Christen, T. Godfraind and R.W. McCallum pp. 153-159. Elsevier, Paris.
27. Bean B.P. (1984): *Proc. Natl. Acad. Sci. USA*, 81: 6388-6392.
28. Bean B.P. (1985): *J. Gen. Physiol.*, 86: 1-30.
29. Bean B.P., Sturek M., Puga A. and Hermsmeyer K. (1986): *Circ. Res.*, 59: 229-235.
30. Bayer R., Kaufmann R., Mannhold R. and Rodenkirchen R. (1982): *Prog. Pharmacol.*, 5: 53-85.

31. Hondeghem L.M. and Katzung B.G. (1984): *Annu. Rev. Pharmacol. Toxicol.*, 24: 387-423.
32. Sanguinetti M. and Kass R.S. (184): *Circ. Res.*, 55: 336-348.
33. Yu J. and Bose R. (1991): *Gastroenterology,* 100: 1448-1460.
34. Hering S., Bolton T.B., Beech D.J. and Lim S.P. (1989): *Circ. Res.*, 64: 928-936.
35. Loirand G., Dacquet C., Pacaud P., Rakotoarisoa L., Sayet I., Mironneau C. and Mironneau J. (1989): *Eur. J. Pharmacol.*, 167: 265-274.
36. Terada K., Kitamura K. and Hirosi K (1987): *Pflugers Arch.*, 408: 552-557.
37. McDonald T.F., Pelzer D. and Trautwein W. (1980): *Pflugers Arch.*, 385: 175-179.
38. Bayer D., Kalusche D., Kaufmann R. and Mannhold R. (1975): *Naunyn-Schmiedeberg's Arch. Pharmacol.*, 290: 81-97.
39. Lee K.S. and Tsien R.W. (1983) *Nature,* 302: 390-394.
40. Beech D.J., MacKenzie I., Bolton T.B. and Christen M.O. (1990): *Brit. J. Pharmacol.*, 99: 374-378.
41. Bolton T.B., MacKenzie I. and Beech D.J. (1989): In: *Calcium Antagonism in Gastrointestinal Motility,* edited by M.O. Christen, T. Godfraind and R.W. McCallum pp. 109-115. Elsevier, Paris.
42. Terada K., Nakao K., Okabe K., Kitamura K. and Kuriyama H. (1987): *Brit. J. Pharmacol.*, 92: 615-625.
43. Jacquot C., Trouvin J.H. and Christen M.O. (1989): In: *Calcium Antagonism in Gastrointestinal Motility,* edited by M.O. Christen, T. Godfraind and R.W. McCallum pp. 141-151. Elsevier, Paris.
44. Reeder D.D., Becker H.D. and Thompson J.C. (1974): *Surg. Gynec. Obst.*, 138: 847-851.
45. Douglas W.W. (1976): In: *Stimulus-Secretion Coupling in the Gastrointestinal Tract,* edited by R.M. Case and H. Goebell pp. 17-19. MTP Press, Lancaster.
46. Szelenyi I. (1980): *Agents Actions,* 10: 187-190.
47. Glavin G.B. (1990): *Ann. NY Acad. Sci.*, 597: 293-297.
48. Deakin C.D., Fagan E.A. and Williams R. (1991): *J. Hepatol.*, 12: 251-255.

THERAPEUTIC APPLICATION OF CALCIUM ANTAGONISTS IN GASTROENTEROLOGY

R. McCallum

University of Virginia, Health Science Center
Charlottesville, Virginia 22908

As a gastroenterologist and a clinician, I look after patients who may benefit from therapy with calcium channel blockers. I would like to share with you some of the experiences we have had over the last few years, which may provide some insight into the clinical role these agents currently have in the gastrointestinal (GI) tract and also allow for some speculation about future directions.

An organ which has already been discussed to some degree by Dr. Christen is the esophagus, and this part of the GI tract has been very receptive to the role of calcium channel blockers. As a review, what follows are some basic concepts of smooth muscle of the gut (fig. 1) physiology. When a bolus is ingested, or when the gastric antrum, small bowel, or colon is distended, there is a distal descending relaxation preceding the bolus. Hence when one swallows, the lower esophageal sphincter (LES) immediately relaxes, waiting to receive the bolus, which descends in a relaxed esophageal body. Behind the bolus, on the orad side, is what is termed an ascending contraction which helps propel the bolus down the esophagus. This contraction as it descends in an orderly fashion is called peristalsis. Clinically in the esophagus, we have to contend with abnormalities in the coordination of a peristaltic wave, and with relaxation of the LES. We will focus on the problems of contractions of very high amplitude, loss of peristalsis (termed as spasm), and impairment or loss of sphincteric relaxation and coordination (termed as achalasia).

Chest pain originating from the esophagus is a major problem in clinical practice. Approximately half a million patients per year in this country have a coronary arteriogram in the name of angina or ischemia of the heart, and that coronary arteriogram is often not indicative of significant coronary artery disease. The chest pain which is angina-like in quality is later attributed to the esophagus. This is a large potential clinical practice of approximately half a million patients annually in the United States, and about 50% of these patients will eventually have the diagnosis of an esophageal motility disorder established. There are two major areas of diagnosis: 1) esophageal spasm, which is discoordination

T. Godfraind et al. (eds.), Calcium Antagonists, 337–347.
© 1993 *Kluwer Academic Publishers and Fondazione Giovanni Lorenzini.*

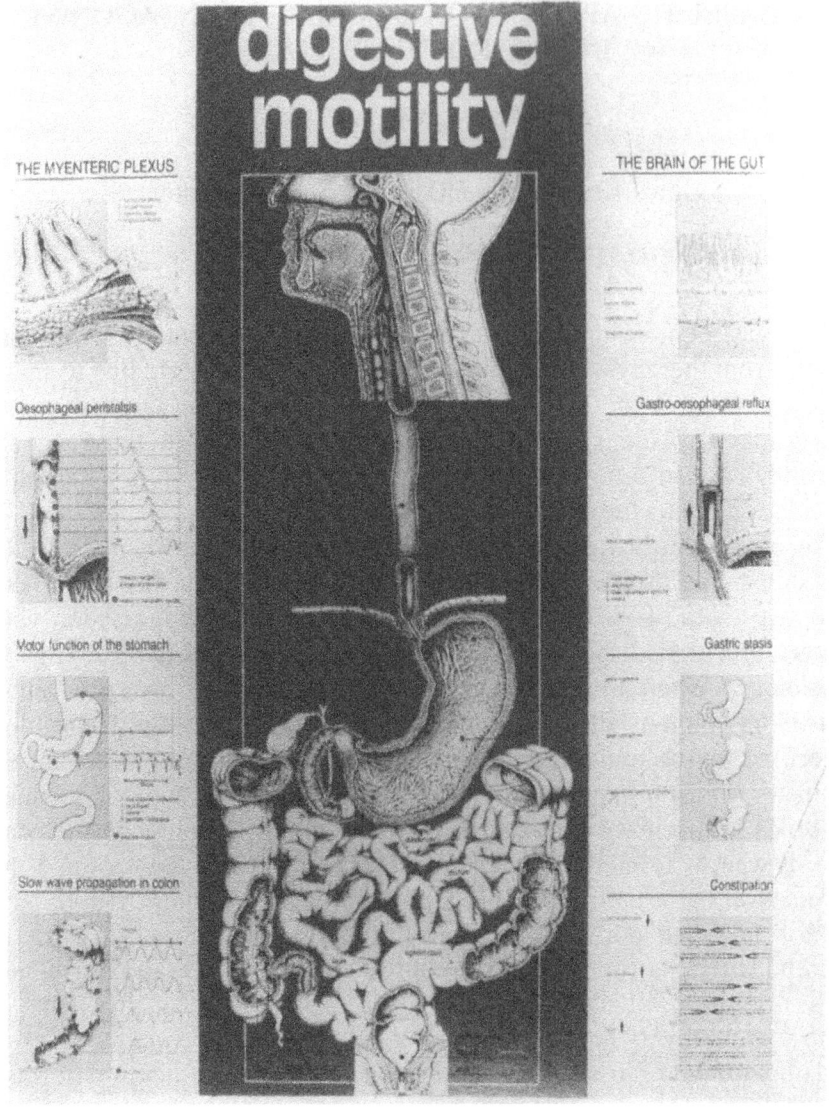

FIG. 1. Overview of neural and myopathic control of gastrointestinal motility, both in the normal physiologic setting and in states of pathophysiology resulting in a clinical syndrome.

and loss of peristaltic motor function in the body of the esophagus and, 2) a nutcracker esophagus or a hypercontracting esophagus which is still peristaltic, but which is so powerfully contracting that it induces pain (1,2).

The medications that are being used to try to relax the smooth muscle of the esophagus include calcium channel blockers, anticholinergics, and combinations of the two. Only nifedipine is predictably able to decrease contraction amplitude in the body of the esophagus and in the lower esophageal sphincter (fig. 2,3). Its effect is equivalent to that of an anticholinergic in the lower esophageal sphincter, but in the case of the body of the esophagus, the two classes can be additive (fig. 4) (4). In open-labelled and double-blind studies calcium channel blockers do decrease chest pain attributed to esophageal motility disturbances such as nutcracker esophagus and diffuse esophageal spasm (5).

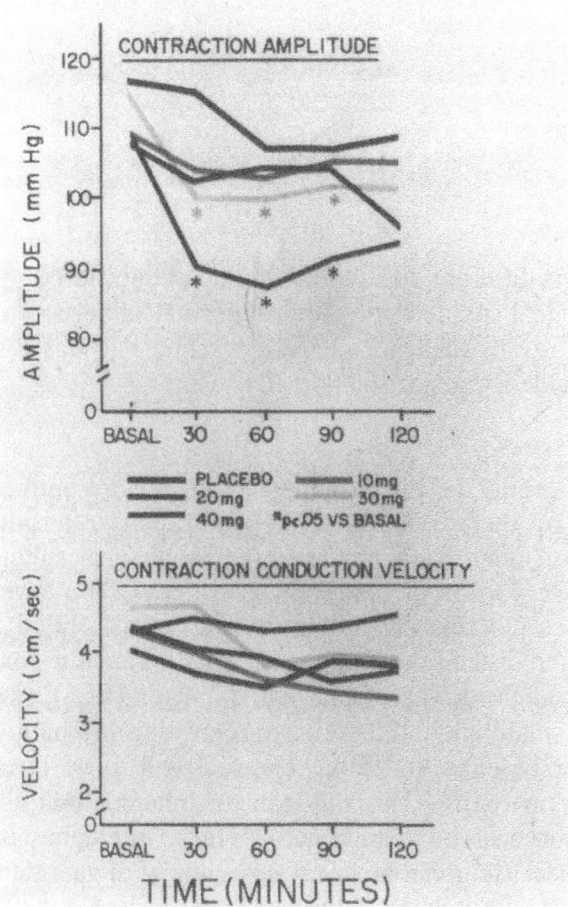

FIG. 2. Effects of sublingual nifedipine on the contraction amplitude of the esophageal body in normal subjects. Doses of 10 to 40 mg were given on separate days. Contraction velocity is also depicted in the lower half of the figure. N = 10

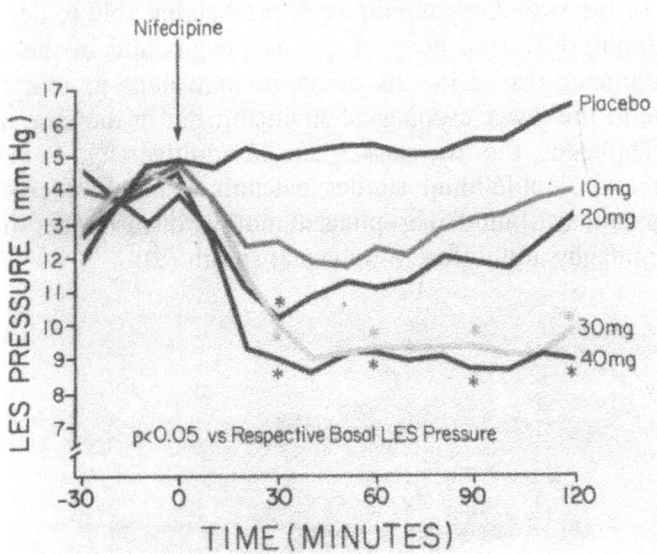

FIG. 3. Effects different doses of nifedipine administered sublingually on lower esophageal sphincter pressure in normal subjects. This study is an extension of results in fig. 2 where a placebo day was also utilized in order to provide a comparison. N = 10

The other entity in the esophagus which is receptive to calcium-channel-blocker therapy is achalasia. Briefly, achalasia is denervation or loss of neural innervation of the lower esophageal sphincter and the smooth muscle portion of the esophageal body. The lower esophageal sphincter does not relax, resulting in aperistalsis. Food and material accumulate and distend the esophagus. The therapeutic goal is to reduce the pressure in the lower esophageal sphincter (LES), so that when it does relax a small amount, its relative relaxation will be greater starting from a lower baseline and more material will move through into the stomach by gravity (6). This will lead to some nutrient absorption and some nutrition will be maintained. Here, nifedipine is particularly interesting, because it can be given sublingually to guarantee absorption and overcome the relative blockage that is occurring in the distal esophagus. It will be quickly absorbed and when given 30 minutes or so before a meal, nifedipine plasma levels will be achieved at the time of eating. It results in decrease in LES pressure and improved emptying (fig. 5) (7). The more sustained slow-release preparations of nifedipine now available will induce a longer inhibition of sphincter pressure.

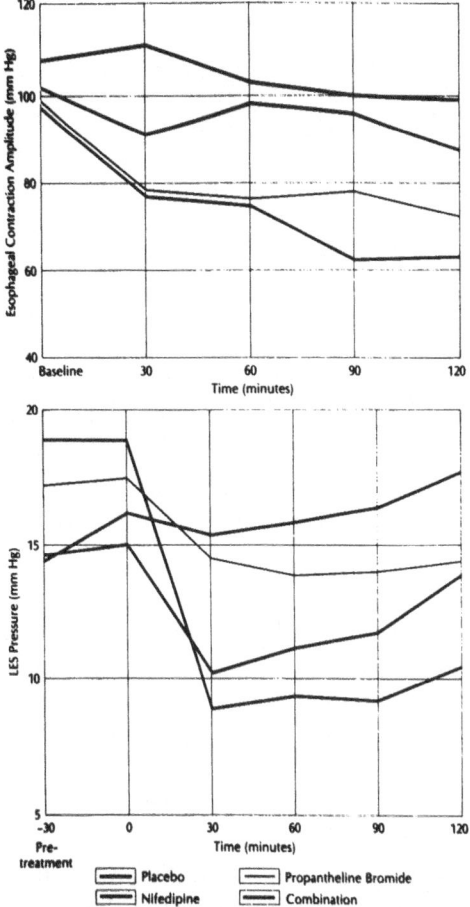

Comparison of the effects of well-tolerated doses of nifedipine (20 mg sublingually) and of propantheline bromide (15 mg orally), alone and in combination in seven normal individuals, indicates that the anticholinergic may be a better treatment for hypertensive esophageal body disorders (greater effect on esophageal contraction, top), while the calcium antagonist may be more effective for hypertensive LES disorders (larger reduction of LES pressure, bottom). The additive effect of the two drugs suggests that combined therapy for sphincter disorders will avoid side effects that occur with higher doses of the two drugs given alone; such combinations might also have merit in esophageal body dysfunction.

FIG. 4. Comparison of the anticholinergic, propanthelene bromide (15 mg orally) compared to nifedipine (20 mg orally) on contraction amplitude in the body of the esophagus following wet swallows in normal subjects.

Let me move now into the stomach, and discuss the clinical effects of calcium channel blockers. The stomach has two major and different functions, emptying solids and liquids (8). The portion dedicated to emptying liquids is the proximal part of the stomach which has tonic contractions, inducing a gastric-duodenal gradient. The antral part of the stomach is involved with the mixing and grinding of solid food together with acid and pepsin to triturate material into a fine particle form which can be passed though the pylorus into the duodenum and maximally

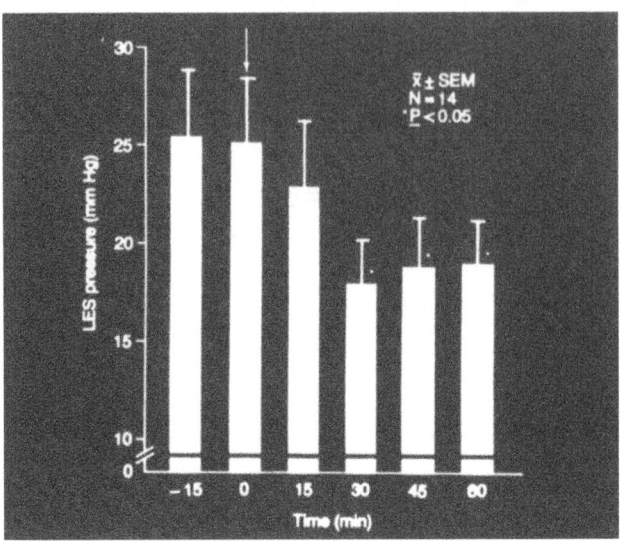

FIG. 5. Effect of sublingual nifedipine on lower esophageal sphincter pressure in patients with achalasia. Lower esophageal sphincter pressure is recorded basally (-15 minutes) and then for 60 minutes following the administration of 20 mg of nifedipine sublingually.

absorbed. The gastric pacemaker is located on the upper part of the greater curvature near the junction of the fundus and the body (fig. 1). The depolarization signal is circumferentially and distally propagated at 3 cycles per minutes in man increasing to 12 in the duodenum. At times when the membrane threshold is exceeded there will be an action potential. The mechanical equivalent is a contraction and this contraction helps propel food into the duodenum. Hence, with postprandial distention of the stomach, vagal innervation, acetylcholine release and gastrointestinal hormone release (e.g., secretin, cholecystokinin, peptide YY, motilin), this milieu at the motor end plate of the gastric smooth muscle, allows for excitation and action potentials to the rate of 3 per minute maximizing the electromechanical coupling and hence emptying the stomach (8).

Does a calcium channel blocker effect this activity? Our group studied the effect of nifedipine (20 mg) on gastric emptying of liquids and solids, found that at that dose it did not interfere with the gastric emptying rate of solid foods or liquids (9). Whether it was working on the pacemaker activity, or whether it was affecting tone, it did not significantly impair gastric emptying.

I will not fully address acid secretion and its role in experimental ulcer disease. It is clear the nifedipine and other calcium channel blockers do decrease gastric acid secretion. However, they are not unique in that there are many agents available which increase acid secretion. Whether calcium channel blockers will find a special niche for being cytoprotective, or be advantageous in peptic ulcer disease, is not clear, but I would predict they will not be useful.

Another very important clinical area is the area of biliary physiology. There are some interesting data in patients with cirrhosis of the liver reporting that nifedipine and other calcium channel blockers, given intravenously or even directly into the superior mesenteric vein, can decrease portal pressure, and can decrease the potential for bleeding from esophageal varices, one of the potentially fatal complications of cirrhosis.

Let us next examine the sphincter of Oddi. In patients whose gallbladder is removed but still have recurrent abdominal pain or pancreatitis, one question is, could the sphincter of Oddi be abnormal; could there be some kind of impairment of the sphincter of Oddi function? A study by Gelrud et al. (10), investigated the individual basal sphincter of Oddi pressures in patients with so-called sphincter of Oddi dysfunction. He administered nifedipine (20 mg) and showed a predictable decrease in sphincter pressure from an abnormally high basal level (fig. 6). Whether this can be translated into a clinical treatment base is not clear, but it is an area of active investigation.

Finally, I would like to discuss the colon (fig. 1). In the practice of gastroenterology, organic disease, peptic ulcer, cholelithiasis, pancreatitis, diverticulitis (definite structural anatomical problems) account for about 50% of the practice referral base. The other 50% involves such upper gastrointestinal symptoms such as dyspepsia, indigestion, esophageal problems, so-called functional disease, and in turn a large part of that overlaps the spectrum of irritable bowel syndrome (IBS) (11). Irritable bowel syndrome (IBS) involves a heterogenous group of entities. Patients can have varying bowel habits, either diarrhea or constipation, sometimes alternating, are exquisitely sensitive to bowel distention, have a very low threshold for pain, and some have an enhanced gastrocolic reflex (12). There may be stress factors, there may be psychological implications in some of these patients, and indeed there may be different sensitivities in the right side of the large bowel compared to the transverse colon, and compared to the left side of the bowel (13). We are just starting to appreciate the wide spectrum of the entity. It also accounts for a large degree of morbidity and absence from work throughout the world. Recently in our laboratory, we have placed catheters into the human colon and investigated simultaneously motility in the sigmoid colon, descending, transverse, and ascending colon. After each meal there is a

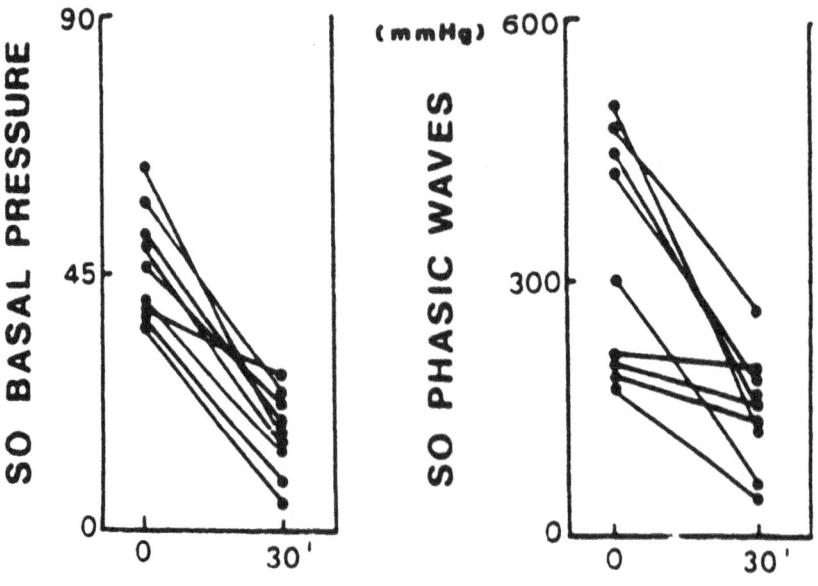

FIG. 6. Effect of nifedipine (20 mg) orally administered to patients with the diagnosis of sphincter of Oddi (SO) dysfunction. The sphincter of Oddi basal pressure is measured through a catheter passed endoscopically and the effect of nifedipine monitored over time. The hypertensive sphincter of Oddi pressure was significantly decreased following the administration of nifedipine. Individual values of basal pressure of the SO before and 30 minutes after administration of nifedipine (20 mg).

predictable gastrocolic reflex or increase in motility throughout the colon (fig. 7). The most pronounced change with the largest increase in motility occurs in the sigmoid colon, which also seems to be the site of most symptoms in IBS patients (14). Each of the segments has a sensitivity to eating, perhaps least so in the ascending colon which is the part of the colon where fecal contents of the stool are rather liquidy and perhaps less distending. Is there a rationale for calcium channel blockers in IBS? Nifedipine has been shown to decrease the rectosigmoid motility index which occurs in response to distention; in other words, it will decrease or minimize the contraction and pain related to distention of the bowel (15,16). Nifedipine will also decrease the postprandial rectosigmoid gastrocolic reflex (table 1).

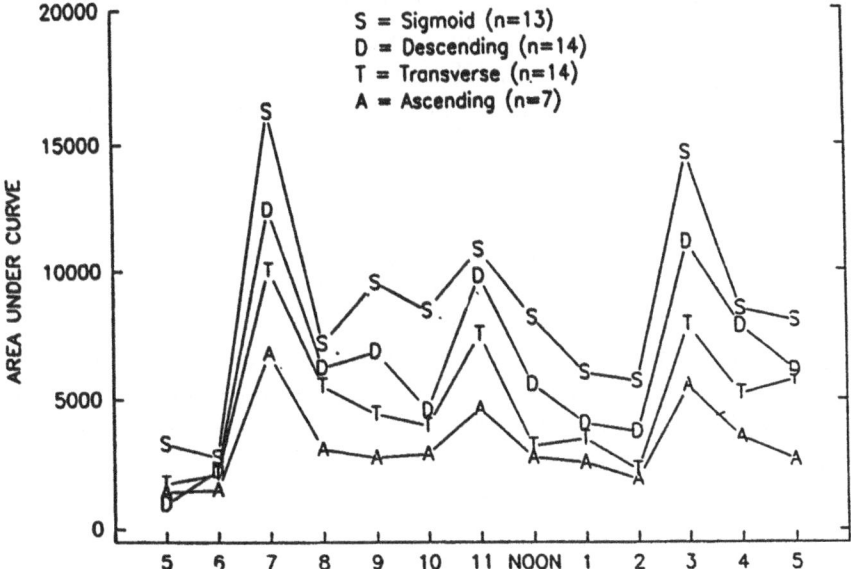

FIG. 7. Comparison of increased motility throughout the colon and specifically in the rectosigmoid region following a meal and baseline or preprandial state. This term in the past has been referred to as the "gastrocolic colic reflex" but can also be induced by directly feeding the duodenum or small bowel and/or with simple distention of the stomach without caloric intake.

TABLE 1.   The effect of nifedipine on the
         increased motility in the rectosigmoid region of the
       colon which predictably occurs immediately after eating.

Calcium channel blockers in IBS

■   rationale

■   nifedipine 20 mg po decreased rectosigmoid MI, contraction #, and % activity time in response to distention (Blume *Gastro.* 1983)

■   nifedipine 20 mg sl decreased postprandial rectosigmoid activity (Narducci *Gastro.* 1983, *AJG* 1985)

■   octylonium bromide reduced the postprandial rectosigmoid MI (Narducci *Dig. Dis. Sci.* 1986)

The agent pinaverium bromide has already been discussed in this symposium. This very interesting new agent has spasmolytic and calcium entry blocking properties, with GI tract selectivity. It inhibits colonic contractile activity in animals and the sigmoid activity in man (table 2). It accelerates left colonic transit in normal and constipated patients. An important observation is that calcium channel blockers, when given intracolonically, not parenterally or orally, but placed into the colon, have a very powerful effect. Studies have shown that they reduce the potassium-chloride-stimulated colonic pressure, whereas the same agents had no effect intra-arterially. In man, there is interest in seeing whether this particular agent, pinaverium bromide and others, might have a similar and selective effect if placed intracolonically. Also, could it reduce the rectosigmoid sensitivity to distention? Whether pinaverium bromide or other agents given intracolonically might somehow interact with serotonin located in the mucosal cells and believed important in mediating pain, is also yet to be determined.

TABLE 2.   The effect of pinaverium bromide on
                manometric aspects of the colon in man.

Pinaverium Bromide
(DICETEL)

■   spasmolytic with calcium entry blocking properties

■   GI tract sensitive

■   inhibits colonic contractile activity in animals and neostigmine-stimulated sigmoid activity in man (Passaretti 1989)

■   accelerates left colonic transit in normals and constipated patients (Froguel *J. of GI Motility* 1990)

In summary, I have reviewed the effects of the established calcium channel blockers, and the agent pinaverium bromide on the motility patterns of the esophagus, stomach, small bowel, and colon, in respect to both contractile activity as well as sphincteric function. I have also reviewed their clinical effects and current status as well as speculated about future directions. This important class of therapeutic agents clearly will have a significant and evolving role in the care of patients with disorders of gastrointestinal motility.

REFERENCES

1. Traube M., Albibi R. and McCallum R.W. (1983): *JAMA*, 250: 2655.
2. Benjamin S.B., Richter J.E., Codova C.M., Knuff T.E. and Castell D.O. (1983): *Gastroenterology*, 84: 893.
3. Hongo M., Traube M., McAllister R.G. and McCallum R.W. (1984): *Gastroenterology*, 86: 8.
4. Hongo M., Traube M. and McCallum R.W. (1984): *Dig. Dis. Sci.*, 29: 300.
5. Siverstein B.D., Kramer C.M. and Pope C.E. (1982): *Gastroenterology*, 82: 1181.
6. Vantrappen G. and Hellesmans J. (1980): *Gastroenterology*, 79: 144.
7. Hongo M., Traube M. and McCallum R.W. (1984): *JAMA*, 252: 1733.
8. Minani M. and McCallum R.W. (1982): *Compr. Ther.*, 8: 50.
9. Traube M., Lange R.C., McAllister R.G. and McCallum R.W. (1985): *Dig. Dis. Sci.*, 30: 710.
10. Guelrud M., Mendoza S., Rossiter G., Ramirez L. and Barkin J. (1988): *Gastroenterology*, 95: 1050.
11. Harvey R.F., Salih S.Y. and Read A.E. (1983): *Lancet*, 1: 632-634.
12. Drossman D.A., Powell D.W. and Sessions J.T. (1977): *Gastroenterology*, 73: 811-822.
13. Whitehead W.E., Engel B.T. and Schuster M.M. (1980): *Dig. Dis. Sci.*, 25: 404-412.
14. Lind C.D. and McCallum R.W. (1989): In: *Pathogenesis of Functional Bowel Disease*, edited by W.J. Snape, Jr. pp. 249-266 Plenum Publishing Corporation, New York.
15. Prior A., Harris S.R. and Whorwell P.J. (1987): *Gut*, 28: 1609-1612.
16. McLeod J. (1983): *Med. J. Aust.*, 2: 119.

CALCIUM ANTAGONISTS AND THE KIDNEY: IMPLICATIONS FOR RENAL PROTECTION

Murray Epstein M.D., F.A.C.P.

Nephrology Section
Veterans Administration Medical Center
Division of Nephrology
University of Miami School of Medicine
1201 NW 16th Street, Miami, Florida 33125

Since the introduction of calcium antagonists over two decades ago, attention has focused on their beneficial effects in the management of symptomatic coronary artery disease, and on their ability to lower blood pressure. During the past decade, it has become apparent that this class of drugs also has beneficial effects on the kidney (1-4). The purpose of this brief review is to consider the potential salutary therapeutic applications of calcium antagonists on renal hemodynamics and on renal excretory function.

We have recently reviewed the pharmacologic effects of the calcium antagonists on renal hemodynamics (1-5). In brief, calcium antagonists do not affect the vasodilated isolated perfused kidney; however, they dramatically alter the response of the kidney to vasoconstrictor agents (1,2). In order to directly characterize the effect of calcium antagonist on the renal microcirculation, our laboratory has developed a model of the isolated perfused hydronephrotic kidney, which facilitates direct observations of the renal microvasculature under defined *in vitro* conditions (6,7). Using this isolated perfused hydronephrotic rat kidney model, which permits direct visualization of afferent and efferent arterioles, we have recently demonstrated that the calcium antagonists preferentially vasodilate preglomerular vessels. Collectively, our studies with diverse agonists indicate that calcium antagonists reverse afferent arteriolar vasoconstriction induced by widely divergent stimuli, including putative mediators of deranged renal hemodynamics such as endothelin (8). Such observations suggest that the activation of potential dependent calcium antagonists (PDCs) constitutes a final, common mechanism of afferent arteriolar vasoconstriction by diverse agonists. In contrast, the efferent arteriole appears to be highly refractory to the vasodilatory effects of calcium antagonists, indicating a remarkable intra-organ heterogeneity of mechanisms which activate smooth muscle within the renal microcirculation (2,8).

T. Godfraind et al. (eds.), Calcium Antagonists, 349–354.
© 1993 *Kluwer Academic Publishers and Fondazione Giovanni Lorenzini.*

Calcium antagonists have additional properties that may contribute to their ability to afford renal protection under diverse experimental conditions and perhaps in clinical disorders (2,3,5). Some of the more prominent mechanisms postulated to mediate renal protection are listed in table 1.

TABLE 1.   Known and Postulated Mechanisms
                  Mediating the Renal Protective Actions
                        of Calcium Antagonists

---

1.   Reduce systemic blood pressure.
2.   Reduce renal hypertrophy.
3.   Modulate mesangial traffic of macromolecules.
4.   Inhibit mitogenic effects of platelet-derived growth factor
      (PDGR) and of thrombin.
5.   Scavenge toxic oxygen-free radicals.
6.   Reduce metabolic activity of remnant kidneys.
7.   Ameliorate uremic nephrocalcinosis.
8.   May block pressure-induced calcium entry.

Modified from reference 5 with permission.

These include the ability of calcium antagonists to lessen injury by retarding renal growth (9,10), to attenuate mesangial entrapment of macromolecules (11,12), to countervail or attenuate the mitogenic effects of growth factors (11), and to act as free radical scavengers.

Possible Applications for Calcium Antagonists in Renal Disease

The above-enumerated pharmacologic effects of calcium antagonists on renal hemodynamics, in concert with their other renal-protective properties suggest potential applications in clinical medicine. First, the striking effects of calcium antagonists on renal hemodynamics and renal sodium handling recommend their use in the treatment of hypertension (5,13). Direct-acting vasodilators such as hydralazine, that reduce peripheral vascular resistance directly, have been used in antihypertensive therapy for many years, but their effectiveness is limited both by reactive stimulation of renal and hormonal responses that counteract their antihypertensive actions (5,13), and by the induction of sodium retention. The consequent volume expansion results in pseudotolerance to the antihypertensive effects of vasodilators such as hydralazine.

In contrast to direct-acting vasodilators such as hydralazine and minoxidil, calcium antagonists attenuate the expected adaptive changes in peripheral vascular resistance (PVR), heart rate, cardiac output, and extracellular fluid volume that eventually countervail the initial reduction in blood pressure. Calcium antagonists interfere with angiotensin II and alpha adrenergic-mediated vasoconstriction. They also countervail the sodium-retaining effects of decreased renal perfusion (2,4) and possibly decreased levels of natriuretic hormones.

Aside from their role in treating hypertension, the salutary effects of calcium antagonists on renal hemodynamics, in concert with their effects on cellular calcium metabolism, suggest a future role in managing certain types of acute renal insufficiency (4,5,14). Table 2 summarizes several such examples. Possibilities include the utilization of their ability to augment renal perfusion in clinical settings in which renal hemodynamics are compromised (for instance, the amelioration of radiocontrast-induced reductions in renal hemodynamics).

TABLE 2.   Current and Potential Applications of
            Calcium Antagonists in Clinical Medicine

---

1.   Amelioration of renal insufficiency from:
      a) Radiocontrast agents.
      b) Cyclosporine nephrotoxicity.
      c) Aminoglycoside nephrotoxicity.
      d) Chemotherapy

2.   Organ preservation during harvesting of kidneys for transplantation.

Modified from reference 5 with permission.

Protection Against Radiocontrast-Induced Acute Renal Failure

Several lines of evidence have demonstrated that the intra-renal administration of radiocontrast medium results in both a prolonged vasoconstrictive response and a reduction in glomerular filtration rate (GFR) (4,5,15).

Recently, Neumayer et al. (16) carried out a prospective randomized double-blind study assessing the effects of three days of nitrendipine administration on radiocontrast-induced nephrotoxicity. They interpreted their data as indicating that prophylactic administration of calcium antagonists ameliorate radiocontrast-induced renal dysfunction. Additional studies are presently being conducted to delineate the possible protective role of calcium antagonists in this clinical setting. In an analogous manner, it has been proposed that calcium antagonists might exert a salutary effect in protecting against other experimental models of acute renal failure (4,5,14).

Role in Transplant-Associated Acute Renal Failure

The prophylactic administration of calcium antagonists to donor kidneys ameliorates posttransplantation renal insufficiency. In a prospective randomized trail, Wagner et al. (17) evaluated the influence of diltiazem on the development of acute renal failure (ATN) following transplantation of cadaveric kidneys. In this initial study, cadaver kidneys were harvested locally at the study center, and diltiazem was added to Eurocollin solution (20 mg/L) at the time of donor nephrectomy. The graft recipient was given a preoperative bolus injection of diltiazem, followed by maintenance diltiazem therapy. In the control group (n = 22), 9 patients (41%) developed ATN, compared with 2 patients (10%) in the diltiazem group (p < 0.05). In the control group, 3.5 ± 0.4 hemodialyses per patient were necessary, compared with 0.6 ± 0.2 in the diltiazem group (p < 0.005). Subsequently, Dawidson and Rooth (18) confirmed and extended these observations utilizing additional calcium antagonists including verapamil. Collectively, these observations with diverse calcium antagonists indicate that the protection afforded is a class effect.

Protection Against Cyclosporine Nephrotoxicity

Cyclosporin A (CyA) is now established as the immunosuppressant of choice in human organ transplantation because it improves graft survival and does not cause myelosuppression. Unfortunately, nephrotoxicity resulting from CyA therapy remains a major hurdle in the wider application of this important drug (19). Within the past several years, evidence indicating that calcium antagonists exert protective actions against cyclosporine nephrotoxicity has emerged. Although the exact mechanisms whereby calcium antagonists ameliorate cyclosporine nephrotoxicity are not clear, it is conceivable that they act by countervailing the effects of thromboxane and/or endothelin. Cyclosporine augments both endothelin and thromboxane A_2, thereby inducing renal vasoconstriction (20,21). Several studies from our laboratory demonstrate that calcium antagonists reverse thromboxane-mediated afferent arteriolarvasoconstriction (22,23). These findings provide a scientific framework for anticipating a protective effect of calcium antagonists on CyA-induced renal vasoconstriction (2).

SUMMARY

Collectively, the above-cited observations of the renal vasodilatory effects of calcium antagonists on the renal microcirculation commend their use in the management of hypertension. They constitute a means of lowering systernic blood pressure while preserving renal perfusion to this vital organ. Furthermore, the present observations raise the possibility that calcium antagonists may ameliorate acute renal insufficiency in clinical settings in patients at increased risk of developing acute renal failure. Although I have focused on the effects of calcium antagonists on renal hemodynarnics, it should be emphasized that the renal protective effects of these agents may also be attributable to other renal protective mechanisms including retardation of renal growth and attenuating the effects of diverse growth factors (table 1). Additional investigation to evaluate these consequences of the renal hemodynamic actions of calcium antagonists is required.

ACKNOWLEDGEMENTS

Portions of this article are adapted with permission from *Calcium Antagonists and the Kidney*, edited by M. Epstein and R. Loutzenhiser, Hanley & Belfus, Philadelphia, 1990.

REFERENCES

1. Loutzenhiser R. and Epstein M. (1985), *Am. J. Physiol.*, 249: F619-F629.
2. Loutzenhiser R. and Epstein M. (1990): In: *Calcium Antagonists and the Kidney,* edited by M. Epstein and R. Loutzenhiser pp. 33-74. Hanley & Belfus, Philadelphia.
3. Epstein M. (1992). In: *Calcium Antagonists in Clinical Medicine,* edited by M. Epstein pp. 309-346. Hanley & Belfus, Philadelphia.
4. Epstein M. and Loutzenhiser R. (1990): *Am. J. Kidney Dis.,* 16(suppl.4)1: 10-14.
5. Epstein M. and Loutzenhiser R. (1990): In: *Calcium Antagonists and the Kidney,* edited by M. Epstein and R. Loutzenhiser pp. 275-298. Hanley & Belfus, Philadelphia.
6. Loutzenhiser R., Hayashi K. and Epstein M. (1988): *J. Pharmacol. Exp. Ther.,* 246: 522-528.
7. Steinhausen M., Snoel H., Parekh N., Baker R. and Johnson P.C. (1983): *Kidney Int.,* 23: 794-806.
8. Loutzenhiser R., Epstein M., Hayashi K. and Horton C. (1990): *Am. J. Physiol.,* 258: F61-F68.
9. Dworkin, L.D. (1990): *J. Am. Soc. Nephrol.,* 1: S21-S27.

10.   Dworkin, L.D., Levin, R.I. and Benstein, J.A. (1990): *Am. J. Physiol.*, 259: F598-F604.

11.   Sweeney C., Schultz P. and Raij L. (1990): *J. Am. Soc. Nephrol.*, 1: S13-S20.

12.   Raij L. and Keane W. (1985): *Am. J. Med.*, 79(suppl.36): 24-30.

13.   Epstein M. and Oster J.R. (1988): In: *Hypertension. Practical Management*, edited by M. Epstein pp. 114-126. Battersea Medical Publications, Miami.

14.   Michael U. and Lee S.M. (1990) In: *Calcium Antagonist and the Kidney*, edited by M. Epstein and R. Loutzenhiser pp. 187-201. Hanley & Belfus, Philadelphia.

15.   Bakris G.L. and Burnett J.C. (1985): *Kidney Int.*, 27: 465-468.

16.   Neumayer H.H., Junge W., Kufner A. and Wenning A. (1989): *Nephrol. Dial. Transplant.*, 4: 1030-1036.

17.   Wagner K., Albrecht S. and Neumayer H. (1987): *Am. J. Nephrol.*, 7: 287-291.

18.   Dawidson I. and Rooth P. (1990): In: *Calcium Antagonist and the Kidney*, edited by M. Epstein and R. Loutzenhiser pp. 233-246. Hanley & Belfus, Philadelphia.

19.   Myers, B.D. (1986): *Kidney Int.*, 30: 964-974.

20.   Perico N., Benigne A., Zoja C. and Remuzzi G. (1986): *Am. J. Physiol.*, 251: F581-F587.

21.   Kon V., Sugiura M., Inagami T., Harvie B.R., Ichikawa I. and Hoover R.L. (1990): *Kidney Int.*, 37: 1487-1491.

22.   Loutzenhiser R., Epstein M., Horton C. and Sonke P. (1986): *Am. J. Physiol.*, 250: F619-F626.

23.   Epstein M., Hayashi K. and Loutzenhiser R. (1989): *Kidney Int.*, 35: 291.

CALCIUM ANTAGONISTS IMPROVE
THE OUTCOME AFTER RENAL TRANSPLANTATION

Ingemar Dawidson, M.D., Ph.D., Biff Palmer, M.D.,
Christopher Lu, M.D., Richard Risser, M.S.,
Pål Rooth, M.D., Ph.D., Arthur Sagalowsky, M.D.,
and Zsolt F. Sandor, M.D.

University of Texas Southwestern Medical Center
Parkland Memorial Hospital
5323 Harry Hines Blvd., General Surgery - E7.126
Dallas, Texas 75235-9031

Calcium antagonists (CATs) have a role in the management of certain types of acute renal insufficiency (1,2). These include prophylaxis against postischemic acute renal failure, radio-contrast-induced acute renal failure, nephrotoxic drugs and post transplant-associated acute renal failure and cyclosporine A (CsA)-induced renal dysfunction. For the transplanted kidney, CATs may be beneficial in several settings. First, the use of a CAT during organ procurement may protect the kidney during the warm and cold ischemic periods (3). Second, the use of CATs perioperatively further protects the kidney during reperfusion and early after transplantation (4). Third, CATs also offer protection against CsA-induced renal dysfunction (5). This chapter summarizes and updates our experimental and clinical experience with CATs in cadaver renal transplantation (CRT).

MATERIAL AND METHODS

Animal Experiments. Subcapsular renal blood flow was studied, in mice using *in vivo* fluorescence microscopy. Blood flow changes were observed after bolus, or continuous infusion of CsA, as well as after various drug treatments (6).

Clinical Studies. Two randomized clinical studies (4,5) and one retrospective study were performed (7). Immunosuppression includes 375 mg methylprednisolone on day 1, tapered to 20 mg/day by postoperative day 10. Azathioprine, initially 100 mg on day 1 decreased to 25 mg/day for 5 days. Antilymphocyte globulin (14 mg/kg bwt) overlapped with CsA on day 6 (7 mg/kg) and day 7 (12 mg/kg). Verapamil (VP) was initiated on day 3 (study 1), or given into the renal artery (study 2) and continued for two weeks in both studies in an oral dose of 120 mg twice daily. Using

355

T. Godfraind et al. (eds.), Calcium Antagonists, 355–361.
© 1993 *Kluwer Academic Publishers and Fondazione Giovanni Lorenzini.*

doppler ultrasonography, diastolic parenchymal blood flow velocities were obtained in the renal subcapsular parenchyma. Kidney function was assessed from serum creatinine and glomerular filtration rate (GFR) on days 1 and 7, using subcutaneous ^{125}I-iothalamate (8).

RESULTS

Graft Survival.
Patients in study 2 have been followed for a mean of 18 months with a current graft survival (GS) for VP patients of 90% (27/30), significantly better than that for the control patients or 68% (18/29) (p < 0.01). These differences were also evident in an actuarial GS analysis for all patients (p < 0.0237) (fig. 1). Since two kidneys lost early in the VP group were never-function kidneys, no immunologically mediated graft losses within one year in the CAT group have been identified. The greatest benefits seem to occur with repeat transplants where 1 of 10 VP-treated patients

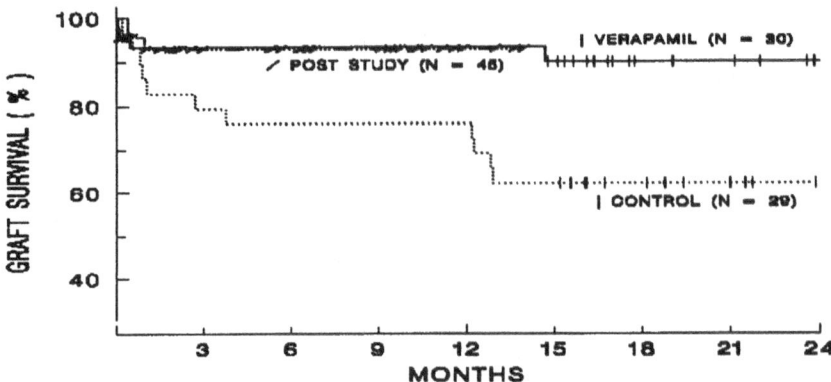

FIG. 1. Actuarial GS was significantly improved in 30 patients treated with VP compared to 29 control recipients translated after the randomized study. All these patients received perioperative VP, with GS identical to that of the VP randomized group.

has lost the graft early. In contrast, 3 of 8 control kidneys were still functioning at one year (p < 0.05). Since the termination of study 2, all CRT at the UT Southwestern Medical Center are receiving perioperative treatment with VP. Fig. 1 also includes the actuarial survival curve (as of 9/16/91) for this group of patients (N = 45). The tick marks on the lines indicate the follow-up time for each patient with a surviving graft. The 93% actuarial one-year graft-survival estimate in these patients is similar to the 95% rate among the study 2 patients randomized to VP treatment.

Fig. 2 demonstrates survival curves for the four groups (first and repeat transplant with or without VP) based on a Cox regression model for survival analysis. Both VP and the transplant number are significant factors related to survival according to this analysis (p = 0.0095 and 0.0055, respectively). This model predicts a one-year graft survival of 96% for first transplants treated with VP. Repeat transplants with a CAT have a similar predicted graft survival (88%) as first transplants without CAT (83%) while the repeat-transplant control patients have a 44% predicted graft survival at one year.

In the retrospective study where 17 patients receiving a CAT for treatment of hypertension, graft survival at one-year was 93% versus 78% for 23 patients who did not receive a CAT. These patients fulfilled the

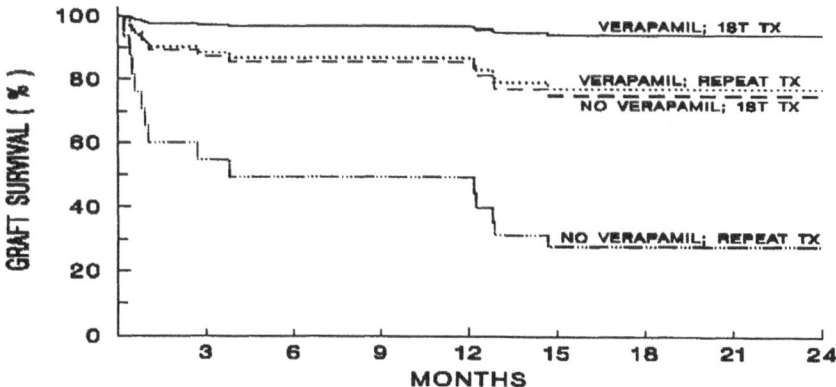

FIG. 2. Cox regression analysis model for survival analysis predicts a 96% one-year GS for first transplants treated with VP. Repeat transplants with VP have the same predicted GS as first transplants without VP while repeat transplants without VP have lost their grafts in half of the cases.

criteria of being on a CAT for at least one year after CRT. The control patients never received a CAT during the same one-year period (7).

Rejection Episodes. In the first randomized study of 40 patients, only 3 of 22 patients randomized to VP were treated for a rejection episode within 4 weeks of transplantation. This was in sharp contrast to 10 of 18 of the control patients treated for rejection within 4 weeks (p < 0.01) (5). In the retrospective study, CAT-treated patients had significantly fewer or 35% first-rejection episodes during the one-year followup in contrast to 83% in patients who did not receive a CAT (p < 0.01) (7). Also there were fewer second rejections with a CAT treatment, or 18% compared to 33% in the control patients (p < 0.01).

CsA Blood Levels and CATs. CsA blood levels were almost twice as high in patients receiving VP compared to controls in both our randomized studies (4,5).

CATs and Protection from Ischemia. When VP was given intra-arterially during surgery (study 2), serum creatinine values on days 1 and 2 after transplantation were significantly lower compared to control patients. With VP, serum creatinine fell by 2.7 mg% between days 1 and 2, in contrast to 1.3 mg% for the control patients. On the second day after transplantation, creatinines were 7.4 and 5.6 mg% for control and VP patients, respectively (p < 0.01) (4). By day 7, the majority of patients (77%) receiving VP had serum creatinines below 2.0 mg% versus only 26% for control patients (p < 0.01). Accordingly, on day 1 GFR was 35 and 19 ml/min for VP and control patients, respectively. By day 7, GFR had increased to 44 and 28 ml/min for VP and control patients (p < 0.01).

CATs and CsA Nephrotoxicity. Despite the higher CsA blood levels during VP treatment (study 1), serum creatinine levels at one week were lower with VP (1.08 ± 0.41 mg%), than those of control patients (1.46 ± 0.46 mg%) (p < 0.008). Also, the increase in GFR from day 1 to day 7 was greater with VP (32 ± 13 ml/min) compared with 18 ± 13 ml/min in control patients (p < 0.002) (5).

Renal Blood Flow and CATs. CsA in mice causes a linear dose-related blood-flow inhibition reaching a maximum at about 40 mg/kg bwt (unpublished data). The blood-flow inhibition induced by 12 mg/kg of CsA was prevented by pretreatment with CATs such as VP and isradipine (6,9). Alpha blocking agents, beta blockers, prostaglandin analogs, and a thromboxane A inhibitor were all ineffective in this experimental model, both as pretreatment and treatment after blood-flow inhibition was established (unpublished data). The CsA-induced blood-flow inhibition in animals was later confirmed in CRT recipients where mean diastolic-blood-flow velocity in 10 patients decreased from 10 to 3 cm/sec. Despite continued CsA administration blood flow returned to pre-CsA levels within 3-4 days (10). Pretreatment with VP prevented this fall in renal-blood-flow characteristics (5). When VP was given intra-arterially, blood flow was significantly better on the first postoperative day (4). Only 8% (2/25) of the VP patients had parenchymal-blood-flow velocity less than 8 cm/sec versus 54% for no CAT (p < 0.01). A CAT, therefore, seems to prevent both the blood-flow inhibition induced by organ procurement procedures, as well as from CsA.

COMMENTS

These studies demonstrate several significant benefits from perioperative use of CATs in cadaver renal transplantation. Most importantly graft survival and kidney function are improved. This is further supported by the fact that these results have been corroborated by a current 96% graft survival in CRT recipients with VP given perioperatively (unpublished data). The beneficial effects from CATs may be due to several actions of CATs occurring separately or in combination. CATs also have direct and indirect effects which may increase the level of immunosuppression.

The decreased incidence of acute rejection episodes may be related to the blockage of cellular calcium influx which inhibits lymphocyte activation and macrophage proliferation, both in animal and human *in vitro* systems (11,12). At least part of the beneficial effect of VP on transplant outcome may be due to the increased CsA immunosuppressive effect without accompanying nephrotoxicity because of the increased blood CsA level. Although CATs have complex and incompletely understood interactions with CsA metabolism, both diltiazem and VP compete with CsA for the cytochrome P-450 pathway (3,13). In contrast to these two CATs, the dihydropyridine CAT, nifedipine, does not increase CsA blood concentration (14).

Previously, we demonstrated by *in vivo* fluorescence microscopy in mice that VP prevents CsA-induced decrease in renal blood flow (6). Subsequently, these data were confirmed in the clinical setting in CRT recipients (4,5). These two clinical studies reconfirm the protective effect of VP against CsA-induced blood-flow decrease, as well as after organ procurement procedures. The relative importance of cytoprotection from CATs and their preferential vasodilatation of the afferent arterioles is hard to distinguish. Experimental and clinical data suggest that both mechanisms are contributing.

The present studies strongly support routine perioperative use of CATs in CRTs to improve renal function and graft survival. Although VP produces higher CsA blood levels, acute nephrotoxicity is less common and CsA doses are not empirically lowered. Better immunosuppression from increased CsA levels without toxicity likely plays a role in the improved results. Some investigators have steadily reduced the CsA dose to minimize cost (15). Routine decreases in CsA dose, based on CsA blood levels, may have played a role in the lack of benefits of a CAT in other studies (16). Based on the results in our two clinical studies, the argument could be made not to reduce the CsA dose, but rather accept higher CsA blood levels without nephrotoxicity and gain from increased immunosuppression. Better renal function and graft survival vastly outweigh the small monetary gain from decreased CsA dosing.

In summary, verapamil restores and maintains renal blood flow and minimizes renal injury associated with organ procurement and cold ischemia. The randomized clinical studies confirm our previous animal research that VP prevents CsA associated deterioration of renal blood flow. VP given intraoperatively into the renal artery also reduces the need for postoperative hemodialysis. VP-treated patients have fewer rejection episodes and most importantly VP is associated with improved graft survival.

The beneficial effect of VP on renal transplant outcome may be related to cytoprotection from ischemia, the preferential vasodilatation of the preglomerular arterioles, elevated blood CsA levels, and inherent immunosuppressive properties.

REFERENCES

1.   Loutzenhiser R. and Epstein M. (1990): In: *Calcium Antagonists and the Kidney,* edited by M. Epstein pp. 33-73. Hanley & Belfus, Philadelphia.
2.   Schrier R.W, Arnold E.D., Van Putten V. and Burke T.J. (1987): *Kidney Int.,* 32: 313-321.
3.   Neumayer H.H. and Wagner K. (1987): *J. Cardiovas. Pharmacol.,* 10: 170-177.
4.   Dawidson I., Rooth P., Lu C., Sagalowsky A., Diller K., Palmer B., Peters P., Risser R., Sandor Z. and Seney F. (1991): *JASN,* (in press).
5.   Dawidson I., Rooth P., Fry W., Sandor Z., Willms C., Coorpender L., Alway C. and Reisch J. (1989): *Transplantation,* 48: 575-580.
6.   Rooth P., Dawidson I., Diller K. and Taljedal I.B. (1988): *Transplantation,* 45: 433-437.
7.   Palmer B., Dawidson I., Sagalowsky A., Sandor Z. and Lu C. (1991): *Transplantation,* (in press).
8.   Israelit A., Long D., White M. and Hull A. (1973): *Kidney Int.,* 4: 346.
9.   Rooth P., Dawidson I., Clothier N. and Diller K. (1988): *Transplantation,* 46(4): 566.
10. Fry W.R., Dawidson I., Alway C.C. and Rooth P. (1988): *Trans. Proc.,* 20(3): 222.
11. McMillen M.A., Lewis T., Jaffe B. and Wait R. (1985): *J. Surg. Res.,* 39: 76-80.
12. Weir M.R., Peppler R., Comolka D. and Handwerger B.S. (1988): *Trans. Proc.,* 20: 240-244.   13.Renton K.W. (1985): *Biochem. Pharmacol.,* 34: 2549-2553.
14. Dy G., Raja R. and Mendez M. (1991). *Trans. Proc.,* (in press).
15. Neumayer H. and Wagner K. (1986): *Lancet,* p. 523.

16. Pirsch J.D., Voss B.J., D'Alessandro A.M., et al. (1990): American Society of Transplant Physicians, 9th Annual Meeting in Chicago, May 29-30, 1990, abstract.

REVERSAL OF P-GLYCOPROTEIN-MEDIATED MULTIDRUG RESISTANCE IN TUMOR CELLS BY CALCIUM CHANNEL ANTAGONISTS

Lee M. Greenberger

Lederle Laboratories, American Cyanamid Company,
Oncology and Immunology Research Section,
Pearl River, NY 10965

Drug resistance to chemotherapeutic agents significantly limits the effective treatment of cancer. In tissue culture, tumor cells can become cross resistant to structurally and functionally unrelated hydrophobic agents (e.g., vinblastine, doxorubicin or taxol), although such cells were selected for resistance to a single anticancer drug (1). This type of multiple drug resistance (MDR) is mediated by overproduction of a membrane protein known a P-glycoprotein (P-gp). Elevated production of P-gp is due to gene amplification and/or mRNA overexpression of the gene encoding P-gp. The protein functions as a drug efflux pump with broad specificity for hydrophobic antitumor agents. This action allows resistant cells to maintain intracellular antitumor agents below cytotoxic levels. Expression of P-gp in tumors from patients can be associated with a poor prognosis and probably accounts, at least in part, for inherent and acquired resistance to cancer chemotherapy (2).

P-gp (140-170 kDa; approximately 1280 residues) is composed of two symmetrical and highly homologous halves. Each half contains a cassette of six transmembrane spanning domains followed by a nucleotide binding consensus site. A large superfamily of transport proteins share the same motif. For example, the cassette appears twice in the cystic fibrosis transmembrane conductance regulator involved in chloride transport, the *Plasmodium falciparum mdr* gene expressed in chloroquine-resistant parasites, the yeast STE 6 gene product involved in export of a-mating factor, and proposed peptide transporters involved in antigen presentation (3,4). Many bacterial transporters contain the entire cassette or elements of the cassette, found within individual polypeptides or subunits of transporter complexes, respectively (5).

A variety of reversal agents, which are also known as chemosensitizers, have been found for MDR (6). These compounds are nontoxic alone, but in combination with a drug in the MDR phenotype, such as bisantrene, resensitize the MDR cell to the chemotherapeutic agent. Verapamil was the first reversal agent identified. However, a variety of calcium

T. Godfraind et al. (eds.), Calcium Antagonists, 363–368.

antagonists, notably the 1,4-dihydropyridine nicardipine, also reverse MDR (fig. 1). In addition, other unrelated molecules also reverse MDR (e.g., reserpine, cyclosporine A, phenothiazines). Verapamil resensitized MDR cells by inhibiting afflux of anticancer agents. This effect may be directly on P-gp since verapamil competitively inhibits vinblastine binding to vesicles containing P-gp (7) and inhibits photoaffinity labeling of P-gp by photoactive vinblastine (8). Furthermore, vinblastine inhibits the binding of photoactive verapamil analogs to P-gp (9,10).

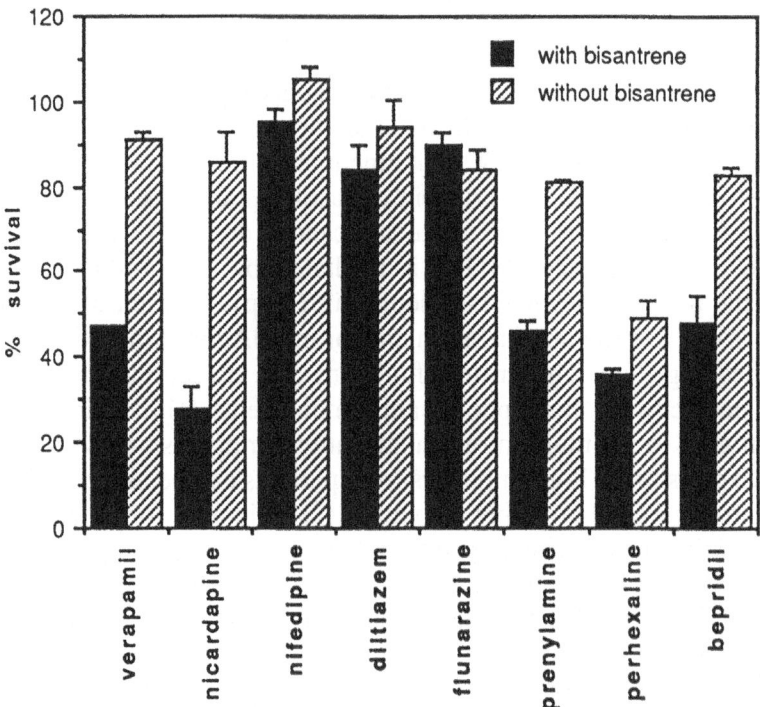

FIG. 1. The effect of calcium channel antagonists on the reversal of resistance to bisantrene. Agents (10 μM) were tested alone ($\square$) or in combination with 20 μM bisantrene ($\blacksquare$). A human OVCAR MDR cell line, which was completely resistant (100% survival) to 20 μM bisantrene, was used. Compounds were incubated with cells for 3 days. Present survival (mean $\pm$ SD) was measured by a colormetic assay that is directly proportional to the cell number.

Photoaffinity analogs of antineoplastic drugs, as well as reversal agents, have been used to understand the interaction of these agents with P-gp. A preferential order of inhibition of photoaffinity labeling has been found regardless of the probe (11). At equal molar concentrations, the order of inhibition is vinblastine > actinomycin D > doxorubicin $\geq$ colchicine. (Hydrophilic drugs such as bleomycin have no effect on photoaffinity labeling.) This suggests that structurally diverse agents interact with a common site in P-gp. Consistent with this, ^{125}I-iodoaryl azidoprazosin (^{125}I-IAP; an α1-adrenergic antagonist) and ^{3}H-azidopine (a 1,4 dihydropyridine) bind to a 6-kDa common domain in P-gp (11).

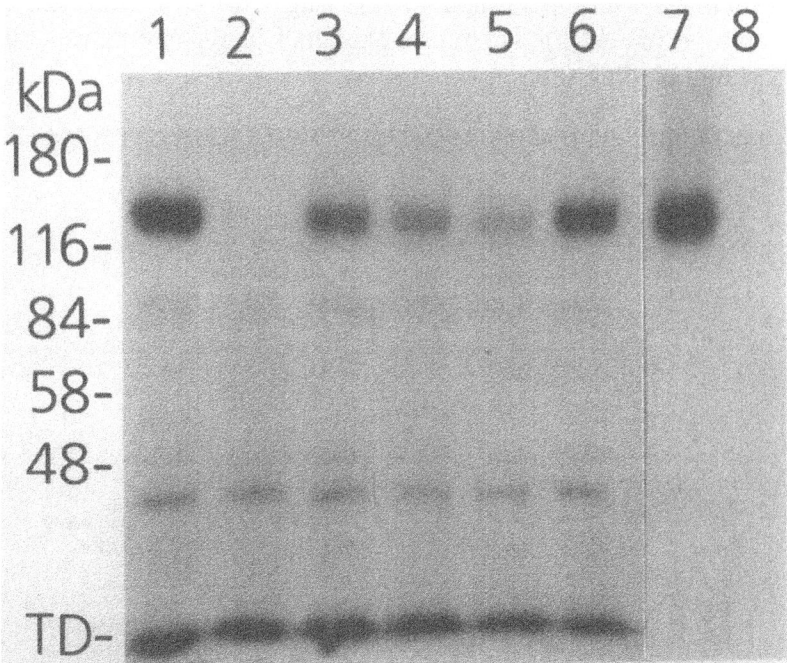

FIG 2. The effect of chemotherapeutic drugs on photoaffinity labeling of P-gp by ^{125}I-iodoaryl azidoprazosin ([25]I-IAP). Membrane (100 μg) from mouse MDR cells was incubated with 2 nM ^{125}I-IAP alone (lane 1) or with 20 μM vinblastine (lane 2), colchicine (lane 3), doxorubicin (lane 4), actinomycin D (lane 5), or bleomycin (lane 6). After UV exposure, samples were resolved by gel electrophoresis (lanes 1-6). Lanes 7 and 8: as in lane 1 but samples were immuno-precipitated with an antibody to P-gp (lane 7) or pre-immune serum (lane 8) before SDS-PAGE. An autoradiogram of the gel is shown. Molecular mass markers are shown. TD denotes tracking dye. (adapted from 11).

An immunological mapping approach was used to identify the photoaffinity drug binding sites in P-gp (12). To do this, a battery of site-directed polyclonal antibodies were generated to synthetic peptides that mimic P-gp. Epitopes were chosen such that the antibodies would have little or no cross-reactivity between each half of P-gp. In these experiments, P-gp within vesicles was photoaffinity labeled with ^{125}I-IAP, and then vesicles were digested with trypsin. The resultant fragments were immuno-precipitated by site-specific antibodies. The results are summarized in fig. 3. Initially, P-gp was digested to 95- and 55-kDa fragments. Both fragments were photoaffinity labeled. After further digestion, a 40-kDa photoaffinity-labeled fragment was derived from the 55-kDa fragment. Cleveland mapping has been used to demonstrate that photolabeling within the 95- and 40-kDa fragments were restricted to a small epitope (6-8 kDa). Therefore, each half of P-gp contains a small photoaffinity binding site.

LOCATION OF PHOTOAFFINITY DRUG BINDING SITES IN P-GLYCOPROTEIN

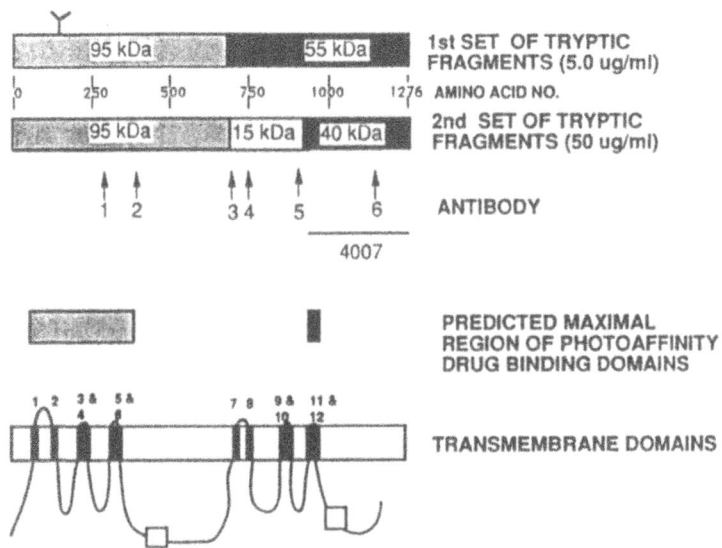

FIG. 3. Location of photoaffinity drug binding sites in P-glycoprotein (from 12, with permission).

Transmembrane domains 11-12 are likely to contain the photoaffinity drug binding site in the C-terminal half of P-gp. The following deductions and experimental evidence are consistent with this prediction:

a) The drug binding pocket in P-gp is expected to be hydrophobic. Photoaffinity probes that bind to P-gp are hydrophobic. The only hydrophobic domains in the 40 kDa fragment are TM 11-12 and the nucleotide binding fold. The latter site is an unlikely drug binding domain since photoaffinity labeling of P-gp by azidopine or azido-ATP are not inhibited by ATP and vinblastine, respectively (13,14).

b) TM 11 is perfectly conserved among two mouse *mdr* genes (*mdr*1a and *mdr*1b) that mediate MDR. However, TM 11 of an inactive mouse *mdr* gene (*mdr*2) is substantially divergent compared to the homologous region in active *mdr* genes (15).

c) Markedly different drug resistant profiles are found in drug-sensitive cells transfected with wild type P-gp compared to cells transfected with a P-gp that contains a point mutation in TM 11 (16).

d) Point mutations in TM 11 of the *Plasmodium falciparum* P-gp homolog are associated with parasites and malaria that are chloroquine resistant (17).

The L-type calcium channel is composed of four cassettes (18). Each cassette contains six transmembrane domains. Striessnig et al. (19) has shown that a photoactive verapamil analog, LU 49888, binds to the $\alpha 1$ subunit of the L-type calcium channel. Binding occurs in, and immediately surrounds, the sixth transmembrane domain of the fourth cassette. LU 49888 also binds to P-glycoprotein (10). If the binding sites for LU 49888 include TM 11-12, then this suggests that verapamil may bind to a homologous region in P-gp and the L-type calcium channel. This hypothesis may help explain why verapamil interacts with P-gp.

ACKNOWLEDGEMENTS

Portions of this work have been done in collaboration with Dr. Susan B. Horwitz. I thank Dr. Dennis Powell for helpful discussions pertaining to this work as well as Ms. Suzanne Carvajal, Ms. Jillian Silva, and Mr. Christopher Lisanti for their excellent technical assistance.

REFERENCES

1. Pastan I. and Gottesman M.M. (1991): *Annu. Rev. Med.*, 42: 277-286.
2. Bellamy W.T., Dalton W.S. and Dorr R.T. (1990): *Cancer Invest.*, 8: 547-562.
3. Greenberger L.M., Hsu S.I.-H., Yang C.P.-H., Cohen D., Lothstein L., Han E.K.-H., Kirschner L.S., Pierkarz R., Yu L. and Horwitz S.B. (1992): In: *Drug Resistance as a Biochemical Target in Cancer Chemotherapy*, edited by T. Tsuruo and M. Owage (in press). Academic Press, California.
4. Parham P. (1990): *Nature*, 348: 674-675.
5. Ames G.F.L. (1986): *Annu. Rev. Biochem.*, 55: 397-425.
6. Yang C.P.-H., Greenberger L.M. and Horwitz S.B. (1991): In: *Synergism and Antagonism in Chemotherapy*, edited by T.-C. Chou and D.C. Rideout pp. 311-338. Academic Press, New York.
7. Naito M. and Tsuruo T. (1989): *Cancer Res.*, 49: 1452-1455.
8. Cornwell M.M., Safa A.R., Felsted R.L., Gottesman M.M. and Pastan I. (1986): *Proc. Natl. Acad. Sci. USA*, 83: 3847-3850.
9. Safa A.R. (1988): *Proc. Natl Acad. Sci. USA*, 85: 7187-7191.
10. Qian X.-D. and Beck W.T. (1990): *Cancer Res.*, 50: 1132-1137.

11. Greenberger L.M., Yang C.P.-H., Gindin E. and Horwitz S.B. (1990): *J. Biol. Chem.*, 265: 4394-4401.
12. Greenberger L.M., Lisanti C.J., Silva J.T. and Horwitz S.B. (1991): *J. Biol. Chem.*, (in press.)
13. Yang C.P.-H. and Horwitz S.B., personal communication.
14. Cornwell M.M., Tsuruo T., Gottesman M.M. and Pastan I. (1987): *FASEB J.*, 1: 51-54.
15. Hsu S., Lothstein L. and Horwitz S.B. (1989): *J. Biol. Chem.*, 264: 12053-12062.
16. Gros P., Dhir R., Croop J. and Talbot F. (1991): *Proc. Natl. Acad. Sci. USA*, 88: 7289-7293.
17. Foote S.J., Kyle D.E., Martin R.K, Oduola A.M.J., Forsyth K., Kemp D.J. and Cowman A.F. (1990): *Nature*, 345: 255-258.
18. Catterall W.A. (1988): *Science*, 242: 50-64.
19. Striessnig J., Glossmann H. and Catterall W.A.(1990): *Proc. Natl. Acad. Sci. USA*, 87: 9108-9112.

REVERSAL OF MULTIPLE DRUG RESISTANCE (MDR) IN MALARIA BY REVERSE ISOMERS OF CALCIUM CHANNEL BLOCKERS AND ITS POSSIBLE APPLICATION TO CANCER AND MDR PARASITIC DISEASES

K. Van Dyke, Z. Ye, and C. Van Dyke

Department of Pharmacology and Toxicology,
West Virginia University, Health Sciences North
Morgantown, West Virginia 26506 and
Cancer Biologics of America
600 Perimeter Drive, Lexington, Kentucky 40517

Several aspects of multiple drug resistance (MDR) have remained unclear since the original experiments demonstrated that toxic, anticancer drugs are pumped out of resistant cancer cells. It was unclear initially how a common mechanism could pump out a diverse series of anticancer drugs including adriamycin, actinomycin D, emetine, vincristine, vinblastine, chloroquine and VP-16. No structural relationship among the affected drugs was apparent.

Tsuruo et al. (1) found that racemic verapamil could reverse the resistance to the above drugs. Previously, a 155 to 170 Kd glycoprotein was found which was responsible for the MDR pumping action (2). This glycoprotein used adenosine triphosphate as an energy source and apparently had ATPase activity. More detailed work indicated that multiple proteins, specifically MDR_1, MDR_2 and sometimes MDR_3, could be separated via electrophoresis, although only MDR_1 was apparently active against the above drugs. The function of the MDR_2 and MDR_3 molecules remained a mystery.

Milhous et al. (3) showed that racemic verapamil was effective in reversing chloroquine resistance in malaria. In malaria, as well as in cancer, reversal proved to be a high-dose effect which was too toxic to be of clinical significance. The dose required to achieve reversal resulted in toxic arrhythmias.

Given that verapamil was a racemic mixture consisting of S(-) and R(+) isomers and that only the S(-) isomer allosterically blocked calcium channels, we decided to investigate MDR reversal using the R(+) isomer. Further, we obtained the R(+) isomers of close congeners gallopamil (D-600) and devapamil and tested all three for reversal activity against the 3- to 5-fold chloroquine resistant *Plasmodium falciparum* malaria. We demonstrated that all three of the above R(+) isomers completely

369

T. Godfraind et al. (eds.), Calcium Antagonists, 369–375.
© 1993 *Kluwer Academic Publishers and Fondazione Giovanni Lorenzini.*

reversed the chloroquine resistance at micromolar doses *in vitro* (see table 1). Since there is little or no binding of these isomers to the calcium channel, resistance reversal must be affecting a channel independent of but similar to the calcium channel. A careful examination of the anticancer drugs, chloroquine and the verapamil congeners reveals that they all form divalent cations in aqueous solution. Since the chemotherapeutic drugs are basically toxins, perhaps this is a detoxifying channel for divalent cations which, unlike the calcium channel, is nonstereospecific.

TABLE 1.   Reversal of Chloroquine-Resistant
Strain of *Plasmodium Falciparum* by R(+)
Verapamil Derivatives

Drugs[b]	IC_{50} OF CQ VS *P. FALCIPARUM* (MEAN ± S.D.[a])		RATIO R/S
	SENSITIVE STRAIN	RESISTANT STRAIN	
CQ	32.3± 10.5[c]	157.5± 59.4	4.9
CQ+V (1 mM)	29.3± 7.3	72.6± 22.4[d]	2.5
CQ+V (2 mM)	26.7± 6.0	39.7± 14.3[d]	1.5
CQ	20.3± 3.4[c]	119.5± 15.0	5.5
CQ+G (1 mM)	25.4± 7.7	35.5± 5.6[d]	1.4
CQ+G (2 mM)	23.1± 9.8	23.9± 8.1[d]	1.0
CQ	24.6± 7.7[c]	137.8± 12.2	5.6
CQ+D (1 mM)	26.6± 6.2[c]	36.5± 3.8[d]	1.5
CQ+D (2 mM)	22.4± 5.6	30.8± 2.8[d,e]	1.4
CQ+D (4 mM)	22.0± 2.1	23.7± 6.2[d,e]	1.1

a All of the data in the table are results from three trials, concentrations in nanomolar.
b CQ,V,G,D are chloroquine, R(+) verapamil, R(+) gallopamil and R(+) devapamil, respectively.
c Significant difference (P<0.05) between sensitive and resistant strains.
d Significant difference (P<0.05) between resistant strain with and without the drug as compared with the data of CQ above.
e Significant difference (P<0.05) between resistant strain with different concentrations of the drug.

Before the advent of agriculture, man consumed a variety of substances containing toxins. This necessitated a system which either metabolized and then excreted the toxins (P450 system) or pumped the toxins from the cells or organs (MDR). Removing toxins by pumping seems more sensible then metabolism because it takes less energy and does not require multiple enzymes for multiple toxins.

D-TETRANDRINE AND MDR IN MALARIA

Since it has been shown that drugs which reverse multiple drug resistance inhibit the exit mechanism for toxins, this is a crucial characteristic of resistance reversing drugs. Such pump inhibitors are known as disease modifying drugs (DMD). A second desirable characteristic of a DMD would be a direct effect against the disease state. Since initial work of Tsuruo et al. pointed to calcium channel inhibitors, we decided to explore weak calcium channel inhibitors which might have direct action against the disease process itself. We tried d-tetrandrine, a bisbenzylisoquinoline, and found it possessed weak calcium channel blocking activity. Next, we tested it for MDR-reversing activity in chloroquine-resistant *P. Falciparum* malaria with astonishing results. D-tetrandrine completely reversed chloroquine-resistance and proved almost as toxic to the resistant malaria strain as chloroquine. While both d-tetrandrine and chloroquine were effective in the nanomolar concentration range against the resistant strain, d-tetrandrine was much less effective against the chloroquine-sensitive strain. These results show that d-tetrandrine has a selective action against chloroquine resistance, possibly against the mechanism of resistance itself.

Based on the above results, we performed a series of *in vitro* experiments to determine the synergistic effect of chloroquine and d-tetrandrine against chloroquine-resistant malaria. Different concentrations of the drugs wee tested and the 50% inhibitory concentrations (IC_{50}) were determined (see table 2). From this data, isobolograms were drawn which revealed a 43-fold synergy between the two drugs. In fact, this combination was so toxic to the resistant strain that the resistant strain was more sensitive to chloroquine than the sensitive strain in the presence of d-tetrandrine. Recently *in vivo* experiments were performed with Aotus monkeys infected with resistant P. falciparum malaria and treated with chloroquine and tetrandrine. The combination completely eliminated all detectable chloroquine-resistant malaria.

TABLE 2. IC_{50} (Nm) of Tetrandrine
                                   (TT) and Chloroquine (CQ) for Each Drug
                                       Alone and in Combination[a]

Malaria[b]	Single drug		Drug combination[c]		
	TT	CQ	TT(1.0μM) CQ(0.3μM)	TT(2.0μM) CQ(0.2μM)	TT(3.0μM) CQ(0.1μM)
S Strain	498.1± 94	26.7± 3.9	54.9± 7.1(TT) 16.5± 2.1(CQ)	114± 23(TT) 11.4± 2.3(CQ)	223± 39(TT) 7.4± 1.3(CQ)
R Strain	197.5± 25	185.8± 4.9	79.5± 14(TT) 23.8± 4.8(CQ)	79.5± 16(TT) 8.0± 1.6(CQ)	125± 9.6(TT) 4.2± 0.3(CQ)

[a] The data in the above table are mean values ± SD (nM) from 3 experiments.
[b] S and R strains represent CQ-sensitive (FCMSU1/Sudan) and CQ-resistant (W2) strains of *P. Falciparum*, respectively.
[c] Ratios of TT/CQ in the drug combinations are 10:3, 10:1 and 30:1 respectively.

Using both photoaffinity labeling with ^3H azidopine and primer specific probes, we have determined that malaria resistance is an overexpression of the MDR glycoprotein, i.e., both chloroquine-sensitive and resistant parasites carry the MDR gene, but the resistant parasite develops 3- to 4-fold more of the protein than does the sensitive parasite (4).

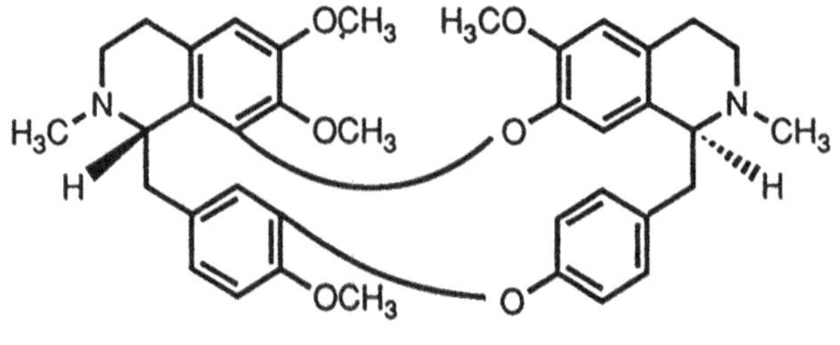

FIG. 1. The structure of d-tetrandrine.

We have done extensive experiments to measure the structure/activity relationship of d-tetrandrine and came to the following conclusions. Tetrandrine is a divalent cation like other MDR_1 inhibitors. For greater activity, the right-hand ring must be in the S configuration while the ring to left may be in either the S or the R configuration (see fig. 1). The placement of OCH_3 groups can be important. The two oxygen bridges are important because they make the molecule rigid as opposed to the more "flexible" molecules of verapamil and congeners.

d-TETRANDRINE AND MDR IN CANCER

Currently, we are directing our attention to MDR in cancer using the

DX-20 cell human uterine sarcoma cell line which demonstrates 75-fold resistance to adriamycin and its parent, the MES-SA cell line. *In vitro* experiments show a 30- to 40-fold synergy between adriamycin and d-tetrandrine as well as complete resistance reversal. Once again, adriamycin in the presence of tetrandrine actually makes the resistant strain more sensitive to the toxin than the sensitive strain.

Next, we directed our studies using the above MDR sarcoma cells in nude mice. These athymic mice were injected with the tumor and left alone for 10 days to allow the tumor to develop untreated. A tetrandrine derivative was then injected daily into appropriate groups for days 11 through 17 to inhibit the MDR exit pump. A single dose of adriamycin was given to the relevant groups on day 17. Groups receiving the tetrandrine derivative continued daily injections for days 18 through 24. Initially, large tumors developed in all four groups: untreated mice, mice receiving adriamycin, mice receiving the tetrandrine derivatives, and mice receiving the tetrandrine derivative and adriamycin. Following treatment, the tumors had all but disappeared in the combination therapy group while extremely large tumors were present in all other groups with no significant difference in the tumor size. After more than 40 days,the combinational therapy group was still surviving while all other groups had uniformly died.

OTHER MDR CHANNELS AND MDR IN OTHER DISEASES

Our recent studies with mefloquine-resistant *P. falciparum* malaria indicate that d-tetrandrine is ineffective in reversing resistance to mefloquine. The fact that mefloquine is a monovalent cation while tetrandrine and chloroquine are divalent cations proves a possible clue about the function of the MDR_1, MDR_2, and MDR_3 glycoproteins. We speculate the existence of a nonstereospecific MDR pump for $+1$ ions which is similar to the $+2$ pump described above, i.e., in order to reverse resistance to a $+1$ ion, a $+1$ cation channel blocker must be used. Therefore, the possibility exists that a whole series of such channels are involved in detoxifying substances charged either negatively or positively. Such a system may be as extensive and as biologically important as the P_{450} system is for metabolizing drugs. Indeed, we conclude that MDR is not a unique system developed under stress by cancers or invading organisms to combat chemotherapeutic agents; rather, it is an intricate mechanism employed by many normal body tissues for flushing out toxins. In the case of chemotherapy, we believe that the development of MDR is the selection of individual cells in the invading population that overexpress the corresponding, pre-existing detoxifying pump and are thus more able to survive.

Furthermore, many other major diseases exist in which the detoxifying system (MDR) may be important. We point in particular to amoebiasis, trypanosomiasis, fungus infections, leprosy, and tuberculosis. When the MDR mechanism is overexpressed, the organism is more vulnerable to these DMD, particularly ones which offset the disease itself. We predict that new combinations of drugs will be the rule of thumb to treat diseases where resistance is due to the overexpression of the MDR pump.

CONCLUSIONS

Based on the above research, we are able to draw the following conclusions:

1. It is important for drug manufacturers to produce pure drugs and not racemic mixtures, so that therapeutic and toxic effects can be separated as much as possible.

2. Multiple drug resistance is misnamed. In reality, it is a series of detoxifying channels which eliminate toxic substances by pumping them out of cells. MDR_1 is the divalent cation pump which excretes a variety of $+2$ compounds out of the cell. Drugs which inhibit the pump are also $+2$ compounds like verapamil, gallopamil (D-600), devapamil and d-tetrandrine.

3. By selecting isomers, e.g., $R(+)$ that do not interfere with calcium channels, we escape the toxicity associated with high-dose effects on the calcium channels, e.g., arrhythmias.

4. Drugs which simply inhibit the exit pump are less effective than drugs which have another action against the invading cell. d-Tetrandrine synergizes with chloroquine and the two drugs are most effective together (dose of chloroquine is decreased 45-fold) in malaria.

5. By blocking the exit pump first and then treating with the toxin, e.g., adriamycin, we can force the toxin to remain inside the cell (cancer or malaria) and treat at lower doses than previously suspected. We have clearly demonstrated this concept with a 75-fold resistant cancer by treating infected nude mice with adriamycin and a tetrandrine derivative.

6. We were the first to show the nonstereospecific nature of the detoxifying, $+2$ MDR channel. Now we have the first evidence of a $+1$ detoxifying channel as well. Resistance to mefloquine, a $+1$ ion has developed and is not inhibited by d-tetrandrine, a $+2$ ion.

7. We speculate that detoxifying channels for negative ions may exist as well. This is logical given that toxic substances come in a variety of forms that cells have had to handle since cellular life began.

8. The detoxifying system is pervasive throughout the animal kingdom. Among others, amoeba, trypanosomes, malaria, and tuberculosis have demonstrated MDR. By modifying the amount of gene product (overexpression), more of the pump is produced which protects the organisms against higher doses of the drug.

9. We have produced a new drug combination (chloroquine and d-tetrandrine) which has been shown to be effective *in vivo* against chloroquine-resistant malaria.

10. Two million people die each year from malaria, 90% of whom are victims of chloroquine-resistant *P. falciparum* malaria. We have found an effective combination for those people.

11. We believe that bisbenzylisoquinolines will be effective in many other diseases including cancer.

REFERENCES

1. Tsuruo T., Iida H., Tsukasohi S. and Sakurai Y. (1981): *Cancer Research,* 41: 1967.
2. Juliano R.L. and Ling V. (1976): *Biochem. Biophys. Acta.,* 455: 152-162.
3. Martin S.K., Oduola A.M. and Milhous W.K. (1987): *Science,* 235: 899-901.
4. Ye Z., Van Dyke K., Spearman T. and Safa A.R. (1989): *Biochem. Biophys. Res. Comm.,* 162: 809-813.

INDEX

Medical Science Symposia Series

1. A. M. Gotto, C. Lenfant, R. Paoletti (eds.) and M. Soma (ass.ed.): *Multiple Risk Factors in Cardiovascular Disease.* 1992   ISBN 0-7923-1938-9
2. A. L. Catapano, A. M. Gotto, Jr., L. C. Smith and R. Paoletti (eds.): *Drugs Affecting Lipid Metabolism.* 1993

          ISBN 0-7923-2232-0
3. T. Godfraind, S. Govoni, R. Paoletti and P. M. Vanhoutte (eds.): *Calcium Antagonists. Pharmacology and Clinical Research.* 1993

          ISBN 0-7923-2259-2

KLUWER ACADEMIC PUBLISHERS – DORDRECHT / BOSTON / LONDON

The manufacturer's authorised representative in the EU is Springer
Nature Customer Service Centre GmbH, Europaplatz 3, 69115 Heidelberg,
Germany. If you have any concerns regarding our products, please
contact ProductSafety@springernature.com

Printed and bound by CPI Group (UK) Ltd, Croydon, CR0 4YY

23/04/2026

02095623-0007